Bioseparation and Bioprocessing

Volume II: Processing,
Quality and Characterization,
Economics, Safety and Hygiene

Edited by
G. Subramanian

WILEY-VCH

Further Reading from WILEY-VCH

A Practical Approach to Chiral Separations by Liquid Chromatography
Edited by G. Subramanian
Fundamentals and Applications, 2nd Edition
1994, 422 pp., hardcover, VCH, ISBN 3-527-28288-2

Process Scale Liquid Chromatography
Edited by G. Subramanian
1994, XVI, 225 pp., hardcover, VCH, ISBN 3-527-28672-1

Bioseparation and Bioprocessing

Volume II: Processing, Quality and Characterization, Economics, Safety and Hygiene

Edited by
G. Subramanian

WILEY-VCH

Weinheim · New York · Chichester · Brisbane · Singapore · Toronto

Ganapathy Subramanian
60 B Jubilee Road
Littlebourne
Canterbury
Kent CT3 1TP, UK

Cover illustration: Three-dimensional model of human choriogonadotropin. The model is based on the crystal structure of deglycosylated hCG[9] (PDB code 1hrp). The protein part of the molecule (ribbon) and the four N-linked carbohydrate chains (spheres) are shown on the same scale. The oligosaccharides are attached to Asn52 (top, right) and Asn78 (bottom) of the α-subunit (green), and to Asn13 and 30 (top, left) of the β-subunit (blue). The binding region is indicated in red. It should be noted that the spatial orientation of the carbohydrate chains is arbitrarily set as they are not present in the crystal structure. The carboxy-terminal peptide of the β-subunit (amino acid residues 131–145) is not depicted because its 3D-structure could not be deduced from the crystal [Figure reproduced by courtesy of Prof. Dr. P. D. J. Grootenhuis (Dept. of Computational Medicinal Chemistry, N. V. Organon, Oss)]. See also Chapter 5 of this volume.

Library of Congress Card No. applied for

British Library Cataloguing-in-Publication Data:
A catalogue record for this book is available from the British Library

Die Deutsche Bibliothek – CIP-Einheitsaufnahme
Bioseparation and bioprocessing / ed. by Ganapathy Subramanian. – Weinheim ; New York ; Chichester ; Brisbane ; Singapore ; Toronto : Wiley-VCH
ISBN 3-527-28876-7 ✗

Vol. II. Processing, quality and characterization, economics, safety and hygiene. – 1998

© WILEY-VCH Verlag GmbH, D-69469 Weinheim (Federal Republic of Germany), 1998

Printed on acid-free and chlorine-free paper

Composition: Hagedorn Kommunikation, D-68519 Viernheim
Printing: strauss offsetdruck GmbH, D-69509 Mörlenbach
Bookbinding: Wilhelm Osswald + Co., D-67433 Neustadt
Printed in the Federal Republic of Germany

Preface

Biotechnology represents the confluence of several disciplines. The European Federation of Biotechnology has defined biotechnology as an integrated use of biochemistry, microbiology and chemical engineering in order to achieve the technological (industrial) application of the capacities of microbes and cultured cells. Thus, to produce purified biologically active components really depends on the effective separetion process. Within this versatile area of separation it would be incorrect to claim that this book covers the entire field of separation technology comprehensively; it does not; nor is it intended to be used as a textbook for a specific course. This book is intended to project an overview on selected techniques that are actively applied in the biotechnology industries.

The book is organised into three parts containing fifteen chapters contributed by experienced scientists. The first eight chapter in part one gives an overview of various aspects in processing methods that are applied in the industries for the production of bioactive compounds. Quality and characterisation are addressed in chapters 9, 10, 11 and 12. (Part two). Part three consisting of chapters 13, 14, and 15 deals with Economics, Safety and Hygiene.

It is my hope that this volume will bring together accumulated knowledge in a way which will promote the advancement of technology, which will continue to grow and develop on the basis of fascinating discoveries in the control and separation of biomolecules to create technologies that are useful to society.

I gratefully acknowledge the authors for their time and motivation in preparing their contributions, without which this volume would not have been possible. I should be most grateful for any suggestions which could serve to improve future editions of this book.

Finally I would like to thank the staff of Wiley-VCH for their help.

Canterbury, Kent
January 1998
G. Subramanian.

Contents

Part One: Processing

5   The Application of Glycobiology for the Generation of Recombinant Glycoprotein Therapeutics

Jan B. L. Damm

6   The Release of Intracellular Bioproducts

Anton P. J. Middelberg

7   Microcarriers in Cell Culture Produktion   165

Björn Lundgren and Gerald Blüml

Part Two: Quality and Characterization

Part Three: Economics, Safety and Hygiene

**14   Process Hygiene in Production Chromatography and
          Bioseparation** 417

Glenwyn D. Kemp

Contributors

Rose M. Baker
Research & Development Department
Bio Products Laboratory
Dagger Lane, Elstree
Hertfordshire, WD6 3BX
United Kingdom
(*Chapter 8*)

Dr. Stephen J. Bayne
Novo Nordisk A/S
Hagedornsvej 1
DK-2820 Gentofte
Denmark
(*Chapter 11*)

A. Binieda
British Biotech Pharmaceuticals Ltd.
Watlington Road
Oxford OX4 5LY
United Kingdom
(*Chapter 10*)

Dr. Gerald Blüml
Microcarriers and Cell Separation Media
Pharmacia Biotech AB
c/o Institute for Applied Microbiology
Muthgasse 18 B
A-1190 Wien
Austria
(*Chapter 7*)

Dr. N. Burns
British Biotech Pharmaceuticals Ltd.
Watlington Road
Oxford OX4 5LY
United Kingdom
(*Chapter 10*)

Dr. George E. Chapman
Research & Development Department
Bio Products Laboratory
Dagger Lane, Elstree
Hertfordshire, WD6 3BX
United Kingdom
(*Chapter 8*)

Dr. Yusuf Chisti
Department of Chemical Engineering
University of Almeria
E-04071 Almeria
Spain
(*Chapter 1 and 13*)

Dr. Jan B. L. Damm
Akzo Nobel
N. V. Organon
Quality Assurance
Analytical Control Labs. B
P.O. Box 20
NL-5340 BH Oss
The Netherlands
(*Chapter 5 and 12*)

Dr. Gerrit J. Gerwig
Bijvoet Center for Biomolecular Research
Department of Bio-Organic Chemistry
Utrecht University
P.O. Box 80075
NL-3508 TB Utrecht
The Netherlands
(*Chapter 12*)

Dr. Kim Hejnaes
Novo Nordisk A/S
Hagedornsvej 1
DK-2820 Gentofte
Denmark
(*Chapter 2*)

Dr. Philip J. Jackson
Perkin Elmer
Applied Biosystems
Kelvin Close
Birchwood Science Park North
Warrington WA3 7PB
United Kingdom
(*Chapter 11*)

Dr. Glenwyn D. Kemp
Therapeutic Antibodies Limited
St. Bartholomew's Medical College
Charterhouse Square
London EC1M 6BQ
United Kingdom
(*Chapter 14*)

Dr. Björn Lundgren
Microcarriers and Cell Separation Media
Amersham Pharmacia Biotech
S-75182 Uppsala
Sweden
(*Chapter 7*)

Dr. Paul Matejtschuk
Research & Development Department
Bio Products Laboratory
Dagger Lane, Elstree
Hertfordshire, WD6 3BX
United Kingdom
(*Chapter 8*)

F. Matthiesen
Novo Nordisk A/S
Hagedornsvej 1
DK-2820 Gentofte
Denmark
(*Chapter 2*)

Dr. Anthony Meager
Division of Immunbiology
The National Institute for
Biological Standards and Control
Blanche Lane
South Mimms
Hertfordshire EN6 3QG
United Kingdom
(*Chapter 9*)

T. J. Meyers
British Biotech Pharmaceuticals Ltd.
Watlington Road
Oxford OX4 5LY
United Kingdom
(*Chapter 10*)

Dr. Anton P. J. Middelberg
Department of Chemical Engineering
The University Cambridge
Pembroke Street
Cambridge CB2 3RA
United Kingdom
(*Chapter 4 and 6*)

Dr. John Noble
Foster Wheeler Energy Limited
Foster Wheeler House
Station Road, Reading
Berkshire RG1 1LX
United Kingdom
(*Chapter 3*)

Dr. Brian K. O'Neill
Department of Chemical Engineering
The University of Adelaide
5005 Adelaide S.A.
Australia
(*Chapter 4*)

J. A. Purvis
British Biotech Pharmaceuticals Ltd.
Watlington Road
Oxford OX4 5LY
United Kingdom
(*Chapter 10*)

L. Skriver
Novo Nordisk A/S
Hagedornsvej 1
DK-2820 Gentofte
Denmark
(*Chapter 2*)

P. G. Varley
British Biotech Pharmaceuticals Ltd.
Watlington Road
Oxford OX4 5LY
United Kingdom
(*Chapter 10*)

Dr. Joachim K. Walter
Process Development
Dept. Biotechnical Production
Boehringer Ingelheim
Pharma KG
D-88397 Biberach
Germany
(*Chapter 15*)

Dr. Gordon Wright
PPL Therapeutics
(*Chapter 3*)

Part One
Processing

1 Strategies in Downstream Processing

Yusuf Chisti

1.1 Introduction

Biological products come from many sources: human and animal tissue (e.g., blood, pancreas, pituitary) and body fluids (e.g., milk of transgenics), plant material (e.g., Taxol® from the bark of *Taxus* species, oils), microbial fermentations, cultures of higher eukaryotes, and raw broths from enzyme bioreactors. Irrespective of the source, crude extracts, fluids and broths invariably undergo separation and purification to recover the product in the desired form, concentration, and purity. Processing beyond the bioreaction step is termed downstream processing. Here the 'bioreaction step' includes producing plants and animals.

A recovery process consists of physicochemical operations such as those listed in Table 1-1. The steps of a properly engineered downstream process are integrated with each other and with the bioreaction stage to yield an optimal recovery scheme [19]. This discussion is limited to factors which must be considered in developing any economically viable product purification and concentration scheme based on a

Table 1-1. Bioseparation operations.

Solid-liquid separations [1–3]
Centrifugation, filtration, flocculation, flotation, sedimentation [4,5].

Membrane separations [1–3,6]
Diafiltration and dialysis, microfiltration, pervaporation [7], reverse osmosis, ultrafiltration.

Extractions
Aqueous liquid–liquid extraction [1,8,9], extraction and leaching of solids [10], reversed miceller extraction, liquid membrane extraction [11,12], solvent extraction [1,13], supercritical extraction.

Chromatographic methods [1–3,14]
Affinity, gel permeation, hydrophobic interaction, ion exchange. See also Volume 1.

Thermal operations
Distillation, drying [1], evaporation, freeze drying or lyophilization [15].

Miscellaneous
Adsorption, cell disruption [16–18], crystallization, electrophoresis and other electrokinetic methods, precipitation [1].

small selection of the many available processing operations. Individual operations are detailed in other sources [1–3] including some chapters of this series.

1.2 Overview of Process Considerations

Anyone faced with designing a bioseparation scheme can take comfort in the variety of available separation processes (Table 1-1); however, the same variety can be a source of much distress. The number N_f of possible recovery flowsheets that can be theoretically devised to completely separate a mixture of C components by using S number of separation operations is given by

$$N_f = \frac{[2(C-1]!S^{(C-1)}}{C!(C-1)!}$$

(1)

Thus, complete separation of a five-component mixture using two separation methods would generate 224 possible flowsheets! Although only one component, the product, is usually wanted, several 'components' need separating. Examples of such components are cells, water, cell debris, nucleic acid polymers, added salts, and the remainder of the proteins. Often, a process must include additional nonseparating steps such as cell disruption, heating, and mixing. Not all possible flowsheets can be exhaustively evaluated; instead, experience and thorough knowledge of individual bioseparations and relevant fermentation must be relied upon to narrow the choices to a few practicable options for detailed evaluations and experimental testing.

Factors that must be considered in designing a downstream processing scheme include the nature, concentration and stability of the product, the desired purity and end use. Because of contamination and supply considerations, there is a distinct trend to move away from direct extraction of human and animal sources to recombinant cells. Thus, for example, microbially produced recombinant human insulin and growth hormone are now available. For vaccines, too, attempts are underway to engineer safer organisms to produce the antigenic material that would otherwise be obtained from pathogens.

The end use of the product may vary – research, *in vitro* diagnostic, food, animal feeds, soil inoculants, pesticides, medicinals, medical device, cosmetics, etc. The specific form of the product may include live human cells for medical purposes; live microorganisms, viruses (e.g., for vaccines), spores (e.g., for biotransformations, insecticides, and solid state culture), and higher organisms (e.g., nematodes); bioactive polymers, proteins and enzymes; inactive polymers (e.g., food protein, xanthan, PHB); smaller organics (e.g., streptomycin, amino acids, citric acid, ethanol, Taxol®); polypeptides (e.g., cyclosporine); and cellular organelles (e.g., nuclei, mitochondria, chloroplasts).

Location of the product, whether extracellular, intracellular, or periplasmic, affects how it is recovered. Physical and chemical properties of the product and contaminants need addressing, and biosafety issues must be given attention (see Chapter 13).

A further consideration is price relative to existing sources and other competing products. When no competing products or alternative sources can be identified, the estimated production costs would need to be compared with what the market can reasonably be expected to pay.

As far as possible, the requisite purification and concentration should be achieved with the fewest processing steps; generally, no more than six to seven steps are used, a situation quite different from that in chemistry and biochemistry laboratories, where the number of individual steps is often not a major consideration, and purity of the product is usually more important than overall yield or costs [19]. The overall yield of an n-step process with step yield of x percent is $(x/100)^n$. Therefore, n must be minimized for a high overall yield. For example, a train of only five steps, each with 90 % step yield, would reduce the overall recovery to less than 60 % [20]. To minimize reduction of the overall yield, high-resolution separations such as chromatography should be utilized as early as possible in the purification scheme in keeping with the processing constraints that these steps require (e.g., clean process streams free of debris, particulates, lipids, etc.).

Separation schemes incorporating unit operations which utilize different physical–chemical interactions as the bases of separation are likely to achieve the greatest performance for a given number of steps. Combining two separation stages based on the same separation principle may not be an effective approach. As an example, when two chromatographic steps in series are selected, e.g., gel filtration which separates based on molecular size, and ion exchange chromatography which separates based on difference in charge on the molecules, may be a suitable combination.

Speed of processing is another factor that significantly affects the design of a recovery scheme. The size of the bioreaction step and the frequency of harvest usually determine the turnaround time for the downstream process train. Sometimes during processing, exposure of material to relatively severe environmental conditions is unavoidable. Very many factors affect stability, including temperature, pH, proteases and other degrading enzymes, mechanical forces, microbial contamination, oxidants, and other denaturing chemicals. In severe environments, the duration of exposure must be minimized and especial precautions (e.g., low temperature; addition of chemicals to reduce oxidation, etc.) are necessary to reduce the impact of exposure. The need for speedy processing constrains equipment choice and capacity. For example, the low pH necessary during extraction of penicillins affects stability, hence rapid extraction is essential, thus mixer–settler type extraction is contraindicated.

Typically, a separation process must operate within the physiological ranges of pH and temperature (pH $\sim$ 7.0; temperature $\leq$ 37 °C), but differences from the norms are not unusual. For example, enzymes such as lysozyme, ribonuclease, and acid proteases are quite stable at low pH values [19]. Some biologically active molecules, particularly proteins, may be sensitive to excessive agitation; however, enzymes, with the exception of multienzyme complexes and membrane-associated enzymes, are not damaged by shear in the absence of gas–liquid interfaces [1,21].

Except for the final few finishing operations, downstream processing is usually conducted under non-sterile, but bioburden-controlled conditions; however, prevention of unwanted contamination and cleaning and sanitization considerations require that the processing machinery be designed to the same high standards as have been described for sterile bioreactors [22,23]. Containment and hygienic processing requirements may severely affect equipment choice (see also Chapters 13 and 14).

A commercial recovery scheme must be reliable and consistent. Process robustness is essential to economic production, process validation, and product quality. Automation assures consistency and rapid turnaround of the process equipment. Operations such as in-place cleaning are often automated [24].

Additional considerations include biosafety and containment. Bioproducts may be potentially allergenic, and they may produce activity associated reactions in process personnel [10]. In addition, process material may be pathogenic, cytotoxic, oncogenic, or otherwise hazardous. Processing of such material requires attention to containment and biosafety both during design and in operation of the bioseparation scheme [10,25]. Certain processing operations are difficult to contain, and may pose peculiar operational problems. For example, gasket failures during high-pressure homogenization could create high-pressure sprays [16] and, unless designed with containment features, operations such as centrifugation may generate aerosols (see also Chapter 13).

Small quantities of multiple products are sometimes produced in the same plant: a series of runs or campaigns of one product is followed by another. The risk of cross-contamination is high and adequate safeguards are essential. Experience suggests that cross-contamination with penicillins and penicillin-containing substances cannot be reasonably prevented in a multi-product facility. Because penicillins may produce adverse reactions in some patients, Good Manufacturing Practices (GMP) regulations demand dedicated penicillin processing facilities that are segregated from non-penicillin products. Separate air handling systems are necessary if a building processes penicillins as well as non-penicillin products.

GMP regulations including the validation requirements [26], affect all aspects of downstream processing. Requirements depend on the kind of product (e.g., food, bulk pharmaceutical, final dosage form, etc.) and the jurisdiction. Willig and Stoker [27] should be consulted for specific guidance.

The final few downstream processing steps include formulation which is highly product specific. How a product is formulated may critically affect its stability, efficacy, and bioavailability. Formulation may involve addition of fillers (e.g., starch, cellulose, sugar, flour), diluants, preservatives, sunlight protectants (e.g., carbon black, dyes, titanium oxide), dispersal aids, emulsifiers, buffers, moisture retainers, adjuvants (e.g., mineral oils and aluminum hydroxide added to improve antigenicity of certain vaccines), flavors, colors, and fragrances. Additional finishing operations may include sterile filtration, vialing, granulation, agglomeration, size reduction, coating, encapsulation, tableting, labeling, and packaging.

1.3 Product Quality and Purity Specifications

The specifications on product purity and concentration should be carefully considered in developing a purification protocol. Concentration or purification to levels beyond those dictated by needs is wasteful. The acceptable level of contamination in a particular bioproduct depends on the dosage, the frequency of use, and the method of application (e.g., food, drug, oral, parenteral), as well as on the nature and toxicity (or perceived risk) associated with the contaminants [19]. Products such as vaccines, which are used only a few times in a lifetime, may be acceptable with relatively high levels of other than the desired biomolecule. In some cases, contaminating protein levels of about 100 ppm may be acceptable. *In vitro* diagnostic proteins (enzymes, monoclonal antibodies) may tolerate greater levels of contaminants so long as the contaminants do not interfere with the analytical performance of the product. With certain diagnostic proteins, such as the blood typing monoclonal antibodies, cross-contamination causing misdiagnosis is an extreme concern because of possibly fatal consequences of mis-typing. Such concerns influence the design and operation of the downstream process, particularly for multi-product plants.

Parenteral therapeutics usually must be purer than 99.99%. A variety of approaches are used to assure quality. Methods typically used with protein therapeutics are summarized in Table 1-2; Anicetti et al. [29] provide additional details. Requirements relating to some specific contaminants are discussed below.

Table 1-2. Methods for quality assurance of protein therapeutics [28].

Impurity or contaminant	Analytical technique
Protein contaminants (e.g., host cell proteins)	SDS–PAGE electrophoresis, HPLC, immuno-assays (ELISA, etc.)
Endotoxin	Rabbit pyrogen test, LAL[a]
DNA	DNA dot–blot hybridization
Proteolytic degradation products	IEF[b], SDS–PAGE, HPLC, N- and C-terminus analysis
Presence of mutants and other residues	Tryptic mapping, amino acid analysis
Deamidated forms	IEF
Microbial contamination	Sterility testing
Virus	Viral susceptibility tests
Mycoplasma	21 CFR[c] method
General safety	As per 21 CFR 610.11

[a] Limulus amoebocyte lysate; [b] Isoelectric focusing; [c] Code of Federal Regulations.

1.3.1 Endotoxins

Products derived from bacteria such as *Escherichia coli* will invariably be contaminated with bacterial cell wall endotoxins which can cause adverse reactions (headaches, vomiting, diarrhea, fevers, etc.) in patients unless reduced to very low levels (e.g., less than 5×10^{-13} kg per kg body weight). Endotoxins are extremely heat-stable lipopolysaccharides that are not easily removed from solutions of macromolecules. Ultrafiltration and reverse osmosis are effective for depyrogenation of water and small solutes. Other pyrogen removal methods are adsorption on activated carbon and barium sulfate, hydrophobic interaction chromatography, and affinity chromatography. Endotoxins bind to polymixin B affinity columns, but this method must be combined with detergent treatment for effectively removing protein-bound endotoxins [28]. Chromatography using LAL affinity matrix also removes endotoxins.

As a guiding principle, processing must aim to minimize endotoxin contamination by controls on process water and other additives. In addition, aseptic and bioburden controlled operation, and frequent cleaning of equipment help to reduce contamination. The equipment cleaning protocol must include procedures proven for depyrogenation. Standard alkali-based cleaning procedures [24] are quite effective in depyrogenation of stainless steel equipment, but other methods are necessary for cleaning chromatographic columns and membrane filters. The depyrogenation step employed during cleaning of membrane filters usually involves a 30-minute, 30–50 °C treatment with sodium hydroxide (0.1 M), hydrochloric acid (0.1 M), phosphoric acid, or hypochlorite (300 ppm free chlorine). Thorough rinsing with pyrogen-free water follows. Similar procedures are used for chromatographic columns.

An endotoxin-free product should be validated using the LAL test. This test is based on endotoxin-induced coagulation of amoebocyte lysate of horseshoe crab (*Limulus polyphemus*) at 37 °C, pH 7.0. Less than 0.3 ng mL^{-1} endotoxin levels are easily detected. Scrupulously clean glassware and water are necessary to prevent false positives. Some known interferences are EDTA, sodium dodecyl sulfate, urea, heparin, and benzyl penicillin.

1.3.2 Residual DNA

Residual DNA from producing cells can potentially contaminate the product. DNA fragments from established animal cells were once believed to be potentially oncogenic, which prompted the U.S. Food and Drug Administration to recommend a contamination level of no more than 10 pg DNA per dose [30]. Less restrictive limits are now accepted because no oncogenic events were observed following injections of large doses of DNA into animals. Nonetheless, DNA is a contaminant and demonstration of its satisfactory clearance is essential to quality assurance of the product [30]. Residual DNA is removed usually by adsorption on strong anion-exchange resins at pH ≥ 4. Hydrophobic interaction chromatography is also effective and so is affinity chromatography under conditions that bind the desired protein but not the DNA.

1.3.3 Microorganisms and Viruses

Parenteral products, other than certain vaccines, must be free of microorganisms and viruses [19]. Products derived from potentially contaminated sources such as human donors, animals, and some cell lines, can be especially problematic. For such products, the purification scheme must demonstrate viral inactivation or removal unless the product is terminally sterilized by validated means (see also Volume 1, Chapter 16). Usually, in-series processing with at least two steps, each capable of six log virus removal or deactivation, would be necessary. Viruses can be removed by ultrafiltration, or deactivated by methods such as heating, treatment with chemicals (e.g., β-propiolactone), solvents and detergents, and ultraviolet or gamma irradiation. In one study with plasma derived human serum albumin, heat treatment at 60 °C for 10 h in the final container produced more than five log reduction of vaccinia, polio-1, vesicular stomatitis, Sindbis and HIV-1 within 10 minutes [31]. In another case, freeze-dried coagulation factors were treated at 80 °C for 72 h in the final vial. For Factor VIII, inactivation of HIV-1 occurred within 24 h, without significant deterioration of the product [31]. For a Factor IX preparation, treatment with solvent/detergent combination of tri-(*n*-butyl) phosphate and Tween-80 for 5 h inactivated a range of typical enveloped viruses within an hour [31]. Up to six log reduction of some typical enveloped viruses such as herpes simplex-1 and Sindbis could be achieved in spiked samples using protein G column chromatography with acid elution; however, only three log reduction was observed for acid tolerant non-enveloped polio virus [31].

1.3.4 Other Contaminants

For many biological products, particularly pharmaceuticals, seemingly minor alterations in downstream processing can have important implications on the performance of the product. For example, penicillins may be recovered by liquid–liquid extraction of either the whole fermentation broth or solids-free broth. The latter scheme requires an additional solid–liquid separation step than the whole broth process. However, the whole-broth extracted product has been known to cause more frequent cases of allergenic reactions in comparison with the other processing alternative. In fact, some pharmaceutical companies now demand of contract suppliers that, in addition to meeting product specifications in terms of measurable contamination, the product they supply must conform to a certain production method, in this case extraction after removal of fungal solids. When raw penicillin is for bulk conversion to semi-synthetic penicillins, whole-broth extraction may be acceptable in view of the security afforded by the additional steps involved in making and purifying 6-aminopenicillanic acid from raw penicillin [19].

1.4 Impact of Fermentation on Recovery

Downstream processing should not be considered in isolation with the bioreaction step. Development of biocatalyst by natural selection, mutation, and recombinant DNA technology is a powerful means of influencing downstream processing [32]. Similarly, modification of fermentation feeding strategies, culture media and conditions profoundly affect the downstream process [32].

1.4.1 Characteristics of Broth and Microorganism

Composition of the fermentation medium affects downstream recovery. Relatively poorly defined complex media components are often acceptable for producing commodity chemicals and bulk antibiotics, but usually not for parentral proteins. Low-serum and protein-free media are commonly employed in animal cell culture to greatly simplify recovery of sparing amounts of proteins produced. Similarly, the type of antifoam and its concentration must accommodate the recovery constraints.

For some processes, alternative microorganisms may be a viable option. Preference should be given to faster growing, easy to process organisms. Selection of a producer must consider the overall productivity of the process, not just that of the fermentation step. Production of recombinant proteins in *Saccharomyces cerevisiae* may have important advantages relative to production in genetically modified bacteria such as *Escherichia coli* [33]. *S. cerevisiae* is generally recognized as safe for food and pharmaceutical use. In addition, unlike bacteria, the yeast does not produce endotoxins, and its broths are much easier to process than those of mycelial fungi and filamentous bacteria [33]. Unlike the DNA-laden homogenates of bacteria such as *E. coli*, yeast lysates are not excessively viscous. In yet other cases, it may be possible to naturally select autoflocculating strains, as has been done with certain brewing yeasts and bacteria. Cells may also be genetically modified into flocculating ones.

Genetic engineering of producing organisms and products provides new opportunities for influencing downstream bioseparations. For example, recombinant fusion proteins with added polypeptide 'affinity tags' have been produced to facilitate purification [34,35]. Affinity tags have been developed for ion exchange, hydrophobic interaction, affinity, immunoaffinity and immobilized metal ion chromatography. Specific cleavage sites between the tag and the protein allow removal of the tag after purification [34]. Some of the available affinity tags and the chromatographic methods applied with those tags are listed in Table 1-3. Reagents and enzymes that have been used to cleave the tags, and the specific cleavage sites, are noted in Table 1-4.

Another strategy for simplifying downstream recovery is genetic manipulation to enable extracellular secretion of the recombinant protein. Failing outright secretion, it may be possible to achieve secretion into the periplasm of microorganisms such as *E. coli*. Relatively mild disruption or extraction conditions can then be used for

Table 1-3. Affinity tags and corresponding chromatographic separations [34].

Affinity tag	Chromatography scheme
Polyarginine	Ion exchange
Polyphenylalanine	Hydrophobic interaction
β-Galactosidase	Affinity
Protein A	Affinity
Antigenic peptides	Immunoaffinity
Polyhistidine	Metal ion chelate

recovery in comparison with products produced in the cytoplasm. Periplasmic secretion has additional advantages: periplasm of *E. coli* contains only seven of the 25 cellular proteases [36], hence, the likelihood of proteolysis is reduced. Moreover, periplasm contains only 100–200 proteins [36], therefore, selective extraction of periplasm yields a less complex, easier to purify mixture. In addition, the oxidative environment of periplasm is more favorable to formation of disulfide bonds than the environment of cytoplasm. Disulfide linkages determine the correct folding of the polypeptide chain and, therefore, its biological activity. Chemicals such as chloroform, Triton X-100, and combinations of lysozyme and EDTA [36] facilitate release of periplasmic proteins. Extraction chemicals should be tested for possible effects on protein stability. In one study, Garrido et al. [33] observed loss of β-galactosidase activity even at 4 °C when the enzyme was extracted with a mixture of chloroform and sparing amounts of sodium dodecyl sulfate (SDS). In larger quantities, SDS is a well-known protein denaturant [37].

Table 1-4. Chemicals and enzymes for specific cleavage of fusion proteins [34].

Cleavage reagent	Cleavage site
Cyanogen bromide	Met ↓
Formic acid	Asp ↓ Pro
Hydroxylamine	Asn ↓ Gly
Collagenase	Pro-Val ↓ Gly-Pro
Factor Xa	Ile-Glu-Gly-Arg ↓
Enterokinase	Asp-Asp-Asp-Lys ↓
Rennin	His-Pro-Phe-His-Leu-Leu ↓
Carboxypeptidase A	C-terminal aromatic amino acids
Carboxypeptidase B	C-terminal basic amino acids

Secretion or extracellular leakage of an otherwise intracellular product is some-times achieved simply by modifying the fermentation conditions. For example, addi-tion of penicillin during growth in certain amino acid fermentations produces cells that leak the amino acid which is recovered by isoelectric precipitation from the extracellular fluid.

1.4.2 Product Concentration

Concentration of the product in the source material affects the cost of recovery. Con-centrations are usually quite low; some values typically seen in culture broths are noted in Table 1-5. In addition to the product, the broth contains many contaminants – proteins, lipids, surfactants, carbohydrates, nucleic acid polymers, salts, compo-nents of the culture medium, pigments, organic acids, alcohols, aldehydes, esters, amino acids, and other metabolic products – some of which may be quite similar to the desired product. Some of the contaminants may be toxic or otherwise hazar-dous (e.g., endotoxins, mycotoxins).

Downstream processing typically represents 60–80 % of the cost of production of fermentation products. Thus, superficially it may appear that process improvement should focus on downstream. This is not so. Even small improvements in the yield or purity of the product in the bioreaction step can have a significant effect on downstream recovery costs. As a rough guide, the selling price P (US\$ kg^{-1}) of a product (i.e., a reflection of cost of production) depends on its concentration C_i in the broth or the starting material. This dependence can be described by the equation

$$P = 528 \cdot C_i^{-1}, \tag{2}$$

which is based on data compiled by Dwyer [38]. The potential for yield improvement at the bioreaction stage is usually high. Major yield enhancements have been fairly commonly achieved by strain selection, medium development, optimization of feed-ing strategies, and environmental controls.

Table 1-5. Typical concentration of various products in raw fermentation broth.

Product	Final concentration (kg m^{-3})
Vitamin B$_{12}$	0.06
Monoclonal antibodies	0.1–0.5
Riboflavin	0.1–7
Antibiotics	0.2–35
Gibberelic acid	1–2
Amino acids	2–100
Yeast	30–60

1.4.3 Combined Fermentation–Recovery Schemes

In keeping with a global approach to process improvement or intensification, schemes that combine the bioreaction stage and parts of downstream processing are potentially attractive [32]. Such schemes include extractive fermentations, fermentationa–distillation, perfusion culture using membranes, inclined settlers or 'spinfilters' to retain the cells in the bioreactor, fermentation–adsorption using chromatographic media, as well as other methods. Combining fermentation and recovery not only reduces the number of individual processing steps, but the productivity of the fermentation may also be substantially enhanced by eliminating or reducing the inhibitory effects of certain products.

A novel scheme for retaining particles, particularly animal cells, in perfusion bioreactors relies on standing sound waves applied perpendicular to a vertically aligned harvest flow channel [39–41]. The sound waves concentrate the suspended cells in bands aligned with the flow [42,43]. Gravity sediments such aggregated particles against the flow once the sound is switched off; hence, a clarified liquor leaves the flow channel whereas the solids are concentrated in the feed vessel. This type of separation in ultrasonic flow fields provides an effective means of retaining cells in continuous flow bioreactors. This technique allows easy maintenance of sterility as no mechanical items penetrate the sedimentation chamber. Moreover, there is nothing to clog, foul, or breakdown. Process-scale implementation of this method is being developed.

1.5 Initial Separations and Concentration

The first few processing operations in a purification train are aimed at volume reduction to minimize processing costs by reducing the size of the downstream machinery. Removal of suspended material and substances which might interfere with further downstream operations are additional requirements of some of the early separation steps. Further, because viscous broths are difficult to handle, viscosity reduction should be achieved as early as possible to simplify pumping, mixing, filtration, sedimentation, etc. Removal of suspended solids, digestion of carbohydrates, or removal of nucleic acids are some of the operations that may be needed to improve broth handling.

Typically, solid–liquid separation would be among the first processing steps for extracellular as well as intracellular products. For the latter, solid–liquid separations are usually a means of concentration of the biomass, or removal of the suspending culture fluid prior to disruption or other downstream treatment. Cell or other solid product washing operations often employ solid–liquid separation steps. The commonly used methods of solid–liquid separation are filtration and centrifugation. Centrifuges are used also to separate difficult to break emulsions and other liquid–liquid systems. Some examples are recovery of cream from milk, recovery of oil drops, fats (e.g. in rendering and meat processing plants) and waxes, and liquid–liquid extraction.

Table 1-6. Types and applications of centrifuges [1,19].

Tubular bowl. Tubular bowl machines are capable of high g-forces, usually up to 20 000 g in industrial devices. Solids accumulate in the bowl and must be removed manually at the end of operation. Bowl capacity limits solids-holding capability. To ensure sufficient interval between bowl cleaning, the solids concentration in the feed should usually be $\leq$ 1 % volume/volume; higher concentrations can be processed with smaller batches, for example, in production of certain vaccines. Good dewatering of solids is obtained.

Multichamber bowl. Similar to tubular bowl machines. Division of bowl into multiple chambers increases solids-holding capacity. Solids must be discharged manually; hence, economic operation is feasible only with feeds with low concentration of solids. Good for polishing of otherwise clarified liquors. Capable of high g-forces. Gradation of g-forces from inner to outer chamber. Smallest particles sediment in the outermost chamber. Good dewatering of solids.

Disc-stack. Lower g-forces than tubular bowl machines. Solids may be retained, or discharged intermittently or continuously by various mechanisms (e.g., periodic ejection of solids by hydraulic separation of upper and lower parts of the bowl; nozzle discharge under pressure; valves; etc.). Not all discharge methods are suitable for all solids. Solids must flow. Poor dewatering. Not suited for mycelial solids; good for slurries of yeasts and certain bacteria. Depending on the mechanism of solids discharge, may handle feeds with up to 30 % (v/v) solids.

Scroll discharge. Scroll discharge decanter centrifuges are suitable for slurries with high concentration of relatively large, dense solids. Feed solids concentrations of 5–80 % (v/v) can be handled. Solids are discharged continuously. The g-forces are low. Suitable for fungal broths and dewatering of sewage sludge.

Perforated bowl or basket centrifuges. Also known as filtering centrifuges. Useful for low-g recovery of relatively large, mostly crystalline solids. The perforated bowl is lined with filter cloth to retain solids, whereas the liquid passes through. Sedimented cake may be washed and recovered as fairly dry material. Not effective for particles below 5 μm, and loadings < 5 % (v/v) [44].

Solid–liquid separations can be implemented in a variety of ways that are best suited to particular applications. Thus, as detailed in Table 1-6, many different designs of centrifuges are available [1,44]. Similarly, filtration may be performed in conventional filter presses, horizontal and vertical leaf-type pressure filters, rotary drum pressure or vacuum filters with or without filter aid (or body feed or admix) and using different means of solids discharge. Production scale rotary drum filters tend to be quite large: 0.9–4.3 m drum diameter and up to 6 m drum width. Sterile operation is usually not feasible, and containment is difficult. Alternatively, solids may be recovered by membrane filtration either in dead end (e.g., in many filter sterilizations) or cross-flow modes; the latter may be implemented in flat plate, hollow fiber or spiral wound static membrane cartridges, as well as in dynamic modes [1]. While the variety of available options helps to ensure that specific needs are met, careful consideration of the problem at hand is required for selection of the optimal processing method. Alternatives should be considered whenever possible. For example, rotary drum filters with string discharge usually perform well in separating mycelial solids from penicillin broths, but this discharge mechanism, without filter aids, causes problems with broths of *Streptomycetes* and other bacteria [19]. Precoat drum filtration may be used with bacterial broths when biomass is not the desired product. A knife blade (or doctor blade) discharge mechanism is used to continuously remove the deposited solids along with a thin layer of the precoat. Knife discharge without precoat or filter aids is suitable for recovering yeast from the filter cloth on drum filters; however, knife blades are not suited to cleanly cutting away a layer of deposited mycelial fungi because of the stringy nature of solids. Similarly, because of the concentration and the morphology of the solids, the disc stack centrifuge is not suitable for fungal fermentation broths, but properly selected scroll discharge machines are effective. Leaf filters are generally batch devices that are inexpensive to install, but labor-intensive to operate. Leaf filters are suitable for broths with little solids, e.g., in polishing of beer [19]. Gravity sedimentation may be employed as a volume reduction step prior to removal of solids by other means, but sedimentation by itself is not common for biomass removal in processing of high value products. Gravity sedimentation in thickeners and clarifiers [4,5] is encountered widely in sludge recovery in biological wastewater treatment. Certain solids may be recovered using hydrocyclones, but this method is little used in bioprocessing.

When more than one processing option its technically feasible, evaluations of the economics of use in terms of capital expenditure on equipment and its operating costs (processing time, yields, labor, cleaning, maintenance, analytical support) is necessary for optimal process selection. Economic evaluations should be performed over the expected lifetime of the equipment [19]. For example, for separation of solids from fermentation broth, centrifugation and microfiltration may be two competing alternatives [1]. In still other applications, for example when very fragile cells are to be separated from suspending liquid, centrifugation may not be an option.

Some other concentration steps, applicable to products in solution, are precipitation [1,45], adsorption, chromatography [14], evaporation, pervaporation [7] and ultrafiltration [1]. Some of these operations are equally capable purification steps (e.g., chromatographic separations). Certain steps (e.g., some chromatographic separations; membrane separations) may require a relatively clean process stream, free of debris, lipids or micelles which may cause fouling of the equipment. Such steps are often used downstream of steps which can handle cruder material [19].

Sometimes the characteristics of fermentation broth or process liquor may be modified by pretreatment to enable processing by a certain method. Major changes in processing characteristics may be achieved by pH and/or temperature treatment, use of additives such as polyelectrolytes, other flocculants and enzymes, and changes in ionic strength [19]. Flocculants (e.g., alum, calcium and iron salts, tannic acid, quaternary ammonium salts, polyacrylamide) can enhance sedimentation rates by thousands of fold relative to unflocculated suspension. Aging of protein precipitates and crystals can substantially improve filtration and sedimentation. Addition of salts is sometimes helpful in dewatering difficult to dewater solids such as protein precipitates. Water is drawn out of the pores of the solid into the salt containing liquid film on the outside. Osmosis or chemical potential difference drives the flow. Among other factors, time of harvest can beneficially alter processing behavior of the broth as well as the stability of the labile product. Culture conditions and methodology influence microbial morphology, product formation and downstream recovery. For example, cells grown in defined media are generally easier to disrupt than ones cultured in complex media [16]. Also, high specific growth rates produce less robust cells.

1.6 Intracellular Products

In general, a biological product is either secreted into the extracellular environment, or it is retained intracellularly. In comparison with the total amount of biochemicals produced by the cell, very little material is usually secreted to the outside; however, this selective secretion is itself a purification step which simplifies the task of the biochemical engineer. Extracellular products, being in a less complex mixture, are relatively easy to recover. On the other hand, because a greater quantity and variety of biochemicals are retained within cells, intracellular substances are bound to eventually become a major source of bioproducts [16]. Among some of the newer intracellular products are recombinant proteins produced as dense inclusion bodies in bacteria and yeasts. Recovery of intracellular products is more expensive as it requires such additional processing as cell disruption [16–18], lysis [16], permeabilization [46], or extraction. Intracellular polymers such as poly-β-hydroxybutyrate (PHB) may be recovered either by cell disruption [17,37] or solvent extraction. In principle, selective release of the desired intracellular products is possible, but in practice it is neither easily achieved nor sufficiently selective. Hence, the desired product must be purified from a relatively complex mixture, complicating processing and adding to the cost [19]. Nevertheless, an increasing number of intracellular products are in production. Economics of production may be improved by recovering several products (intracellular and extracellular) from the same fermentation batch [21].

As for other separations, many options exist for the disruption of cells (Table 1-7). Of these, high-pressure homogenization is apparently the most suitable for bacterial broths, whereas bead mills are more widely used for fungal cultures [1,16]. For dissolved products, cell disruption conditions (e.g., pressure, number of passes) must be selected to prevent excessive micronization of debris because micronization complicates solid–liquid separation further downstream [16]. However, when the product is an intracellular solid that is undamaged by homogenization, micronization of debris actually favors product recovery. This strategy is useful with protein inclusion bodies, certain cellular organelles, and sometimes with granules of bioplastics such as polyhydroxyalkanoates. Nonetheless, overzealous disruption conditions should be avoided in view of the recently published evidence that suggests loss of intracellular solids by micronization [37].

Disruption of bacterial cells releases large amounts of nucleic acids which increase the viscosity of the broth, often producing viscoelastic behavior. To ease further purification, the nucleic acids are usually removed by precipitation (e.g., with manganous sulfate, streptomycin or polyethyleneimine) [1]; alternatively, viscosity may be reduced by enzymatic digestion of nucleic acids or high-shear processing in high-pressure homogenizers [19]. Another alternative for eliminating nucleic acid polymers is heat shock treatment prior to disrupting the cells. Heat shock treatment

Table 1-7. Cell disruption options [16–18].

High-pressure homogenization. Frequently used for large-scale disruption of yeasts and non-filamentous bacteria. Generally not suitable for mycelial broths. Broth must be free of large suspended solids, tight cell clumps and flocs. Maximum acceptable particle size is about 20 µm, but a lower size is preferred. Slurry viscosity should not normally exceed 1 Pa s [1,16]. Optimal viscosity and solids concentration ranges are narrower than for bead mills.

Bead milling. Bead mills come in vertical and horizontal configurations with different mechanisms for retention of grinding media, and different types of agitators. Agitators that reduce back-mixing are preferred. Vertical mills are susceptible to fluidization and accompanying loss in performance. Typically three to six passes should achieve complete disruption. Useful for yeasts, mycelial fungi, algae; less efficient with bacteria. Grinding bead size affects disruption. Smaller the microbial cell, smaller the optimal bead size [16,33].

Autolysis. Under suitable conditions certain cultures would autolyse in the stationary phase upon completion of fermentation. Baker's yeast can autolyse.

Osmotic shock. Useful for animal cells and in specific cases for bacteria. Large dilutions may be necessary.

Thermolysis. Sufficiently heat-stable products may be released by heat shocking the cells. Microbial susceptibility to heat shock treatment varies widely. Monovalent metal ions such as Na^+ and K^+ may aid thermolysis. Suited to specific cases.

Enzymes and chemicals. Detergents, EDTA, solvents (e.g., toluene), antibiotics, and lytic enzymes may be used. Sometimes enzymes and chemical additives are used in combination with homogenization or bead milling to reduce the severity of mechanical treatment. Treatment with acids and alkalis may be useful in specific cases. Especially useful for extraction from periplasm.

Others. Ultrasonication, desiccation, freeze–thaw, extrusion of frozen paste. Applicable only to laboratory scale.

would typically require rapid heating to at least 64 °C and a holding time of 20–30 minutes. This treatment should digest almost all DNA/RNA. Shorter holding times may be satisfactory if complete degradation is not necessary for processability. Rapid temperature rise preferentially destroys proteases relative to RNA-hydrolyzing enzymes. Thermal treatment may be feasible for heat stable products [37] as well as for those produced as denatured inclusion bodies.

Processing considerations relevant to some specific bioseparations are discussed in the following section.

1.7 Some Specific Bioseparations

1.7.1 Precipitation

Proteins are easily concentrated by precipitation with organic solvents (e.g., ethanol, acetone), polymers (e.g., poly(ethylene glycol), poly(propylene glycol), dextran), and salts. Fractional precipitation allows for a degree of separation [1]. Fractionation with ammonium sulfate is commonly used. Organic solvents produce a denaturing environment making low temperature processing necessary [1]. Alcohol precipitation is frequently used in recovering biologically inactive dissolved polymers such as polysaccharides. Examples include precipitation of xanthan and gellan with isopropanol. Precipitation methods can handle large amounts of crude material, are easily scaled up, and can be implemented in continuous processing modes [1,47]. However, precipitation is generally not useful for recovery from very dilute animal cell culture fluids. Ammonium sulfate precipitation for recovery of recombinant β-galactosidase from *S. cerevisiae* has been detailed by Zhang et al. [47].

1.7.2 Foam Fractionation

Foam fractionation, microflotation or froth flotation is potentially useful for concentrating particles (cells, organelles, other small solids such as granules of PHB) and proteins into a foam phase for further recovery. The technique involves gentle bubbling of air (or other inert gas) at the base of a column of broth or solution. Hydrophobic solids and surface active molecules accumulate at the gas–liquid interface and rise with the bubbles. Collector surfactants and other promotors are often added to improve attachment. Additives such as frothing agents and stabilizers may be necessary. Enrichment in the foam depends on physical collection efficiency of bubbles (i.e., on bubble size, hydrodynamics, bubbling rate, concentration of particles) and adsorption chemistry. Empirical investigation is essential for selecting suitable additives, concentrations, hydrodynamic regimes, and for assessing performance, including recovery from the foam phase. Culture conditions may be used to influence adsorption behavior. Froth flotation is encountered only occasionally in bioprocessing. Potentially, fermenters used in batch cultivation could subsequently be employed for froth flotation. Airlift bioreactors with gas–liquid separators [48] and added means of skimming the gas-floated biomass are used in activated sludge treatment of wastewater. Part of the harvested sludge is returned to the reactor as inoculum.

1.7.3 Solvent Extraction

Rapid solvent extraction can be carried out in centrifugal extractors such as the Podbielniak and the Alfa Laval machines that are commonly used in antibiotics processing [1,13]. These devices were originally designed to handle solids-free liquids, but have been adapted to media containing limited amounts of small particles. Other more conventional extractors are banks of mixer–settlers, York–Scheibel column (suitable for solids-free liquids), and the reciprocating plate Karr column (suitable for whole broths). Supercritical extraction of solids and liquids with carbon dioxide or other solvents (e.g., pentane) may be useful for small organic solutes. In these cases a concentrated solute is obtained easily by boiling off the solvent. Recently, serum albumin has been extracted into aqueous reverse micelles formed in carbon dioxide using a perfluoropolyether surfactant [49]. This opens up new opportunities for purification of proteins and other large molecules.

1.7.4 Aqueous Liquid–Liquid Extraction and its Variants

Conventional liquid–liquid extraction based on partitioning between an aqueous phase and a water-immiscible organic solvent is not suitable for proteins and protein-based cellular organelles because of low protein stability in organic solvents. A suitable alternative is partitioning between two immiscible aqueous phases [1,8,9]. Such phases are obtained by adding two incompatible polymers – for example, poly(ethylene glycol) and dextran – to water, or by mixing a relatively hydrophobic polymer solution with salts. Examples of such systems are aqueous mixtures of PEG-PVA, PPG-dextran, PPG-potassium phosphate, PEG-ammonium sulfate, as well as others. Partitioning of solutes is brought about by differences in net charge and hydrophobicity. Higher-polarity molecules solubilize preferentially in the salt-rich phase, whereas the relatively hydrophobic molecules concentrate in the polymer-rich phase. Polymers with attached affinity ligands – hydrophobic and ionizable functional groups – can improve partitioning behavior. Partitioning is strongly affected by pH, composition and type of phases (e.g., molecular weight of polymer, ionic strength, salt, polymer). In addition, the volume ratio of the phase mixture to that of the protein solution should be such that neither phase approaches saturation with protein. Aqueous two-phase systems have been successfully employed for enrichment of proteins, cells, organelles, and small molecules. Proteins that extract into the polymer phase are back extracted into the salt phase for recovery. Phase separation can be slow because of high viscosity and small density differences. Gravity separation is generally satisfactory for PEG-salt systems, but centrifugal separation may be necessary for PEG-dextran. Aqueous two-phase extraction is commercially employed, but it is relatively uncommon.

Among relatively new developments in liquid–liquid extraction is reversed miceller extraction [12] also known as liquid membrane emulsion extraction. Reversed micelles are surfactant stabilized microdroplets of an aqueous phase suspended in a water-immiscible solvent. Contacting the reversed micelle-laden organic phase with an aqueous mixture of proteins or other solutes results in preferential transfer of one or more species from the aqueous phase to the organic phase, and from there to the aqueous core of the reversed micelles. The intervening organic phase constitutes a liquid 'membrane.' Extraction is influenced by pH and ionic strength of the bulk aqueous phase, and the nature of the reversed miceller core. Usually, a protein solubilizes in the reverse miceller phase at pH values below its isoelectric pH when the ionic strength is low. Once a component has been extracted, reversed micelles can be back-extracted with buffers to yield a solution rich in the desired substance. Back-extraction is favored by altering the pH and ionic strength. Factors such as hydrophobicity of the protein also contribute to partitioning behavior.

A variation of the liquid membrane emulsion extraction is the supported liquid membrane extraction [11; see also Volume 1, Chapter 11]. No stabilizing surfactant is necessary in this case; instead, the liquid membrane-forming organic phase is supported in the pores of a porous solid that separates the two aqueous phases. Additives may be employed to enhance mass transfer through the organic phase [11]. Reversed micelles and liquid membranes are not widely used at present.

1.7.5 Membrane Separations

Cross-flow membrane filtration flux typically ranges over 10–120 $L \cdot m^{-2} \cdot h^{-1}$; the exact value depends on the membrane pore size and the viscosity of the suspending fluid. Microfiltration of animal cells and microbial homogenates is done best at transmembrane pressures less than 1.38×10^4 Pa. Higher pressures, typically 6.9–34.5×10^4 Pa, are used in recovering microbial cells. Because of the small pore size, ultrafiltration membranes invariably require high transmembrane pressures (13.8–27.6×10^4 Pa) for reasonable flux.

Polymer membranes predominate in bioprocessing, but ceramic and sintered metal membranes are used occasionally. Hydrophilic membranes are preferred for liquids. Hydrophobic polymer membranes are easily fouled by silicone antifoams which may cause as much as 50 % decline in flux. Low-molecular weight poly(propylene glycol) or poly(ethylene glycol) based antifoams are usually better. Mechanical foam control [24,50] during fermentation is sometimes helpful in eliminating or reducing antifoam consumption.

Even without antifoams, membrane performance deteriorates over time, making periodic replacement necessary. Prior experience or experimentation are the only reliable predictors of membrane life [6]. Membranes are not easily cleaned; detectable residues of bioactive material may remain after any reasonable cleaning. Such situations require product-dedicated filters to prevent cross-contamination. Furthermore, polymer-based membrane filters cannot usually be heat sterilized; chemical sanitization and atmospheric steaming are the only options. Chemical cleaning, sanitization, and steaming lower membrane life; hence the choice of chemicals and cleaning conditions need to be carefully assessed.

The major costs associated with ultrafiltration and microfiltration are the initial capital expense and the cost of membrane replacement; energy is not a major expense. The frequency of membrane replacement determines feasibility of membrane separations. In contrast, in reverse osmosis where the high transmembrane pressure is unavoidable, pumping expense and membrane replacement costs are major contributors to operating costs. As with centrifuges, membrane filter selection requires experimental evaluations [1,6].

Even in cross-flow operation, membrane filters experience performance loss due to concentration polarization or accumulation of a solute layer at the surface of the membrane. Small amounts of relatively large, dense inert solids such as cellulose fibers or polymer beads added to the feed are known to reduce concentration polarization by disturbing the fluid boundary layer on the membrane surface. Cross-flow channels are sometimes also inserted with static turbulence enhancers such as wire screens, but such filter modules are not suitable for mycelial or filamentous biomass especially at high concentration of solids. Mechanical methods of increasing turbulence are employed in dynamic filters, but few such devices have gained any commercial acceptance. One dynamic configuration utilized two porous concentric cylinders with microfiltration membranes supported on the surfaces of the annulus. The inner cylinder rotated at high speed; differences in angular velocities of the fluid elements along the width of the annular gap produced Taylor vortices that substantially enhanced filtrate flux relative to static cross-flow operation [51]. Nonetheless, limited scale-up potential prevented further development. A variation on the concentric cylinder theme has recently been introduced by Pall Filters. This design consists of a stack of supported circular microfilter membranes with mechanically agitated circular steel discs mounted inbetween. Rotation of discs dramatically enhances filtrate flux [52]. The stack supports up to 1.5 m^2 membrane surface, but this may be substantially increased in future designs simply by increasing the overall height of the stack. The device is suited to recovering yeasts and non-filamentous bacteria from relatively less viscous broths.

Membrane filters are used also in the diafiltration mode for buffer exchange, washing of solids, desalting, and removing other small molecules from solution of macromolecules.

Pervaporation is another membrane separation that is particularly useful for low energy recovery of relatively volatile liquids (e.g., ethanol) from fermentation broths [7]. Permselective membranes separating the broth from a vapor phase allow only selective permeation of the desired solvent to the other side, where hot air or heat supplied to the membrane continuously evaporates the solvent, hence maintaining a mass transfer driving force. Membrane chemistry determines permselectivity.

1.7.6 Electrically Enhanced Bioseparations

Electric fields may be used to enhance bioseparations [53,54], but commercial use is limited at present because of the damaging effects of ohmic heating that accompanies current flow. Electrolysis can be another problem. Nevertheless, electrokinetic forces on charged particles have been demonstrated to reduce concentration polarization and membrane fouling during microfiltration and ultrafiltration, thereby enhancing filtration rates [53]. Up to sevenfold enhancement of transmembrane flow has been recorded during microfiltration with direct current (DC) electric field strengths of 100–120 V cm^{-1} [53]. Some of the problems associated with electric fields may be reduced by replacing the steady DC fields with pulsed direct current fields [53]. Electric discharges have been used also to break foams instantaneously during processing.

The separation potential of electric fields is best illustrated by electrophoresis, which is a well-established extremely high resolution method for separation of proteins. Differences in molecular charge and weight are the bases of separation. However, despite attempts to scale-up [55], electrophoresis remains confined mostly to laboratory use. Except for small volumes, rapid removal of heat generated has proven difficult without convective mixing that would destroy any separation.

1.7.7 Chromatographic Separations

Enhancing speed has been a major preoccupation with chromatographic processing. Except for bed height dependent gel permeation, the speed of most chromatographic processes can be enhanced by replacing the usual high-resistance packed vertical columns with radial flow devices [55]. Adsorption media used in conventional columns can still be utilized, but the medium is packed in the annulus between two porous concentric cylinders. Radial flow columns attain 10- to 50-fold greater flow rates than conventional columns [55]. Industrial-scale simulated moving bed chromatographic systems are now available [56].

Among other improvements, better, more rigid yet porous chromatographic media that are less susceptible to bed compression have been developed [14]. Other novel media have enabled extremely high speed or perfusion chromatography. Unlike conventional media, perfusion media contain throughpores for bulk flow of fluid through the particle. Diffusional pores as in conventional media are also present. Throughpores allow high flow rates – up to 100-fold greater than in diffusive media [55]. Resulting convection within the particles reduces diffusive transport limitations.

Another high-rate chromatographic system is expanded or fluidized bed chromatography. The medium bed is expanded or fluidized during loading by upflow of unclarified fermentation fluid or cell homogenate [55]. There is little pressure drop through the expanded bed. Plug flow of fluid is desired and easily attained. After adsorption, the microbial solids are washed away by upflow of water or buffer. The adsorbed product is recovered as in conventional chromatography by downward elution of settled, packed bed. Because this method handles unclarified fluids, some solid-liquid separation steps are eliminated. Fluid bed chromatography has been demonstrated with numerous fluids including broths *E. coli,* yeast, mammalian cells [55], autolysed yeast, and blood plasma.

A further rapid chromatographic method that may potentially handle solids-laden fluids is membrane chromatography. This technique employs ion exchange groups or other high-specificity adsorption ligands attached to inner surfaces of pores of conventional microfiltration membranes. Rapid flow through pores reduces diffusion limitations, hence speeding adsorption, and, later, desorption. Hollow fiber membrane modules that allow compact packing of large membrane areas have been used for membrane chromatography [57].

Some especially high-resolution chromatographic separations include HPLC and bioaffinity-based methods. Process-scale HPLC continues to be useful for small batches [38], but this method is expensive, slow, and the high-pressure columns appear to have reached an upper limit of about 0.3 m diameter and 2.4–3.0 m height. Bioaffinity chromatography with affinity ligands – receptors, antibodies, enzymes, and other active proteins – immobilized onto the support media has been used for quite some time, but it remains expensive. Other problems are often poor stability of the affinity matrix, and ligand leakage into the product (see also Volume 1, Chapter 17). With few exceptions (e.g., protein A affinity columns can be cleaned with the strong denaturant guanidine hydrochloride (6 M) which solubilizes adsorbed proteins without affecting the ligand), ligand stability limits the column cleaning regimen. Because of those factors, a trend toward replacing labile bioaffinity ligands with inexpensive and robust alternatives (e.g., dyes, metal ions) is apparent.

Note that some of the speed-enhancing techniques used with chromatography are equally applicable to non-chromatographic adsorptions. Adsorption using columns or slurries of activated carbon is commonly encountered in bioprocessing, particularly for removing pigments.

1.8 Recombinant and other Proteins

Many of the newer recombinant biotechnology products are proteins [30,58]. While the general features of a bioseparation scheme for these products are the same as for other proteins, there are some unique constraints. Genetically modified microorganisms and cells of higher life forms are often more fragile than the corresponding wild strains [59,60]. This has implications for the design of cell–liquid separation stages. Also, recombinant proteins formed in bacteria and yeasts frequently precipitate inside the cell as dense, insoluble, denatured inclusion bodies. In this form proteins which may otherwise be toxic to the cell may be overproduced and remain protected against proteolytic activity within the cell.

Most bacteria and fungi used in producing recombinant proteins also produce a variety of proteases that may degrade some of the desired protein within the cell and during recovery, soluble, non-inclusion body proteins being particularly susceptible to degradation. Degradation by acid proteases with a pH optimum of 2–4 may be minimized by processing at higher pH and low temperatures. Neutral proteases are not particularly thermostable and may be inactivated by heating to 60–70 °C for 10 minutes [19]. Many proteases are metalloproteins and require a divalent metal ion for proteolytic activity; chelating agents such as ethylenediaminetetraacetic acid (EDTA) or citric acid may be used to inactivate such proteases by binding the metal ions. Alkaline proteases of *Bacillus* sp., such as subtilisin, contain serine at the active site and are not affected by EDTA, but are inhibited by diisopropylfluorophosphate. The short-lived reagent phenylmethylsulfonyl fluoride protects against serine proteases. Antioxidants such as vitamin E and ascorbic acid protect against oxidation [19].

Proteins tend to be more stable in concentrated solutions. Addition of poly(ethylene glycols) and other proteins such as albumins may have a stabilizing effect. Glycerol, sucrose, glucose, lactose, and sorbitol are often used as stabilizers in concentrations of 1–30 %. Enzyme substrates usually have an stabilizing effect, as do high concentration of salts such as ammonium sulfate and potassium phosphate. Metalloproteins may be stabilized by addition of metal salts. Divalent metal ions such as Ca^{2+}, Cd^{2+}, Mn^{2+} and Zn^{2+} stabilize various enzymes [19].

Some commonly used sequences of protein purification methods have been outlined by Bonnerjea et al. [45] and by Wheelwright [3]. Chromatographic procedures are indispensable to producing high-purity proteins. Typically, the mean recovery or yield of separation steps such as those listed in Table 1-1 is ~ 60–80 % [45]. Average and high values of purification factors associated with some protein purification operations are shown in Table 1-8 which is based on data compiled by Bonnerjea et al. [45]. Clearly, affinity chromatography far outperforms other methods, but compared with operations such as ion-exchange chromatography, the scope for further improving performance is small because many affinity separations already operate close to theoretical maximum [45].

Changes in processing volume, product yield, and total and specific activities occur during processing as illustrated in Table 1-9 for a relatively simple purification of brain tumor plasminogen activator (PA) from supernatants of cultured, anchorage dependent rat cells [61]. The purification in Table 1-9 was done at 4 °C. The serum-free conditioned medium used for recovery had an initial plasminogen activator activity of only 9 IU mL^{-1} [61]. Zinc chelate-agarose chromatography was used as the first concentration/purification step. The culture fluid (6 L) was applied to the column (5 × 8 cm) at a flow rate of 200 mL h^{-1}. The column was washed with Tris–HCl buffer (0.02 M, pH 7.5, 1 L) that contained 1 M sodium chloride, aprotinin and Tween-80 (0.01 %, vol/vol). Aprotinin, a protease inhibitor, and Tween-80 (poly(oxyethylene sorbitane monooleate)), a surfactant, are generally added at all stages of PA processing to, respectively, suppress proteolysis and overcome the surface adherent tendency of plasminogen activators [30]. After the wash, the column was eluted with a linear gradient of imidazole (0–0.05 M) in the wash buffer (1 L, 120 mL h^{-1}). Pooled PA fractions were further purified on a concanavalin A-agarose affinity chromatography column. Dialysis was used to concentrate the pooled fractions, and a final gel filtration step (Sephadex G-150 superfine) was employed. The overall yield was 39 % [61]. This figure is fairly typical of large-scale protein

Table 1-8. Approximate values of purification factors observed during protein purifications. Based on Bonnerjea et al. [45].

Operation	Purification factor	
	Average	High
Affinity chromatography	100	3000
Dye–ligand affinity	17	–
Inorganic adsorption	12	100
Size-exclusion chromatography	6	100
Hydrophobic interaction chromatography	15	60
Ion-exchange chromatography	8	50
Detergent extraction	4	12
Precipitation	3	12

Table 1-9. Purification of tumor plasminogen activator [61].

	Volume (mL)	Total protein (mg)	Total activity (IU)	Volumetric activity (IU mL^{-1})	Specific activity (IU mg^{-1})	Yield (%)	Purification factor
Clarified medium	6000	270	53 000	8.8	196	100	1
Zinc chelate-agarose	100	138	50 000	500	362	94	1.9
Concanavalin A-agarose	52	2.4	19 400	373	22 750	37	116
Gel filtration	7.5	0.53	20 800	2773	39 000	39	199

recovery. For example, overall recoveries of 23–47 % were noted for a variety of processes (e.g., recombinant BST, recombinant human α-interferon, L-leucine dehydrogenase for use in chiral syntheses) reviewed by Wheelwright [3]. One exception was a somewhat impractical process for tissue-type plasminogen activator (tPA) for which the overall yield was only 6 % [3]. Other methods for large-scale tPA recovery have been presented by Rouf et al. [30].

1.8.1 Inclusion Body Proteins

When possible, production of recombinant proteins as inclusion bodies has important advantages. Some proteins that form inclusion bodies are listed in Table 1-10. Inclusion bodies are easy to isolate, highly concentrated forms of the desired recombinant protein. Typically, inclusion bodies are spheroidal particles, 0.2–2.0×10^{-6} m in diameter and 1100–1300 kg m^{-3} density. The sequence of steps in recovery of inclusion body proteins is cell disruption, centrifugal separation of the inclusion body, washing, solubilization of the protein, and renaturation [19]. Cell disruption by homogenization is the preferred technique in large-scale processing. Disruption by high-pressure homogenization has been detailed by Chisti and Moo-Young [16]. Inclusion bodies are not affected by homogenization. Cell homogenates are centrifuged to sediment the dense inclusion body fraction. Centrifugation at 1000–12 000 g for 3–5 minutes is sufficient. Sedimentation of cell debris can be minimized by

Table 1-10. Some proteins produced as inclusion bodies [19].

Bovine pancreatic ribonuclease	Human interleukin-2	Lysozyme
Bovine somatotropin (BST)	Human interleukin-4	Porcine phospholipase
Epidermal growth factor	Human macrophage-colony stimulating factor	Prochymosin
Human insulin		Pro-urokinase
Human γ-interferon	Human serum albumin	Tissue-type plasminogen activator
	Immunoglobulins	

increasing the density and viscosity of the homogenate with additives such as 30 % sucrose or 50 % glycerol. The inclusion body fraction is washed with buffers containing 1 M sucrose, 1–5 % Triton X-100 surfactant [62] and, in some cases, low concentrations of proteolytic enzymes and denaturants. The wash steps remove soluble contaminants, membrane proteins, lipids and nucleic acids. At this stage the remaining solids fraction is > 90 % recombinant protein. The protein solids are solubilized in highly denaturing chaotropic media. Typically, 6–8 M guanidine hydrochloride or 8 M urea are used for solubilization at pH 8–9, 25–37 °C for 1–2 h [62]. Reducing agents are added to the solubilization media to break any inter- and intra-molecular disulfide bonds to fully solubilize the protein. Some reducing agents are 2-mercaptoethanol, dithiothreitol, dithioerythritol, glutathione, and 3-mercaptopropionate. Some typical concentrations are 0.1 M 2-mercaptoethanol, or 10 mM dithiothreitol [62]. The latter has a shorter half-life than 2-mercaptoethanol, but does not have the odor of 2-mercaptoethanol. Stability of thiol compounds in solution is dependent on pH, temperature, and the presence of metal ions such as Cu^{2+}, which lower stability, and of stability enhancers such as EDTA. Good yields of some proteins can be obtained by solubilization without the reducing reagents, but for others reducing agents are essential. Of the denaturants, guanidine hydrochloride is preferable to urea, which may contain cyanate causing carbamylation of the free amino groups on the protein, particularly during long incubation periods in alkaline environments. Note though, that for some proteins, one denaturant may produce significantly higher overall yield than if solubilization with the other is used [19]. Performance has to be empirically evaluated.

For refolding of solubilized protein into active entities, concentration of the denaturant and the reducing agent are reduced by dilution with a refolding buffer. Denaturants can be completely removed by ultrafiltration with addition of renaturing buffer, dialysis, or gel filtration. Renaturation from concentrated protein solutions produces lower yields of the active protein because of intermolecular aggregation in these solutions. Thus, renaturation is done at low protein concentrations, typically $1–20 \times 10^{-3}$ kg m^{-3} protein [62]. Yield of the active protein is enhanced by refolding in the presence of small, non-denaturing amounts (1–2 M) of urea or guanidine hydrochloride [62]. Presence of high-molecular weight polymers such as poly(ethylene glycol) may also improve yield [19].

During refolding, formation of the disulfide bonds is achieved in one of three ways. The air oxidation method uses dissolved oxygen for oxidation of the cystine residues. The refolding buffer containing solubilized protein is aerated or exposed to atmosphere. Oxidation is accelerated by Cu^{2+} ions at approximately 10^{-6} M. Typical reaction conditions are pH 8–9, 4–37 °C for up to 24 h [62]. Traces of 2-mercaptoethanol may enhance yield. Air oxidation is difficult to control [19].

The glutathione reoxidation method typically uses a 10:1 mixture of reduced and oxidized forms of glutathione at a concentration of 10^{-3} M reduced glutathione [62]. Air oxidation is suppressed by using deaerated buffers held under a nitrogen atmosphere. The ratio of the reduced and oxidized forms of glutathione, the ratio of the glutathione and the cystine residues on the protein, the reoxidation temperature (4–37 °C) and time (1–150 h) provide flexibility to this method [62]. Low-molecular weight thiols other than glutathione may also be used. The third method of disulfide bond formation, the mixed disulfide interchange technique, has been detailed by Fischer [62].

The inclusion body production stage should be optimized to rapidly form relatively pure, large and dense inclusion bodies which are easy to recover and solubilize. Production of proteolytic activity should be suppressed as far as possible. Purification and concentration are greatly simplified because of the already high starting protein concentration and purity in the inclusion bodies which are easy to separate from the bulk of the soluble proteins by centrifugation. The recovery of active protein from inclusion bodies is variable, but can approach 100 %. In general smaller polypeptides are easier to refold into active forms [19]. Because of added processing, and the need to refold in dilute solutions, inclusion body-produced proteins tend to be expensive. With certain proteins such as tPA production as an inclusion body in bacteria is technically feasible but is not competitive with animal cell culture-derived product [30], even though the latter is a fairly expensive production method.

1.9 Conclusions

The variety of bioseparations is vast, but usually a small selection of the available methods is sufficient to achieve the requisite purity. The aim always is to employ the fewest possible process steps consistent with the product quality specifications. In-depth knowledge of individual separations must be combined with insight into the bioreaction step to design an efficient, consistent and integrated overall production process. Whereas the scientific understanding of bioseparations continues to improve and several new capable separations have been introduced, downstream processing of biologicals remains an empirical art. Invariably, experimentation must be relied upon to aid process selection, implementation, and scale-up.

Abbreviations and Symbols

BST	bovine somatotropin
C	number of components
C_i	concentration of product in broth or starting material, kg m^{-3}
DC	direct current
DNA	deoxyribonucleic acid
EDTA	ethylenediaminetetra-acetic acid
g	gravitational acceleration, m s^{-2}
GMP	Good Manufacturing Practice
HIV	human immunodeficiency virus
LAL	Limulus amoebocyte lysate
n	number of steps
N_f	number of possible flowsheets
P	selling price, U.S. \$$_{1984}$ kg^{-1}

PA	plasminogen activator
PEG	poly(ethylene glycol)
PHB	poly-β-hydroxybutyrate
PPG	poly(propylene glycol)
PVA	poly(vinyl alcohol)
RNA	ribonucleic acids
S	number of separation operations
SDS	sodium dodecyl sulfate
x	step yield

References

[1] Chisti, Y., Moo-Young, M., in: *Biotechnology: The Science and the Business:* Moses, V., Cape, R.E., (Eds.), New York: Harwood Academic Publishers, 1991, pp. 167–209. Fermentation technology, bioprocessing, scale-up and manufacture

[2] Belter, P.A., Cussler, E.L., Hu, W.-S., *Bioseparations: Downstream Processing for Biotechnology.* New York: John Wiley, 1988.

[3] Wheelwright, S.M., *Protein Purification: Design and Scale up of Downstream Processing.* New York: Hanser Publishers, 1991.

[4] Christian, J.B., *Chem Eng Prog,* 1994, *90*(7), 50–56. Improve clarifier and thickener design and operation.

[5] Tiller, F.M., Tarng, D., *Chem Eng Prog,* 1995, *91* (3), 75–80. Try deep thickeners and clarifiers.

[6] Gyure, D.C., *Chem Eng Prog,* 1992, *88*(11), 60–66. Set realistic goals for cross-flow filtration.

[7] Fleming, H.L., *Chem Eng Prog,* 1992, *88*(7), 46–52. Consider membrane pervaporation.

[8] Abbott, N.L., Hatton, T.A., *Chem Eng Prog,* 1988, *84*(8), 31–41. Liquid–liquid extraction for protein separations.

[9] Raghavarao, K.S.M.S., Rastogi, N.K., Gowthaman, M.K., Karanth, N.G., *Adv Appl Microbiol,* 1995, *41,* 97–171. Aqueous two-phase extraction for downstream processing of enzymes/proteins.

[10] Chisti, Y., in: *Encyclopedia of Bioprocess Technology;* Flickinger, M.C., Drew, S.W., (Eds.), New York: John Wiley, 1999; in press. Solid substrate fermentations, enzyme production, food enrichment.

[11] Patnaik, P.R., *Biotechnol Adv,* 1995, *13,* 175–208. Liquid emulsion membranes: principles, problems and applications in fermentation processes.

[12] Pyle, D.L., *J Chem Technol Biotechnol,* 1994, *59,* 107–108. Protein separation using reverse micelles.

[13] Schügerl, K., *Solvent Extraction in Biotechnology,* New York: Springer-Verlag, 1994.

[14] Chisti, Y., Moo-Young, M., *Biotechnol Adv,* 1990, *8,* 699–708. Large scale protein separations: engineering aspects of chromatography.

[15] Snowman, J.W., *Adv Biotechnol Process,* 1988, *8,* 315–351. Lyophilization techniques, equipment, and practice.

[16] Chisti, Y., Moo-Young, M., *Enzyme Microb Technol,* 1986, *8,* 194–204. Disruption of microbial cells for intracellular products.

[17] Harrison, S.T.L., *Biotechnol Adv,* 1991, *9,* 217–240. Bacterial cell disruption: A key unit operation in the recovery of intracellular products.

[18] Middelberg, A.P.J., *Biotechnol Adv,* 1995, *13,* 491–551. Process-scale disruption of microorganisms.

[19] Chisti, Y., Moo-Young, M., *I Chem E Symp Ser,* 1994, *137,* 135–146. Separation techniques in industrial bioprocessing.

[20] Fish, N.M., Lilly, M.D., *Biotechnology,* 1984, *2,* 623–627. The interactions between fermentation and protein recovery.

[21] Dunnill, P., *Process Brochem,* 1983, *18*(5), 9–13. Trends in downstream processing of proteins and enzymes.

[22] Chisti, Y., *Chem Eng Prog,* 1992, *88*(1), 55–58. Build better industrial bioreactors.

[23] Chisti, Y., *Chem Eng Prog,* 1992, *88*(9), 80–85 Assure bioreactor sterility.

[24] Chisti, Y., Moo-Young, M., *J Ind Microbiol,* 1994, *13,* 201–207. Clean-in-place systems for industrial bioreactors: design, validation and operation.

[25] Flickinger, M.C., Sansone, E.B., *Biotechnol Bioeng,* 1984, *26,* 860–870. Pilot- and production-scale containment of cytotoxic and oncogenic fermentation processes.

[26] Lubiniecki, A.S., Wiebe, M.E., Builder, S.E., in: *Large-Scale Mammalian Cell Culture Technology;* Lubiniecki, A.S., (Ed.) New York: Marcel Dekker, 1990; pp. 515–541 . Process validation for cell culture-derived pharmaceutical proteins.

[27] Willig, S.H., Stoker, J.R., *Good Manufacturing Practices for Pharmaceuticals: A Plan for Total Quality Control,* 3rd edition, New York: Marcel Dekker, 1992.

[28] Garg, V.K., Costello, M.A.C., Czuba, B.A., in: *Purification and Analysis of Recombinant Proteins;* Seetharam, S., Sharma, S.K., (Eds.), New York: Marcel Dekker, 1991; pp. 29–54. Purification and production of therapeutic grade proteins.

[29] Anicetti, V.R., Keyt, B.A., Hancock, W.S., *Trends Biotechnol,* 1989, *7*(12), 342–349. Purity analysis of protein pharmaceuticals produced by recombinant DNA technology.

[30] Rouf, S.A., Moo-Young, M., Chisti, Y., *Biotechnol Adv,* 1996, *14,* 239–266. Tissue-type plasminogen activator: characteristics, applications and production technology.

[31] Roberts, P., *J Chem Technol Biotechnol,* 1994, *59,* 110–111. Virus safety in bioproducts.

[32] Chisti, Y., Moo-Young, M., *Trans I Chem E,* 1996, *74A,* 575–583. Bioprocess intensification through bioreactor engineering.

[33] Garrido, F., Banerjee, U.C., Chisti, Y., Moo-Young, M., *Bioseparation,* 1994, *4,* 319–328. Disruption of a recombinant yeast for the release of β-galactosidase.

[34] Hochuli, E., *Pure Appl Chem,* 1992, *64,* 169–184. Purification techniques for biological products.

[35] Beitle, R.R., Ataai, M.M., *Biotechnol Progress,* 1993, *9,* 64–69. One-step purification of a model periplasmic protein from inclusion bodies by its fusion to an effective metal-binding peptide.

[36] French, C., Ward, J.M., *J Chem Technol Biotechnol,* 1992, *54,* 301. Production and release of recombinant periplasmic enzymes from *Escherichia coli* fermentations.

[37] Tamer, I.M., Moo-Young, M., Chisti, Y., *Ind Eng Chem Res,* 1998, *37,* 1807–1814. Disruption of *Alcaligenes latus* for recovery of poly-β(hydroxybutyric acid): comparison of high-pressure homogenization, bead milling, and chemically induced lysis.

[38] Dwyer, J.L., *Biotechnology,* 1984, *2,* 957–964. Scaling up bioproduct separation with high performance liquid chromatography.

[39] Baker, N.V., *Nature,* 1972, *239,* 398–399. Segregation and sedimentation of red blood cells in ultrasonic standing waves.

[40] Kilburn, D.G., Clarke, D.J., Coakley, W.T., Bardsley, D.W., *Biotechnol Bioeng,* 1989, *34,* 559–562. Enhanced sedimentation of mammalian cells following acoustic aggregation.

[41] Whitworth, G., Grundy, M.A., Coakley, W.T., *Ultrasonics,* 1991, *29,* 439–444. Transport and harvesting of suspended particles using modulated ultrasound.

[42] Mandralis, Z.I., Feke, D.L., *AIChEJ,* 1993, *39,* 197–206. Fractionation of suspensions using synchronized ultrasonic and flow fields.

[43] Weiser, M.A.H., Apfel, R.E., *Acustica,* 1984, *56,* 114–119. Interparticle forces on red cells in a standing wave field.

[44] De Loggio, T., Letki, A., *Chemical Engineering,* 1994, *101*(1), 70–76. New directions in centrifuging.

[45] Bonnerjea, J., Oh, S., Hoare, M., Dunnill, P., *Biotechnology,* 1986, *4*, 954–958. Protein purification: the right step at the right time.

[46] Dörnenburg, H., Knorr, D., *Process Biochem,* 1992, *27*, 161–166. Release of intracellularly
stored anthraquinones by enzymatic permeabilization of viable plant cells.

[47] Zhang, Z., Chisti, Y., Moo-Young, M., *Bioseparation,* 1995, *5*, 329–337. Isolation of a
recombinant intracellular β-galactosidase by ammonium sulfate fractionation of cell homogenates.

[48] Chisti, Y., Moo-Young, M., *Chem Eng Prog,* 1993, *89*(6), 38–45. Improve the performance
of airlift reactors.

[49] Brennecke, J.F., *Chem Ind (Lond.),* 1995, *21*, 831–834. New applications of supercritical
fluids.

[50] Chisti, Y., *Bioproc Eng,* 1993, *9*, 191–196. Animal cell culture in stirred bioreactors: observations on scale-up.

[51] Kroner, K.H., Nissinen, V., Ziegler, H., *Biotechnology,* 1987, *5*, 921–926. Improved dynamic
filtration of microbial suspensions.

[52] Lee, S.S., Burt, A., Russotti, G., Buckland, B., *Biotechnol Bioeng,* 1995, *48*, 386–400.
Microfiltration of recombinant yeast cells using a rotating disk dynamic filtration system.

[53] Brors, A., Kroner, K.H., in: *Harnessing Biotechnology for the 21ˢᵗ Century:* Ladisch, M.R.,
Bose, A., (Eds.), Washington, DC: American Chemical Society, 1992; pp. 254–257. Electri-
cally enhanced cross-flow filtration of biosuspensions.

[54] Rudge, S.R., Todd, P., *ACS Symp Ser,* 1990, *427*, 244–270. Applied electric fields for downstream processing.

[55] Shanley, A., Parkinson, G., Fouhy, K., *Chemical Engineering,* 1993, *100*(1), 28–33. Biotech
in the scaleup era.

[56] Kim, I., *Chemical Engineering,* 1997, *104*(1), 28–33. Biotech's new mandate: more,
cheaper, and faster.

[57] Brandt, S., Goffe, R.A., Kessler, S.B., O'Connor, J.L., Zale, S.E., *Biotechnology,* 1988, *6*,
779–782. Membrane-based affinity technology for commercial scale purifications.

[58] Zhang, Z., Moo-Young, M., Chisti, Y., *Biotechnol Adv,* 1996, *14*, 401–435. Plasmid stability
in recombinant *Saccharomyces cerevisiae.*

[59] Dunnill, P., *Chem Eng Res Des,* 1987, *65*, 211–217. Biochemical engineering and biotechnology.

[60] Moo-Young, M., Chisti, Y., *Biotechnology,* 1988, *6*, 1291–1296. Considerations for designing bioreactors for shear-sensitive culture.

[61] Bykowska, K., Rijken, D.C., Collen, D., *Thromb Haemost,* 1981, *46*, 642–644. Purification
and characterization of the plasminogen activator secreted by a rat brain tumor cell line in
culture.

[62] Fischer, B.E., *Biotechnol Adv,* 1994, *12*, 89–101. Renaturation of recombinant proteins produced as inclusion bodies.

2 Protein Stability in Downstream Processing

Kim Hejnaes, Finn Matthiesen and Lars Skriver

2.1 Introduction

To the protein chemist two issues are of major concern in downstream processing, protein purity and protein stability. This chapter deals with the latter subject, trying to answer some of the commonly asked questions: Is there any measurable property that would predict the stability of a protein during downstream processing? Which factors influence protein stability? Can extreme conditions be accepted during purification? Can proteins be stabilized by choosing optimal conditions during downstream processing to ensure a correct tertiary structure?

With the introduction of recombinant technology and thereby expression of proteins in micro-organisms, the issue of protein stability in downstream processing has gained much attention. First of all, the protein of interest is often expressed in a foreign host organism to levels far exceeding the *in vivo* concentrations, resulting in chemical and physical instability illustrated by the formation of inclusion bodies in *Escherichia coli*. Secondly, a great number of proteins have found their way to industrial applications such as in biopharmaceuticals with strict demands for purity, stability, and process reproducibility. Thirdly, when purified, many of these proteins have been shown to unfold rapidly under formation of aggregates.

Several parameters including protein concentration, pH, temperature, redox potential, and co-solvents influence the stability of proteins. These factors must be taken seriously into consideration in the design of the downstream process, in which the proteins are subject to environments far from what can be found *in vivo*.

This chapter focuses on the physical and chemical factors affecting the stability of small globular proteins in solution. It is our aim to describe the effect of the above-mentioned parameters on protein stability, and to suggest improved strategies to ensure a correct specific biological activity of the final bulk material.

2.2 Protein Stability

The term protein stability often refers to the preservation of the unique three-dimensional structure of a given protein. The globular proteins in aqueous solutions are not very stable. The difference in Gibbs' free energy between the native and unfolded state is in the range of 20–60 kJ mol^{-1} under physiological conditions [1,2], and the stability of even very different proteins does not differ greatly [2]. The native three-dimensional structure of a globular protein is maintained by the sum of interacting forces, which observed isolated contribute very little to protein stability.

2.2.1 The Native State

One of the major contributions to protein stability in aqueous solutions arises from burial of non-polar amino acid residues into the hydrophobic core of the globule in order to avoid the solvation entropy decrease of the hydrophobic residue [3]. The effect contributes 3.4–7.6 kJ mol^{-1} to the conformational stability for each -CH$_2$- group buried into the interior of the molecule [4].

The packing of amino acid residues into the hydrophobic interior results in both formation and optimization of a large number of interactions including ion pairs, hydrogen bonds, weakly polar interactions, and short-range repulsive electron cloud overlap [5]. Potentially stabilizing features include the tight packing of amino acid residues in the protein interior, the α-helix dipole, helix caps, and weakly polar interactions between aromatic groups [6].

The van der Waals forces are weak interactions having a bond energy of 4 kJ mol^{-1} per atom pair. The contact distance between two atoms are from 0.3–0.4 nm, and the attractive force decreases with distance. Though weak, the van der Waals force is of significance for protein stabilization given the large number of interactions in the folded state. Electrostatic forces arise from the interaction between positively charged amino or guanidinium groups and negatively charged carboxyl groups. The effect is strongly pH-dependent. At low or high pH the protein becomes unstable, as the overall charge increases, leading to increased charge repulsion. Interactions among solvent-exposed amino acids on the protein surface are usually weak and contribute little to protein stability [7,8]. In contrast, interactions with α-helix dipoles are consistently seen to contribute to protein stability [9]. Intra-molecular hydrogen bonds are essential to the structure and stability of globular proteins. A hydrogen bond is an attractive interaction between two or more neighboring atoms mediated by a hydrogen atom. The hydrogen atom is shared between two generally electronegative atoms, the hydrogen donor (-NH- and -OH) and the hydrogen acceptor (nitrogen and oxygen atoms). The >C=O·····HN< hydrogen bond is the most prevalent with >C=O·····side-chain, >HN·····side-chain, and side-chain·····side-chain hydrogen bonds accounting for the remainder [10].

Despite the fact that donor and acceptor atoms can form hydrogen bonds with water in the unfolded protein, a net contribution to stability of 2.9–8.0 kJ mol^{-1} is observed in the folded state [11]. Hydrogen bonds are believed to play a major role in protein stability due to the dipolar nature of amino acids, and to the fact that the helix–coil transition is largely driven by hydrogen bonding [12].

Other energetically favorable interactions involving phenylalanine, tyrosine, and tryptophan residues are suggested to comprise an important class (aromatic interactions) of stabilization with energies between 2.5 and 5.5 kJ mol^{-1} [13].

Thus, it is the effective co-operation of different stabilizing forces that make the total contribution to protein stability much greater than the sum of the individual interactions [14–16].

In its native state, a globular protein comprises a tight packed hydrophobic core of density comparable with that of amino acid crystals [17] or small organic molecules [18]. An extensive formation of hydrogen bonds is observed in the interior [19], while a large majority of charged side-chains are located at the surface in contact with water [20]. The solvent-exposed residues, whether polar or non-polar, contribute little to the stability of the native state [21]. Within the interior, charged or polar groups not paired in hydrogen bonds are rarely observed [20,22]. Studies of point-mutated proteins reveal that the most deleterious effects on protein stability usually occur when buried or rigid sites are modified, suggesting that these sites are most important for the stability of the macromolecule [23].

A globular protein should only to some extent be regarded as a rigid molecule. More likely, the native state of a protein is a dynamic system fluctuating around a limited number of preferred conformations. Within limits, these conformational changes are in dynamic equilibrium, creating an intricate balance between rigidity and flexibility [24].

Protein–water interactions play an important role in maintaining the native structure and biological function of the protein. The dense packing of the hydrophobic core of the protein is a result of the polar nature of the water molecule, and although the influence of water on van der Waals forces, hydrogen bonds, and salt bridges is very complex, it is apparent that a change of the water concentration affects protein stability. Therefore, the stability of the protein into solution depends on parameters like pH, redox potential, ionic strength, and co-solvents [25,26].

2.2.2 The Molten Globule State

Recent data strongly indicate that under mild denaturing conditions, protein unfolding is a first-order phase transition between the native and the molten globule state of the protein. The protein molecule undergoes this transition in a highly co-operative manner, exhibiting the all-or-none behavior typical of a two-state denaturation mechanism within a narrow range of parameters [11]. The molten globule state of proteins was recently claimed to be a third thermodynamic state in addition to the native and unfolded state [27].

The molten globule maintains much of the architecture of the native state, although the detailed tertiary structure of the native state is lost in the transition. However, the folding pattern is similar to the native state, and a high content of secondary structure is maintained, as is the compactness of the molecule. The core remains packed while the outer shell is more expanded with increased side-chain fluctuations. Non-polar surfaces can be exposed to the solvent, increasing the protein's susceptibility to proteases and its ability to aggregate. Even small changes of pH, temperature and buffer composition can induce denaturation [28], and it is fair to assume that the protein occasionally will exhibit properties of a molten globule state in downstream processing.

2.2.3 The Unfolded State

The unfolded state of a protein is defined as a random coil representing an ensemble of readily inter-convertible conformers with equal or closely similar energies. Within certain limits each bond can freely rotate along the covalent chain independently of the rotation of all its neighbors [29]. No changes in covalent structure can be observed between the native and the denatured state with notable exceptions such as bovine pancreatic trypsin inhibitor, where folding is driven by the formation of disulfide bonds [30]. For a significant majority of proteins, large amounts of residual structure can persist under denaturing conditions depending on the solution parameters imposed on the polypeptide chain [31,32]. However, the co-operative interactions stabilizing folded proteins would not be expected to persist in the unfolded state [33].

2.3 Chemical and Physical Instability

Changes in proteins can be divided into two distinct classes involving chemical and physical instability. The first class comprises reactions involving cleavage or formation of a covalent bond resulting in a new chemical entity. The derivatives formed as a result of chemical instability may have properties almost similar to those of the product. Separation is possible, but highly specialized methods are often needed.

The second class refers to changes of secondary and tertiary structures, and does not involve covalent bond modification. Physical instability may lead to polymerization of the more or less unfolded molecules in a rapid aggregation reaction [34]. Irreversible aggregation results in great losses of product, and the initiation of the reaction can be very difficult to predict.

2.3.1 Proteolytic Degradation

Enzymatic proteolysis within the cell is closely controlled due to a combination of factors including compartmentalization, inhibition, and structural stabilization. When the cell dies or is disrupted by mechanical or chemical means, the endogenous enzymes are more or less released to the medium. This event dramatically changes the turn-over rate of the cellular proteins. The initial part of the purification process is therefore very much affected by the presence of proteolytic enzymes resulting in partial or total loss of product.

The result of the proteolytic attack may vary from complete hydrolysis, single breaks within the peptide chain, or loss of a few N- or C-terminal amino acid residues [35].

The fact that the biological activity may be maintained following partial proteolysis emphasizes the need for detailed analysis of the isolated protein. Several symptoms can be indicative of a proteolytic problem.

In the early purification phase, one-dimensional Sodium dodecyl Sulfalte (SDS) electrophoresis can be of much help as poor resolution, a high background staining, and loss of molecular weight bands are often indicative of enzymatic degradation, which also can be visualized by Western blotting techniques. Absence of biological or immunological activity may be due to proteolysis, although the decreased activity could also be related to structural changes of the molecule. During intermediary purification and polishing a variety of analytical tools (mass spectroscopy, capillary electrophoresis, peptide mapping, isoelectric focusing) are available to identify even minor changes of the primary structure. Heterogeneity in partly purified preparations constitutes a major problem in the polishing steps, as only the most sophisticated chromatographic methods will be able to separate the closely related compounds. The optimization of such steps is often very time-consuming and much effort should be made to avoid proteolytic cleavage early in the process.

Most enzymes have a pH at which the proteolytic activity is optimal. At a slightly alkaline pH, the highly active enzymes of the vacuoles and lysozomes will be minimally active, and relatively strong buffers in the pH interval of 7–9 are recommended for extraction to ensure that the correct pH is maintained, when the cell content is released. Low temperature decreases the enzymatic activity. In combination with short processing times this is the most effective procedure in large-scale downstream processing, where the use of enzyme inhibitors is often omitted for economical and bio-safety reasons. The addition of protein-stabilizing agents such as glycerol or dimethylsulfoxide may lower the proteolytic damage by stabilization of the tertiary structure. Low-molecular weight substances such as co-factors may also stabilize the native structure.

In many small-scale applications the use of proteinase inhibitors is widely accepted. For obvious reasons most inhibitors are toxic (some even highly toxic), and great care should be taken when handling these agents. Recommended inhibitor cocktails are given in Table 2-1. Finally, it should be mentioned that careful disruption followed by subcellular fractionation is an efficient method of avoiding proteolytic cleavage. Followed by selective removal of specific proteinases by immobilized inhibitors or substrates (such as benzamidine-Sepharose[R] or aprotinin-Sepharose[R]), the proteolytic activity can be substantially reduced. This subject has been reviewed by North [35].

Table 2-1. Suggestions for inhibitor cocktails. The inhibitors mentioned are potentially harmful. Care should be taken in handling, and if the protein is to be used for biological experiments. Some reagents are only slightly soluble in aqueous media, and stock solutions must be prepared in organic solvents.

Tissue or microorganism	Inhibitor cocktail
Animal tissue	1 mM PMSF
	1 mM EDTA
	1 mM benzamidine
	10 µg/ml leupeptin
	10 µg/ml pepstatin
	1 µg/ml apronitin
Yeast	1 mM PMSF
	5 mM phenanthroline
	15 µg/ml pepstatin
Bacteria	1 mM PMSF
	1 mM EDTA

2.3.2 N-terminal Degradation

N-terminal degradation of recombinant human growth hormone (rhGH) via diketopiperazine formation to des-Phe-des-Pro-rhGH has recently been reported [36]. The mechanism occurs without enzymatic catalysis. The degradation will probably take place in other proteins having proline as the second N-terminal residue.

2.3.3 Non-Enzymatic Hydrolysis

In dilute acid, where the carboxyl group of aspartic acid residue is not dissociated, the aspartic acid peptide bonds are cleaved 100 times faster than other bonds [37]. Under conditions where other aspartyl bonds are stable, the Asp-Pro peptide bond is unusually labile at low pH values [38].

At alkaline pH, the peptide bond is difficult to hydrolyse [39]. Thus, little evidence of peptide bond hydrolysis was found when treating ribonuclease A and lysozyme with 0.2 M NaOH at 40 °C for up to 48 h [40].

2.3.4 Deamidation

One of the major events leading to structural deterioration in globular proteins is the deamidation reaction of asparagyl residues and to some extent, glutamyl residues. As the deamidized protein may exhibit reduced biological activity, the maximal content of des-amido forms in bulk materials and in biopharmaceutical preparations is constantly being debated. The deamidation may also affect the stability of the protein by altering its tertiary structure. The list of proteins undergoing *in vitro* deamidation is comprehensive and includes well-known proteins such as insulin [41,42], human growth hormone [43], and cytochrome *c* [44].

The asparagyl and glutamyl residues are involved in the deamidation reaction. In model peptides the deamidation rate is more rapid with asparagyl residues than with glutamyl residues [45,46].

The instability of the groups in proteins correlates strongly with polypeptide chain flexibility [47], and the deamidation reaction depends on the character of the neighboring side-chains. In insulin, six amino acid residues are prone to deamidation: Gln^{A5}, Gln^{A15}, Asn^{A18}, Asn^{A21}, Asn^{B3}, and Gln^{B4} of which the three Asp residues are the most labile sites. Extensive deamidation of residue Asn^{A21} is observed in acidic solution, a reaction catalyzed by the terminal carboxyl group. At neutral pH deamidation takes place at residue Asn^{B3} under formation of a cyclic succinimide intermediary, resulting in a mixture of iso-Asp and Asp derivatives. Increasing formation of Asp relative to iso-Asp derivative was observed with decreasing flexibility of the insulin molecule [42,48].

The deamidation reaction is strongly sequence-specific. The Asn-Pro sequence has a half-life 100 times greater than that of the Asn-Gly [46,49]. To some extent these observations can also be used on proteins taking the structural steric factors into consideration [45,46]. Deamidation in the two Asn-Gly sequences in triosephosphate isomerase was found to be tenfold slower than in Asn-Gly model peptides [50].

In the pH interval from 5–12 the non-enzymatic deamidation of asparagyl (and glutamyl) residues proceeds via a succinimide intermediary as outlined in Figure 2-1. The peptide bond nitrogen attacks the side-chain carbonyl carbon atom, resulting in the formation of a succinimidyl derivative. This reaction is relatively slow, and is followed by a rapid hydrolysis at either the α- or β-carbonyl group to generate iso-aspartyl or aspartyl residues [49]. Alternate reactions can occur, resulting in fragmentation of asparagyl residues of the peptide chain or by deamidation as a result of formation of an isoimidyl intermediate similar to the succinimidyl intermediate [49]. At acidic pH (1–2) direct hydrolysis of the side-chain amide generates aspartic acid as the sole product.

These reaction patterns have been confirmed by studies of the deamidation of ACTH, which under neutral and alkaline conditions deamidates through the formation of a cyclic imide intermediate and at acidic pH hydrolyses to form Asp-ACTH [51]. Short peptides have been used to study the deamidation reaction as a function of temperature, pH, ionic strength, solvent, and effect of solvent dielectric [49,52]. The deamidation of the asparagyl residue through succinimide intermediates increases with temperature, elevated pH, and high ionic strength. In general the buffer species will influence the rate of deamidation as will the buffer concentration [49,53]. Protein deamidation is influenced by the same parameters as above, in a similar pattern, and by the identity of nearby amino acid residues [54,55]. Proteins are, of course, much more complex in structure, and some asparagyl residues may be buried into the interior of the molecule or be part of rigid domains. In biosynthetic human growth hormone nine Asn and 13 Gln residues are prone for deamidation

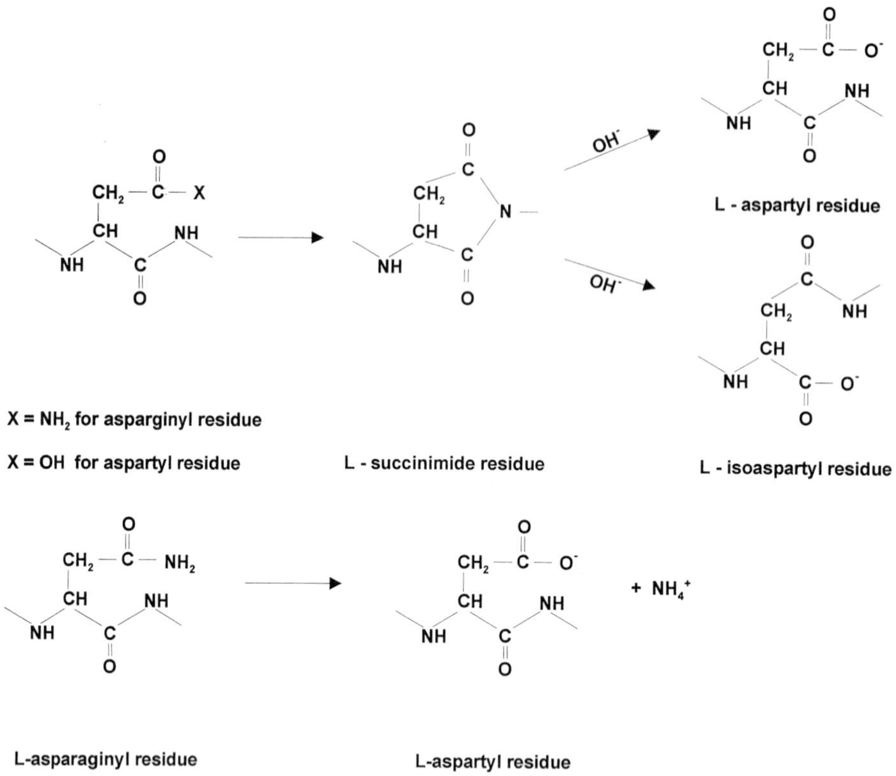

X = NH₂ for asparginyl residue

X = OH for aspartyl residue

L - succinimide residue

L - aspartyl residue

L - isoaspartyl residue

L-asparaginyl residue

L-aspartyl residue

Fig. 2-1. Deamidation. (a) Degradation of aspargyl and aspartyl residues in peptides and proteins at neutral or alkaline pH via intermediate succinimide formation, resulting in a mixture of aspartyl and iso-aspartyl residues. A similar reaction may occur for glutamyl residues. Adapted from [49]. (b) Direct hydrolysis of aspargyl residues in peptides and proteins under acidic conditions. In contrast to the intra-molecular reaction shown in (a), the rate of this reaction is much slower at neutral or alkaline pH. Adapted from [46].

with a major deamidation site at Asn[149] and a minor site at Asn[152] [56]. High ionic strength was found to increase the deamidation of cytochrome c [44,49]. Addition of organic solvent decreased the rate of deamidation by lowering the dielectric constant of the medium [49].

In conclusion, the rate of deamidation of asparagyl and glutamyl residues in proteins can be minimized by working at neutral or slightly acidic pH [55,57], in buffers of low ionic strength, and at low temperatures.

2.3.5 β-Elimination

One of the most frequent degradation reactions of proteins in alkaline solution is caused by the abstraction of a β-hydrogen from cystinyl, seryl, and threonyl residues under formation of a carbanion. Depending on the nature of the side-chain, the carbanion can rearrange to form an unsaturated derivative (dehydroalanine or β-methyl-dehydroalanine) or add a proton to give the L- and D- amino acid residue (racemization). The reaction is outlined in Figure 2-2.

Several studies indicate that the rate of β-elimination is proportional to the hydroxide ion concentration, where the formation of the carbanion by removal of the β-hydrogen is the rate-determining step [39]. Consequently the OH⁻ concentration should be kept low to prevent the β-elimination reaction. Further, the reaction is influenced by temperature and presence of divalent metal ions [58–60]. The dehydroalanine and β-methyldehydroalanine formed are quite reactive with a number of nucleophilic groups of proteins [39].

Heat-induced β-elimination of cystinyl residues has been investigated at 100 °C in the pH range of 4–8. It was concluded that the first-order rate constant of β-elimination for different proteins was remarkably similar, and the reaction, therefore, was independent of the primary structure [61]. The sulfides formed in the degradation of cystinyl residues are themselves reactive, and they strongly influence the redox potential of the solution.

Threonyl and seryl residues are lost through β-elimination at about 3–7 % of the rate of loss of cystinyl residues [39], making the protein disulphide bond the far most sensitive with respect to β-elimination. However, derivatization of the hydroxyl groups of threonyl and seryl residues markedly enhanced the rate of loss [39].

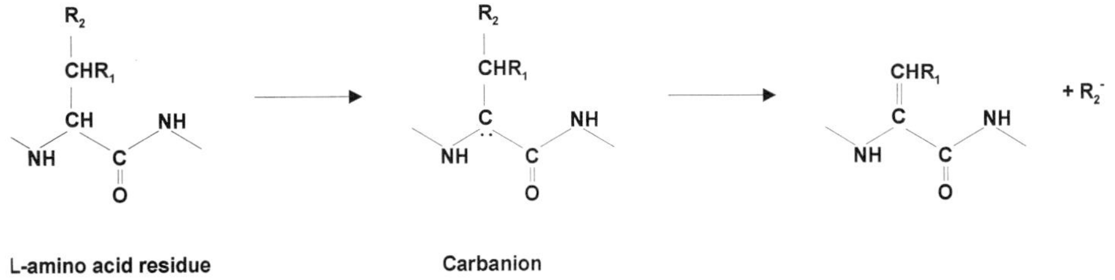

L-amino acid residue   Carbanion

Fig. 2-2. The β-elimination reaction. The reaction is initiated by abstraction of a β-hydrogen from cystinyl, seryl, and threonyl residues under formation of a carbanion. The carbanion can rearrange to form an unsaturated derivative or add a proton to give the L- or D-amino acid residue (racemization, see Fig. 2-3). Adapted from [39].

2.3.6 Racemization

All amino acid residues except glycine are subject to racemization at alkaline pH resulting in formation of the D-enantiomers of the residue.

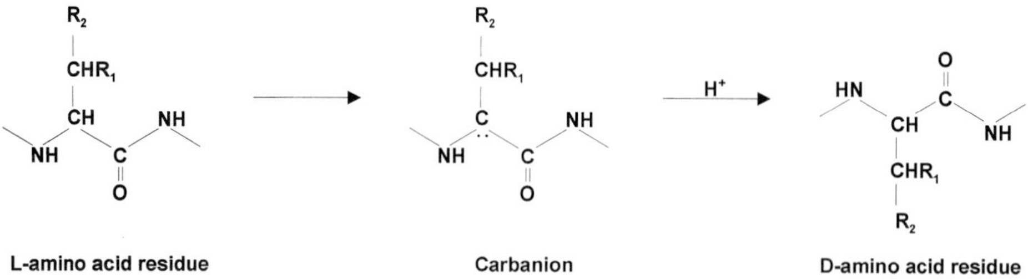

Fig. 2-3. The racemization reaction. The initial step of the reaction is abstraction of the β-hydrogen by the hydroxide ion present in alkaline solution. by uptake of a proton this will result in either the L- or D-amino acid residue. The carbanion formed may also undergo β-elimination (see Fig. 2-2). Adapted from [39].

The initial step of the racemization reaction of amino acid residues is abstraction of the β-hydrogen by the hydroxide ion present in alkaline solutions. By uptake of a proton this will result in either the L- or D-amino acid residue. The rate of racemization will depend upon the ease of abstraction of the β-hydrogen, the stability of the carbanion, and whether the β-elimination reaction is favored [39]. At pH 5–12 Asn, Asp, Gln, and Glu amino acid residues undergo a non-enzymatic modification into a succinimidyl intermediate. The five-membered ring rapidly hydrolyses to Asp or Glu or to the corresponding iso-forms in a variety of reactions resulting in racemization, generating a mixture of D- and L-derivatives [49]. The reaction is outlined in Fig. 2-3.

Several reports have dealt with the racemization of proteins in alkaline solutions [62,63]. In one study the average extent of racemization of the amino acid residues of ribonuclease A, lysozyme, soybean protein, and casein was about 6 % in 0.2 M NaOH after 4 h at 40 °C [63]. Thus, utmost care must be taken at pH above 10 (low temperature, selected buffers, short exposure time), as racemization is inevitably associated with conformational changes and thereby loss of function.

2.3.7 Conversion of Arginine to Ornithine

The guanidinium group of arginine is hydrolyzed by OH⁻ to give ornithine and possibly some citrulline, depending on the nature of the protein. The rate of arginine degradation is generally smaller than for cystine, lysine, threonine, and serine under the same conditions. The reaction is outlined in Fig. 2-4. The extent of loss of the amino acids affected by alkali treatment was proportional to the hydroxide concentration [39].

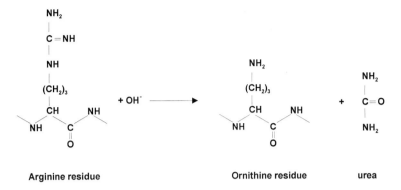

Arginine residue   Ornithine residue   urea

Fig. 2-4. Conversion of arginyl residues to ornithyl residues. The hydroxide ion-catalyzed hydro-lysis of arginyl to ornithyl residues generates ornithyl and urea. Some citrullinyl residues may be formed as well. Adapted from [39].

2.3.8 Oxidation

Oxidation has been observed in many peptides and proteins during their isolation, often resulting in loss of biological and immunological activity [64–66].

At neutral or slightly alkaline pH conditions, the histidyl, methionyl, cysteinyl, tryptophanyl, and tyrosinyl residues are potential oxidation sites, while under acidic conditions the primary reaction is the oxidation of methionine to methionine sulfox-ide [67].

Under mild oxidizing conditions the methionyl residue is oxidized to the corre-sponding sulfoxide as outlined in Fig. 2-5. Oxidation to methionine sulfone requires strongly oxidizing conditions rarely met in downstream processing. Methionine is very susceptible to auto-oxidation and photo-oxidation, the latter being an example of the influence of light rather than oxygen [53].

Cysteine is also easily oxidized to cystine at alkaline pH (where the thiol group is de-protoneated) in the presence of a mercapto-reagent. In another set of reactions, oxidation of cysteine to cystine is catalyzed by divalent metal ions, especially Cu^{2+}. The reaction can be inhibited by addition of EDTA. In the absence of a thiol reagent or a nearby thiol, the cysteine residue may instead oxidize to sulfenic acid. A comprehensive summary of protein oxidation is given in [68].

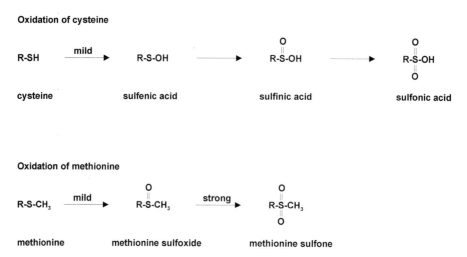

Fig. 2-5. Oxidation of cysteine and methionine residues.

2.3.9 Cysteinyl and Cystinyl Residues

The ability of the cysteinyl residue to form intra- and inter-molecular disulfide bonds makes this amino acid residue unique in terms of protein stability. A large number of extracellular proteins are stabilized via disulfide bonds, and in many cases the covalent cross-link is essential for maintaining their tertiary structure and their biological activity. Ribonuclease, for example, loses almost all activity when the four disulfide bonds are reduced [69]. Insulin and chymotrypsin are completely inactivated by the reduction of a single disulfide bond [70,71].

In addition to the static function described, reactive disulfide bonds sometimes perform a dynamic function in proteins [72]. A disulfide bond remains stable indefinitely in an isolated protein, its stability being entirely dependent on the environment. Major factors are redox potential, pH, and temperature, which must be carefully controlled to prevent bond cleavage. The disulfide bond is degraded in a variety of mechanisms of which some of the most important will be summarized below, (see Fig. 2-6).

In neutral or alkaline media a nucleophile attack on a sulfur atom of the disulfide initiates the cleavage. The reactivity of some nucleophile reagents is $HS^- > RS^- > CN^- > SO_3^{2-} > OH^-$. The thiol–disulfide reaction is perhaps of the most practical importance. This reaction consists of two steps of nucleophile substitution with the formation of a mixed disulfide at the intermediate stage. Both rate and equilibrium constants for the thiol–disulfide exchange exhibits common variation [73], and the reaction rate is influenced by the pK_a of the proton sulfhydryl group (typical value is 8.5), the leaving group, steric hindrance, charge, and entropy [69]. The interaction of -SH groups with aliphatic disulfides proceeds mainly in alkaline media (usually at pH 9.5), although shuffling reactions at pH 7 to 8 have been reported [74].

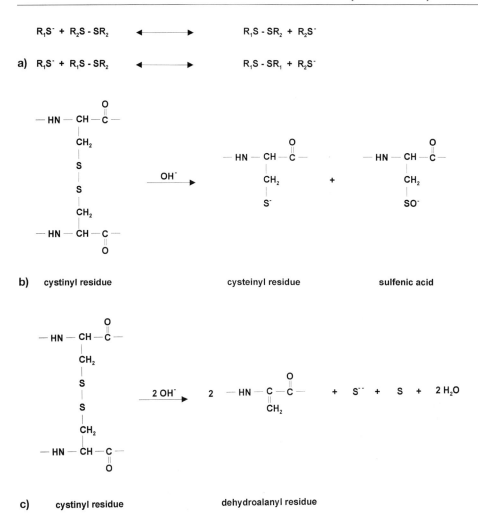

Fig. 2-6. Reactions involving cysteinyl and cystinyl residues. (a) The two-step nucleophilic substitution reaction between the peptide or protein (RS⁻) and a thiol reagent (R₂S-SR₂) in alkaline media. (b) The hydrolysis of the cystinyl residue in alkaline media to the cysteinyl residue and a sulfenic acid residue. Adapted from [39]. (c) The proposed reaction for the β-elimination of protein-bound cystine in alkaline solution. Adapted from [39].

In strongly acidic media the exchange reaction is proposed to take place via a sulfenium cation, which is formed by an attack of the disulfide bond by an electrophil (H^+, Hg^{2+}, Ag^+) displacement on a sulfur atom of the disulfide [75].

OH⁻ is a poor nucleophile towards the disulfide bond, but solvent-accessible bonds were shown to dissociate in 0.2 M NaOH [76]. Three mechanisms were investigated: hydrolysis, α-elimination, and β-elimination. The result of the study points to elimination of β-hydrogen and formation of a persulfide intermediate as being the first step of the degradation by dilute alkali. Only rarely may the α-elimination reaction occur in polypeptides [61,76]. However, in strongly alkaline media the α-carbon proton may also be removed [39,77]. One should be aware that catalytic quantities of HS^- may arise from the β-elimination reaction thus initiating disulfide bond rearrangement.

Presence of divalent metal ions (typically Cu^{2+}) in combination with OH^- and O_2 may result in oxidation of cysteine residues by an ill-defined reaction mechanism [78]. Therefore, stock solutions are often deareated and made 1–2 mM with metal chelators such as EDTA.

The lability of the disulfide bond in proteins may result in the formation of structurally altered forms of the protein with reduced biological activity and/or immunogenic properties. In proteins comprising both free -SH groups and disulfide bonds, rearrangement can happen without the presence of low-molecular weight thiol reagent. However, the number of proteins containing both -SH groups and disulfide bonds are relatively small. Among these proteins are serum albumin, papain, ficin, and β-lactoglobulin [72].

A novel protein derivative has been found during downstream development of biosynthetic human growth hormone. In this derivative a Cys^{182}-Cys^{189} trisulfide bridge was identified using electrospray mass spectroscopy [79].

2.3.10 Denaturation

The term denaturation is used to denote the process in which the tertiary structure of the molecule is changed from the one typical of the native structure to a more disordered arrangement but without alteration of the amino acid sequence. In the process, the hydrogen bonds and intra-molecular interactions resulting in the co-operative stabilization of the native state are disrupted. The precise orientations are lost and regions of 'the inner core' are exposed to solvent water and co-solvent molecules [3,80,81]. For small globular proteins denaturation is an almost all-or-none process approximated rather well by the two-state transition [28]. The co-operativity of the denaturation process in many proteins results in an abrupt transition from the native to the unfolded state within a narrow range of temperature, pH, redox potential, or denaturant concentration. The different phases cannot be transformed gradually into each other, and there is no critical point at which they are indistinguishable [15]. Thermodynamically, the denaturation process can be observed by an increase of molar heat capacity, and a rapid enthalpy increase with increasing temperature [2,82].

In many cases denaturation is a reversible process in which the native structure can be re-established by careful adjustment of solvent composition or of pH, temperature, or redox potential. However, proteins may undergo irreversible denaturation in which the native structure cannot easily be obtained. An example is the group of proteins in which disulfide bonds have been destroyed, and where *in vitro* folding under very restricted conditions must be applied to regain activity. Another example is the hydrophobic aggregation observed among intermediates or unfolded molecules [34,83]. A third example is the formation of stable intermediates as observed during renaturation of AE-IGF-1 [84].

Denaturation is a function of pH, temperature, redox potential, solvent composition, ionic strength, and other parameters of downstream processing. The marginal stability of the native state, and the loss of tertiary structure, make denaturation very difficult to predict. Within milli-seconds the entire protein solution can be transformed into an insoluble gel or precipitated aggregates, resulting in major loss of product.

2.3.11 Aggregation

One of the serious consequences of exposing the protein to denaturing conditions is the exposure of hydrophobic residues to the aqueous solvent. This leads to disorganization of the water molecules, thus increasing the entropy of the system. In order to avoid this change, the water molecules try to maintain organization by aggregation, thereby decreasing the solubility of the protein. The aggregation reaction can be very fast, and will, in severe cases, lead to the formation of insoluble polymers.

Folding of denatured or partially folded intermediates is a first-order reaction resulting in a constant folding rate at increasing concentration of the denatured or partially folded forms. Aggregation is assumed to be controlled by the initial dimerization step in a second-order reaction dominating over the first-order folding reaction. Consequently, an increase of concentration of denatured or partially denatured polypeptide will favor the aggregation reaction and thus form biologically inactive precipitates [85].

It has been stated that aggregates derive from partially folded intermediates rather than from unfolded or native protein [24,86] exemplified in the formation of insulin fibrils from partially unfolded insulin molecules [34] and the aggregation/association behavior of the protonated barnase intermediary [87] (see Fig. 2-7).

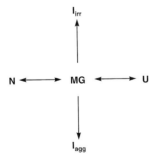

Fig. 2-7. Unfolding and aggregation. The transition from the native state (N) to the molten globule state (MG) and from MG to the unfolded state (U) are reactions of first order (i.e., the reaction rate is independent of the protein concentration). Some proteins do not readily fold back to the native state, either because of stabilization of the intermediary compound I_{irr} or due to the slow and complicated re-establishment of disulfide bonds. The aggregation reaction, in contrast, is a reaction of second order (i.e., the reaction rate depends on the protein concentration), resulting in aggregation of the molten globule to I_{agg}.

Recent data strongly indicate that aggregation occurs by specific interaction of certain conformations of intermediates rather than by non-specific co-aggregation [88]. However, insulin molecules with nearly complete native-like structure, aggregate in a linear fashion [34]. Studies on the aggregation of apomyoglobulin in aqueous urea solutions showed that aggregation also may involve the association of unfolded molecules [89].

Proteins comprising free -SH groups may form inter-molecular disulfide bonds, leading to aggregation of the protein. Purification under reducing conditions erases this problem as demonstrated in the purification of recombinant glutamic acid decarboxylase [90]. Inactivation via aggregation has been reported for several proteins [89,91–98].

Expression of proteins in *Escherichia coli* often results in the formation of insoluble aggregates called inclusion bodies, probably comprising fully or partially denatured protein [99].

2.3.12 Precipitation

One of the most widely used procedures in protein purification is precipitation by altering the solvent properties and thereby lowering the solubility of the protein. Being part of a purification scheme, mild precipitation procedures have been used to limit irreversible aggregation. In iso-precipitation, molecules are precipitated near their isoelectric point, where the electrostatic repulsion is minimal due to the zero net charge of the protein. Inclusion of a solute such as ethanol, which reduces the protein solubility further, will often improve the precipitation process. At low salt concentrations favorable interactions between salt ions and charged groups on the protein surface will result in an increase of the solubility of the protein (salting-in). The effect is identical for all proteins. High concentration of salts will in most cases lead to precipitation of the protein (salting-out). One of the most commonly used procedures is addition of ammonium sulfate to a 2–3 M concentration. The process is largely dependent on the hydrophobicity of the protein, and the optimal salts are those favoring dehydration of the non-polar regions without binding to the protein [100]. The salting-out effectiveness of cations and anions is shown in Table 2-2

Table 2-2. The Hoffmeister series. The first ions in the series are known for their stabilizing effect on proteins. They markedly increase the surface tension of water and are preferentially excluded from the protein surface. The solubility of non-polar molecules is decreased in solvents like ammonium sulfate (salting-out). The last ions of each series may bind to the protein. They have little effect on the surface tension of water, and increase the solubility on non-polar molecules (salting-in).

Cations
$NH_4^+ > K^+ > Na^+ > Li^+ > Mg^{2+} > Ca^{2+} > Gdn^+$

Anions
$SO_4^{2-} > HPO_4^{2-} > CH_3COO^- > Cl^- > NO_3^- > SCN^-$

reflecting the order discovered by Hoffmeister [101]. The effects of the individual ions are additive. Thus GdnCl is a strong denaturant, while Gdn_2SO_4 normally stabilizes proteins.

Organic solvents such as acetone and ethanol have been widely used to precipitate hydrophilic proteins, which are less soluble in non-polar solvents. The water molecules are partially immobilized by hydration of the solvent reducing the water activity. The water molecules around hydrophobic areas will be displaced by the organic solvent reducing the hydrophobic attraction. The principal forces leading to precipitation are, therefore, likely to be electrostatic forces and dipolar van der Waals forces. The technique has its limitations, however, as many proteins tend to denature in the presence of organic solvents. Therefore, it is important to keep the temperature low in order to decrease the conformational flexibility and to prevent the organic solvent from penetrating the internal structure of the protein. A similar effect can be obtained with water-soluble organic polymers, provided they are not too viscous. Polyethylene glycol of a molecular weight between 6000 and 20 000 Da is commonly being used.

2.4 Essential Parameters

The term downstream processing refers to isolation of a given protein from a complex, heterogeneous mixture comprising other proteins, lipids, cell debris, DNA, RNA, Malliard compounds, etc., which results from the fermentation and/or extraction procedure. The ultimate goal is to isolate the protein in its purest form while retaining the specific biological activity of the molecule. Suppliers of protein biopharmaceuticals have met this challenge for years, facing increasing demands for higher purity and specific analytical methods for characterization.

It is, therefore, of interest to focus on the various parameters characterizing each unit operation and to determine their influence on protein stability. These parameters include pH, temperature, redox potential, presence of co-solvents, protein concentration, and pressure.

2.4.1 The Effect of pH

Conventional wisdom states that protein purification is best carried out near the physiological pH, at which the protein is believed to obtain maximal stability. The view can be justified for two reasons. The isoelectric point of many proteins is in the range of pH 5–8 [102], not far from the physiological pH. Further, many proteins are stable around pH 6 regardless of their isoelectric point, reflecting the ionization of histidine in the unfolded state [103]. However, the optimal pH can be very different from neutral. Pepsin is stable at pH 1–2, but denatures rapidly at pH 7 or more [100].

The structural stability of globular proteins depends very much on protonation and deprotonation of potentially titrable groups like carboxyl (pK_a 3.0–4.7), imidazolium (pK$_a$ 8.0–8.5), sulfhydryl (pK$_a$ 8.0–8.5), amino (pK$_a$ 7.6–10.6), and phenolic hydroxyl (pK$_a$ 9.4–10.4) [39]. The effect of pH on secondary structure is reflected in the structural changes of two synthetic polypeptides, poly-L-glutamic acid and poly-L-lysine. At a low pH poly-L-glutamic acid spontaneously forms a helical structure, whereas poly-L-lysine is a random coil. As pH increases, the situation reverses, and at pH 10, poly-L-lysine spontaneously folds into a helix, whereas poly-L-glutamic acid forms a random coil [104]. The same pH dependency is observed in proteins. As pH is increased or decreased from pI, titrable groups are protonated or deprotonated and electrostatic interactions are broken, resulting in structural changes of the protein due to electrostatic repulsion [105]. By a change in pH, interior hydrophilic residues (Asp, Glu, Lys, Arg, His) may be positively or negatively charged, making the residue more solvent-accessible, again resulting in a structural change of the molecule. Although a non-buffered protein solution should automatically adopt a pH very close to pI, the buffer capacity is low. Therefore, even minor changes of the environment may move the system towards pH-extremes. In the pH range 4.5–8.0, the melting temperature, T_m is independent of the pH provided pI of the protein is in the same range [106].

A variety of chemical protein modifications takes place in alkaline solutions, the reaction rate often being directly proportional to the hydroxide ion concentration. In alkaline solutions proteins may undergo degradation by deamidation, β-elimination, hydrolysis, racemization, or breakage of disulfide bonds. Deamidation, hydrolysis, or breakage of disulfide bonds may also take place in acidic environment. Thus, the extent of chemical protein modification in downstream processing as a function of pH can be difficult to predict. Even in mildly acidic or alkaline solutions, deamidation of proteins is often observed, as is degradation of sulfur-rich proteins.

Physical and chemical protein modification is, therefore, not only a phenomenon observed at extremes of pH; it should rather be regarded as a potential risk and possibility anywhere on the pH scale. Protein destabilization as a function of pH is illustrated by the following examples. At acidic pH insulin deamidates in position A^{21} [42], and between pH 1–2 quickly loses six amide groups by hydrolysis [107]. Around neutral pH, insulin deamidation is much slower and takes place exclusively at residue Asn^{B3} [42]. Interferon-γ was reported unstable upon acidic treatment at pH 4 [108,109]. At pH 3.0 L-asparaginase loses 98 % of its biological activity within 50 minutes; the antibody-protected enzyme retained 40 % activity under the same conditions [110]. Degradation of lysozyme and α-lactalbumin at the cystinyl residues was observed by treatment with 0.1 M NaOH for 24 h at 50 °C [60]. A 30 % racemization of the seryl residues was observed after treating lysozyme with 0.5 M NaOH for 2.5 h at 22 °C [111]. The average content of racemization of the amino acid residues of lysozyme, soyabean protein, ribonuclease A, and casein was approximately 6 % in 0.2 M NaOH after 4 h at 40 °C [63]. However, little evidence of hydrolysis of peptide bonds in ribonuclease A and lysozyme was found when the proteins were treated with 0.2 M NaOH at 40 °C for up to 48 h [40]. Cytochrome *c* unfolds in alkaline solution [112]. Degradation of chymotrypsin, trypsinogen, ribonuclease S, and lysozyme in 0.2 M NaOH was reported [76]. In contrast, the activity of low-molecular weight urokinase was nearly independent of pH in the range 2–11 [113].

2.4.2 The Effect of Temperature

The difference in Gibbs' free energy between the native and unfolded state (ΔG_u) of a protein results from a balance between large and opposing entropic and enthalpic effects, ΔS_u and ΔH_u. The ΔH_u and ΔS_u are highly temperature-dependent [82,114] being linearly increasing functions of temperature [115], with a slope equal to the unfolding heat capacity of the protein [2]. The heat capacity (C_P) of both the folded and unfolded states of a protein, is substantial and ΔC_P between the folded and unfolded state is in the order of 8 kJ/(degree Kelvin $\times$ mol) [116]. The significant heat capacity difference results in a non-linear temperature dependence of ΔG as illustrated in Fig. 2-8. Expressions taking the temperature dependence of the thermodynamic parameters of the unfolding into account have been derived [106,117].

ΔG_u is a function of temperature with a maximum in the temperature range of 10 to 30 °C for all globular proteins investigated to date. At this temperature the denaturation entropy change is zero and $\Delta G(T_s) = \Delta H(T_s)$. Thus, the point of maximum thermodynamic stability of the native structure is entirely due to enthalpic contribution.

ΔG_u decreases to zero both at higher and lower temperatures as illustrated in Fig. 2-8 [115]. This results in heat- and cold-denaturation of proteins, respectively. On the assumption of the two-state model for small globular proteins, the native state will collapse in an 'all-or-none' transition upon denaturation.

$\Delta G_u = -RT\ln K$
$\Delta G_u = -RT\ln[N]/[U]$
$\Delta G_u = 0$ when $[N] = [U]$

The temperature at which $\Delta G_u = 0$ is called 'the melting temperature', T_m, where $T_m = \Delta H_m/\Delta S_m$. This shows that the heat denaturation occurs when ΔS_m (chain entropy difference) exceeds the enthalpy from inter-atomic interactions of the native state. Denaturation at low temperature can be explained by the decrease in the strength of hydrophobic interactions as the temperature is lowered [118]. The hydro-

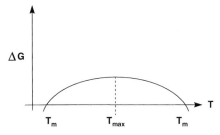

Fig. 2-8. The free energy, ΔG, as a function of temperature. T_m is the transition temperature at which the Gibbs' free energy difference between the native and unfolded state is equal to zero. T_{max} is the temperature at which the stability of the native state is at maximum and where $\Delta S(T_{max}) = 0$. Heat denaturation proceeds with an increase of enthalpy and entropy. Cold denaturation proceeds with a decrease in enthalpy and entropy. Adapted from [106].

phobic interactions are strengthened with increasing temperature, while hydrogen bonds are weakened, resulting in loss of co-operativity [3]. The stability of the native state is therefore limited to the temperature interval where hydrophobic interactions are sufficiently strong to counteract the weaker dissipative forces [118].

The stability of a native protein in aqueous solution is determined not only by the interactions discussed so far, but also by the interactions between residues and the surrounding water molecules. The factors contributing to destabilization of the native state are the increase of the unfolding chain entropy and the decrease of the unfolding hydration enthalpy.

The critical dependence of T_m on solution conditions is illustrated by the pH dependency of T_m for ribonuclease. T_m increases from 46 to 65 °C, as the pH of the solution varies from 3 to 7 [119]. Thermal denaturation is also a function of presence and type of neutral salts. The T_m of ribonuclease decreases from 64 °C to 40 °C as the guanidinium salt is varied from sulfate to thiocyanate [120].

Measurement of T_m(near pI) and ΔC_p leads to evaluation of ΔH, TΔS and ΔG for the unfolding process of a given protein. T_m can be determined by circular dichroism, fluorescence spectroscopy, or UV-difference spectra, which will provide information about the structural content at a given temperature. From these data the transition curve for thermal unfolding in a given solvent can be constructed [121]. For small globular proteins denaturing according to the two-state model, the analysis is straightforward; for large proteins the two-state model may be tested by measuring the transition by at least two different methods [121].

A number of destructive covalent reactions of thermolabile amino acid residues have been observed upon heating. They include deamidation of asparagyl and glutamyl residues in peptides at 100 °C at pH 7.4 [45], cysteine oxidation at pH 8 in α-amylase [122], β-elimination of cystine residues at 100 °C at different pH [61], hydrolysis of peptide bonds at aspartyl residues in proteins [123], and thiol-catalyzed disulfide interchange at 100 °C [61].

The temperature of maximal stability is in the range of 10–30 °C for all proteins investigated, a temperature range well suited for downstream processing. Therefore, thermal inactivation during purification should not be expected provided solvent conditions are kept optimal, i.e. close to pI, low content of buffer additives, low protein concentration, and correct redox potential. However, these optimal conditions are rarely met in purification of proteins, and solvent conditions can suddenly and unexpectedly change, resulting in a shift of T_m into the temperature interval of the downstream process.

Storage of protein solutions at low temperature may constitute a problem, as well, either because of the freezing–thawing procedure, or because of lowered freezing point due to additional co-solvents, at the risk of cold denaturation.

2.4.3 The Effect of Redox Potential

The formation and cleavage of disulfide bonds in proteins is a function of the redox potential of the solution. A decrease in the redox potential towards more reducing conditions will ultimately result in cleavage of the disulfide bond to free cysteinyl residues. Alternatively, an increase of the redox potential towards oxidizing conditions will result in disulfide bond formation, as seen in the *in vitro* renaturation of recombinantly expressed proteins.

Most buffers in downstream processing are designed for maintaining pH in a narrow, well-controlled interval. However, little or no attention has been paid to the very low redox buffer capacity of these same buffers. Consequently, even minor changes of the environment may lead to significant changes of the redox potential of the solution.

For a complete reaction, oxidation as well as reduction must occur:

$$Red_1 + Ox_2 = Ox_1 + Red_2$$

From the free energy of the reaction

$$\Delta G = \Delta G^0 + RT \times \ln K_{eq}$$

and the Gibbs' free energy

$$\Delta G = -nF(E_2 - E_1)$$
$$E_2 - E_1 = E_2{}^{m0} - E_1{}^{m0} - RT/nF \times \ln K_{eq},$$

where E is the redox potential, E^{m0} is the redox potential at $[Ox] = [Red]$ at which $[Ox]$ and $[Red]$ are maintained at unit activities at pH = 0, R is the gas constant, T is the absolute temperature in degree Kelvin, n is the number of electrons transferred, and F is the Faraday constant, and K_{eq} is the equilibrium constant.

Since there is no absolute value for redox potentials, the standard hydrogen half cell is given the value 0 V at any temperature. Thus, for the reaction

$$Ox_2 + ne^- = Red_2$$
$$E_h = E^{m0} - RT/nF \times \ln([Red_2]/[Ox_2])$$

which is the well-known Nernst equation, ignoring transfer of protons. E_h is the redox potential relative to the standard hydrogen electrode.

At 24 °C

$$E_h = E^{m0} - 0.059/n \times \log([Red_2]/[Ox_2]).$$

Obviously the proton in redox reactions involving amino acids, peptides, and proteins cannot be ignored:

$$Ox_2 + e^- + H^+ = Red_2H \ (n = 1)$$
$$E_h = E^{m0} - 0.059 \times \log([Red_2H]/(Ox_2 \times [H^+])) \ (n=1)$$
$$E_h = E^{m0} - 0.059 \times \log([Red_2]/[Ox_2]) - 0.059 \times pH \ (n=1)$$

At $[Ox]/[Red] = 1$, the redox potential (E_h) will change 0.059 V for every pH unit increase. Thus, raising pH from 5 to 12 decreases E_h with 0.42 V creating a much more reducing environment.

Let us consider the two reactions

$$Red_1 = Ox_1 + ne^-$$
$$Ox_2 + ne^- = Red_2$$

with the redox potentials E_1 and E_2. If $E_2 > E_1$, system 2 will oxidize system 1. In most cases the term RT/nF $\times$ ln([Red_2]/[Ox_2]) contributes little to E, and the standard redox potentials can be used to compare the relative reducing or oxidizing power (an exception is $E_1^{m0} \approx E_2^{m0}$).

The standard redox potential of selected half-cell reactions is given in Table 2-3. From the table it is seen that DTT, cysteine, and 2-mercaptoethanol are excellent reducing agents for proteins and that H_2O_2 or potassium ferricyanide are well-suited oxidants.

In practice, a Pt-standard calomel electrode system is used for measuring the redox potential. The potential of the calomel electrode relative to the standard hydrogen electrode is 0.244 V at 25 °C [127]. The system is standardized against a quinhydrone redox buffer solution using $E_c = 0.219$ V at pH 4.00 and $E_c = 0.042$ V at pH 7.00 at 25 °C, where E_c is the redox potential measured with the calomel electrode system.

In order to investigate the correlation between the redox potential of the solution and the stability of the disulfide bonds in proteins, the stability of human recombinant insulin was measured in solutions of increasing DTT concentration. The A- and B-chains of human insulin are held together by the two disulfide bonds only, as outlined in Fig. 2-9. Upon reduction the two chains will separate when the cystinyl residues are reduced to cysteinyl residues. The degradation can be monitored by RP-HPLC analysis of samples taken at different redox potentials. The result is shown in Fig. 2-10.

Table 2-3. Standard redox potentials. The standard redox potentials at $[Ox] = [Red]$ at pH 7 (E^{7m}) of selected half-cell reactions is shown for reducing agents (DTT, cysteine, glutathione) and for oxidizing agents ($[Fe(CN)^6]^{3-}$, O_2 (alkaline solution), H_2O_2). A mixture of reduced and oxidized glutathione is often used in refolding experiments to ensure correct redox potential and mercapto reagent concentration.

Agent	E^{7m}	Reagents
DTT	–0.33 V	Cleland [124]
Cysteine	–0.21 V	Cleland [124]
Glutathione	–0.23 V	Scott et al. [125]
$[Fe(CN)_6]^{3-}$	0.36 V	*CRC Handbook of Chemistry and Physics* [126]
Oxygen (alkaline solution)	0.40 V	*CRC Handbook of Chemistry and Physics* [126]
Hydrogen peroxide	1.78 V	*CRC Handbook of Chemistry and Physics* [126]

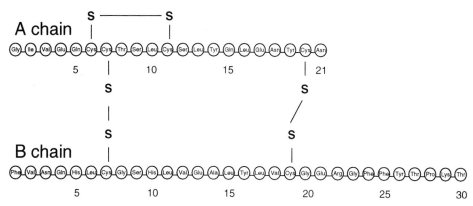

Fig. 2-9. The primary structure of human insulin.

Human insulin remains stable at $E_h > 0.1$ V. At lower redox potential the disulfide bonds are reduced, resulting in loss of intact molecules. At approximately 0 V, about 1 % of the insulin molecules remains intact (corresponding to 4 mM DTT in the said solution). The low buffer capacity is illustrated by the minute amounts of DTT needed to reduce the two disulfide bonds.

A protein comprising cysteinyl residues may be kinetically or thermodynamically trapped in various ways. In the presence of thiols or other nucleophiles (other cysteinyl residues), rearranged disulfide bonds may result in irreversible formation of scrambled forms [84]. Under reducing conditions cysteinyl residues in the reduced form could be trapped, so that an unfavorable structural rearrangement must precede chemical oxidation. The presence of thiol reagents could result in formation of mixed

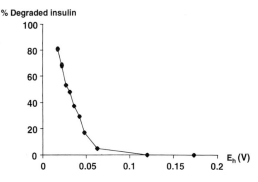

Fig. 2-10. Reduction of human insulin. A solution of 1 mg/ml of human insulin in 0.1 M Tris, 2 mM EDTA pH 8.5 was used. DTT was added stepwise up to 4 mM. At each step the reaction was allowed to proceed for 15 min. Then aliquots of 200 µl were added to 800 µl of 2 M acetic acid to stop the reaction. The amount of intact human insulin was calculated from RP-HPLC analytical data. The redox potential was measured using a calomel reference electrode Ref. 201 (Radiometer Denmark), calibrated against 0.005 M quinhydrone solution adjusted to pH 4.0 (0.1 M acetate) and to pH 7.0 (0.1 M phosphate) at 25 °C. $E_c = 0.219$ V and $E_c = 0.042$ V were used, respectively.

disulfides with available cysteine residues preventing re-establishment of the disulfide bond(s). Further, the cleavage of one or several disulfide bonds may alter the
structure of the protein resulting in other covalent modifications such as proteolysis,
deamidation, or aggregation.

Mercapto reagents have been used carefully in the renaturation of IGF-1, rhodanese, lysozyme, and in stabilization of glutamic acid decarboxylase (GAD$_{65}$) during
downstream processing as outlined in Table 2-4.

The intra- and inter-cellular reactions between cysteinyl residues depend on the
solution pH. The pK$_a$ of the typical cysteinyl sulfhydryl group is approximately
pH 8.6. This value varies from protein to protein and among different positioned
cysteinyl residues. The presence of a thiolate anion is essential for the thiol/disulfide
reaction to occur, and the reaction is strongly inhibited at pH below 7. Maintaining a
mildly acidic environment thus stabilizes proteins with cysteinyl residues.

It appears that the redox potential of the solution is an important parameter in
terms of stability for proteins containing cysteinyl residues. Aqueous buffers will
normally exhibit a redox potential between 0.2 V and 0.5 V ensuring the stability
of the disulfide bond. However, changes in environment (exemplified by shift of
pH) may result in dramatic shifts of the redox potential towards destabilizing conditions. It is recommended that the redox potential be monitored using a reliable and

Table 2-4. Examples of co-solvents used for stabilization of proteins. The buffers indicated for
IGF-1, rhodanese, and lysozyme were used in refolding of the molecules. GAD$_{65}$ were purified
under reducing conditions to prevent inter-molecular disulfide bond formation and aggregation.

Protein	Buffer	Purpose	Reference
IGF-1	50 mM Tris, 2 mM cysteine, 2 mM EDTA, 25 % (v/v) ethanol pH 9.0	Renaturation	Hejnaes et al. [84]
Rhodanese	10 mM Na-phosphate, 200 mM 2-mercaptoethanol, 0.3 M GdnCl, 0.5 % (w/v) lauryl maltoside pH 7.4	Renaturation	Tandon and Horowitz [98]
Lysozyme	0.1 M Tris–HCl 1 mM EDTA 0.695 mM oxidized glutathione 4 M sarcosine pH 8.0	Renaturation	Maeda et al. [128]
GAD$_{65}$	50 mM glycine, 10 mM DTT, 1 % n-octylglycoside, 1 mM glutamate, 1 mM PLP pH 9.5	Stabilization	Moody et al. [90]

stable system such as the Pt-calomel electrode system, and that redox buffers such as mixtures of thiol reagents are introduced. One should, however, be aware of the limitations of such buffers as illustrated by the aggregation of serum albumin in the presence of low levels of thiol reagent [129].

2.4.4 The Effect of Co-solvents

A great number of co-solvents are used in downstream processing to ensure optimal purification conditions. Any of these additives will influence the stability of the native conformation. The choice of co-solvent at a given set of parameters (pH, temperature, ionic strength, protein concentration, and redox potential) is, therefore, of great importance for the net stability of the protein.

The mechanism of stabilization is now well understood following a series of careful interaction measurements between proteins and the complex solvent environment, carried out by Serge Timasheff and co-workers. Intensive studies for more than three decades have shown that if a protein is dissolved into a mixture of two solvents, generally one of them will form more favorable interactions with the protein than the other. The preferred solvent is most often water and hence the expression 'preferential hydration'. Consequently, the co-solvent will be excluded from the protein/solvent interaction layer. The effect of the co-solvent will be stabilization of the native state of the protein since the system tends to move towards minimizing the area of the water–protein interface. Destabilizing agents are able to bind to exposed groups on the protein surface, either by charge–charge or hydrophobic interactions. They will preferentially interact with the unfolded protein and favor its unfolding [130].

Small molecules and ions also contribute to protein stabilization by raising the water surface tension. The mechanism, based on the pertubation of the water surface tension, is a function of the physical chemistry of the water–small molecule interaction [130].

The different mechanisms of exclusion and binding have allowed Timasheff and co-workers [130] to suggest the existence of three classes of co-solvents. The first class comprises sugars, some amino acids, some salts, and certain polyols. Co-solvents belonging to this class (sucrose, glucose, mannose, glycine, alanine, glutamine, glutamic acid, and Na_2SO_4) increase the surface tension with no or weak binding to the protein. Glycerol, mannitol, and sorbitol also belong to this class; they have, however, affinity for polar regions. In general the stabilizing effect is concentration–dependent. The effect becomes apparent only at relatively high co-solvent concentration [131].

The second group comprises weakly interacting salts ($MgSO_4$, $(Gdn)_2SO_4$, NaCl, Arg-HCl, and Lys-HCl). They increase the surface tension and bind to charged groups or to peptide bonds. The stabilizing effect will depend on protein charge and salt concentration. Valine binding to hydrophobic regions belongs to this group.

Polyethylene glycol (PEG) and 2-methyl-2,4-pentanediol (MPD) belong to the third class, where stabilization is caused by steric exclusion and repulsion from charged groups. Solvents able to bind to the surface of the protein make protein stabilization a balance between co-solvent exclusion and binding. Although PEG is strongly excluded from the protein surface domain, its ability to bind to non-polar amino acid residues [132] will lower the unfolding transition temperature [133] and thus destabilize the native conformation. MPD, another strong binder to hydrophobic regions is strongly repelled from charged groups [133] and is, therefore, excluded from the protein domain, stabilizing the native protein. However, a partially unfolded protein will expose non-polar groups to the surface, where MPD will bind and the net result will be destabilization.

Co-solvents able to bind to exposed residues of proteins have been widely used in downstream processing to prevent aggregation of intermediary compounds. Aggregation has been suppressed at well-defined concentrations of denaturants [134,135] or other additives [136]. Use of denaturants in downstream processing is a balance between the denaturing effect of the co-solvent and its power to prevent aggregation. Examples of additives used to reduce aggregation are 0.2 M arginine [137], PEG [97], and lauryl maltoside [138].

The effect of neutral salts (defined as strong electrolytes which are significantly soluble in water without bringing about a major change of solution pH [105]) depends on the individual contribution from the cation and the anion as outlined in the Hoffmeister series (Table 2-2). Direct interaction between the salt and protein can be electrostatic in nature, and this effect may be dominating at low salt concentrations. Ions also react with dipolar groups such as amino, carboxyl, and hydroxyl groups. Other salts have particular non-polar or hydrophobic character and exhibit a solubilizing effect on non-polar residues or surfaces, leading to destabilization or denaturation. Once the electrostatic effects have been saturated, the transition temperature is generally found to be a linear function of the salt concentration. The anion and cation affects T_m in a roughly additive fashion [139]. At higher salt concentrations, protein stability depends on the ability of the salts to interfere with the hydration sphere of the protein. Salts excluded from the hydration sphere generally tend to stabilize the protein by preferential hydration and to decrease the solubility (salting-out). In contrast, salts that tend to disorganize the water structure destabilize the protein and increase the solubility (salting-in). In general, these effects follow the Hoffmeister series [139] making ions like NH_4^+ and SO_4^{2-} excellent stabilizers, while SCN^{2-} and ClO_4^- are de-stabilizers. The same destabilizing ions are also effective in dissociating non-covalent aggregates of globular proteins [105]. The impact of neutral salts on protein stability requires a high concentration (1–8 M) of the added salt, indicating that in general protein–salt interactions can be neither strong nor specific [140].

Detergents are amphiphilic surface active agents (surfactants) which are soluble in aqueous solvents on account of their hydrophilic matrices. The widespread use of detergents in protein chemistry arises from their extraordinary solubilizing effect on membrane proteins and inclusion bodies. The detergent intercedes between the hydrophobic surface of the protein and the bulk aqueous medium. The most effective detergents tend to be the most effective denaturants and the least likely to maintain biological activity. However, many detergents can be used without decreasing the biological activity. A recent interesting development is the synthesis of sodium oligooxyethylene dodecyl ether sulfates.

Protein aggregation may be inhibited by surfactants (1 % SDS, 0.1 % dodecylpoly(oxyethyleneglycolether)$_n$, 0.01 % poly(oxyethylene)$_n$-sorbitane-monooleate, or 0.01 % octylphenolpoly(ethyleneglycolether)$_n$) as shown for insulin [141]. Complete reversibility of rhodanese folding can be achieved with the use of the non-ionic detergent dodecyl-β-D-maltoside (lauryl maltoside). The detergent stabilizes an intermediate with exposed hydrophobic surfaces and prevents aggregation [142]. Not only this detergent, but other non-ionic, as well as zwitterionic, detergents also have favorable effects in activating the denatured state [143]. Anionic detergents such as SDS are known to increase the α-helix content of several proteins at low protein concentrations [144,145]. Increase of helix content to levels above that of the native protein has been shown for lysozyme, Bence Jones protein, W floribunda lectin, and Histone H2B in presence of SDS. The newly formed helices were stabilized by hydrophobic shielding [146].

Detergents bind to proteins and their removal may be extremely difficult. Another drawback is the restriction of purification techniques available when detergents are used in downstream processing. Ionic detergents will rule out ion-exchange chromatography, and most detergents will interfere with hydrophobic interaction and affinity chromatography.

Another specific group of compounds binding to proteins is molecular chaperones that mediate the correct assembly of other polypeptides. Chaperones bind to aggregation-prone conformations of intermediates, thereby stabilizing the unfavorable conformation. The positive effect of protein disulfide isomerase and chaperones were shown in the production of a functional scFv fragment in a *E. coli* cell-free translation system [147].

Preferential interaction measurements have been carried out on a variety of co-solvents [148]. Without exception, all are preferentially excluded from the native globular protein, and preferential hydration always induces salting-out, regardless of the mechanism of the preferential interaction [149]. Thus, a strong precipitant is not necessarily a good stabilizer because of the possible interaction with the protein surface, while a good stabilizer will also be a good precipitant. It is, a priori, impossible to state whether a co-solvent will stabilize or destabilize a protein. The effect will depend on possible interactions with the protein surface. The subject is reviewed by Timasheff [130,148,150].

2.4.5 The Effect of Protein Concentration

Protein denaturation is a reaction of first order [86] and therefore independent of protein concentration. In contrast, protein aggregation is a reaction of second order where the rate of reaction increases with protein concentration [85]. Therefore, formation of aggregates can be suppressed by working at low protein concentrations.

The rate of insulin fibrillation was shown to be a function of the insulin concentration. In acidic solution (pH 2.5, 21 °C) the increase of solution viscosity was much faster in solutions of relatively high insulin concentration [34].

Glucagon was shown to aggregate at concentrations above 1.5 mg/ml in 0.01 M hydrochloric acid at 30 °C. At 1 mg/ml the reaction proceeds very slowly, confirming the strong concentration dependence of the fibrillation reaction [151].

2.4.6 The Effect of Pressure

Pressure does affect protein stability, but at levels far exceeding those met in downstream processing [2,152–155]. At pressures above 570 MPa at 30 °C, secondary structure elements of ribonuclease A co-operatively disrupted to a fully unfolded state without any residual structure [156].

2.5 Summary

Proteins are only marginally stable in aqueous solutions, the predominant stabilizing forces being the hydrophobic effect and hydrogen bonding. Mutual environmental factors affect the chemical and physical stability of proteins, ranging from presence of stabilizing proteins to parameters such as temperature and pH. A protein undergoing unfolding may not easily regain its native structure, either because of covalent modifications, formation of stable intermediates, or because of aggregation. It is, therefore, an important part of the downstream processing strategy to maintain the native structure of the protein throughout purification. In other words, to maintain the specific biological activity at a level comparable with the *in vivo* activity, assuming that the native structure is that of highest biological activity.

Two issues are of concern before a purification strategy can be laid out. The physical and chemical properties of the protein are generally unknown prior to purification, leaving few data available to the protein chemist. Secondly, the influence of pH, redox potential, and co-solvents on protein stability is virtually unknown, increasing the possibility of unintended destabilization of the protein during purification.

A protein's stability in the initial extract can be investigated provided that a bioassay is at hand. Under such circumstances, the bioactivity can be measured as a function of pH, temperature, redox potential, concentration of co-solvents, etc. which can give an idea of the parameter intervals at which the protein is stable, and which co-solvent will be of use during purification. It will also be possible to investigate whether the native structure can be re-established after destabilization, and thereby one can partly characterize the possible upper an lower limits of the said parameters.

A downstream process comprises a set of different unit operations with the purpose of increasing the purity of the protein. In order to obtain the highest possible purification factor of each step, parameters like pH, temperature, ionic strength, and protein concentration are often optimized at the expense of protein stability and recovery. In chromatography, each unit operation consists of an application, a binding procedure, and an elution procedure characterized by a set of (different) parameters. Thus, the application sample, the immobilized protein, or the eluted fraction makes up systems well defined by the parameter intervals. Fortunately, most of the parameters influencing protein stability can be measured in order to determine the lower and upper limits of each parameter which influence the protein stability in a given system.

Once the system is defined, the protein stability can be tested by investigating the transition from the native to the unfolded state as a function of temperature and thereby making it possible to determine T_m.

The robustness of the system and the identification of major interactions between parameters can be studied by employing fractional factorial statistical designs allowing efficient means of testing many variables using a reduced number of data [157]. A well-defined operating space is of course a prerequisite in the production of bulk materials meant for biopharmaceuticals. The notorious instability of many proteins, leading to uncontrolled aggregation and degradation during purification, seriously raises the question whether protein stability should be paid as much attention as protein purity during downstream processing.

A downstream process is normally divided into three parts: capture, intermediary purification, and polishing. The purpose of the capture procedure is to separate the protein from host cell proteins, lipoproteins, lipids, DNA or RNA, cell debris, Maillard compounds, and to remove excess water. In this early purification phase, proteolytic degradation may constitute a major problem, as could expression of misfolded forms of the molecule due to the foreign host used, and to over-expression of the protein. The incorrectly folded forms may initiate aggregation, resulting in severe loss of product. Therefore, very gentle methods are recommended for purification at that stage. The most brilliant solution to the problem is to make use of highly selective affinity matrices with ligands recognizing only the native structure. Not only does this concept assure a high purification factor, but the mild application and elution conditions characterizing affinity chromatography favor the stability of the native protein. The lack of effective cleaning procedures for immobilized antibodies and their high production costs have severely restricted the use of affinity chromatography in large-scale downstream processing. However, with the introduction of mimetic triazine-based organic ligands [158,159] costs can be reduced to levels acceptable for large-scale operations. The triazine-based ligands can even be cleaned in 0.5 to 1.0 M NaOH without disrupting the matrix.

The polishing steps are mainly introduced to isolate the product from derivatives having close to similar physical and chemical properties. Proteins with perhaps a stabilizing effect on the protein have been removed, and the protein is surely present in much higher concentrations than in the initial extract. At this stage of the downstream process, the choice of co-solvents and combinations of parameters may be of primary importance for protein stability, while contaminants like proteolytic enzymes no longer constitute a problem.

With this, hopefully, the questions raised in the introduction to this chapter have been at least partly answered. Factors influencing protein stability have been identified, extreme conditions can be dealt with provided that the parameters influencing protein stability have been carefully adjusted, and a strategy has been laid out to assure the stability of the native structure during downstream processing. It is our belief that improved knowledge of protein stability is an important tool in the design

of the downstream process. By mastering the different unit operations even denatured proteins can be brought back to their stabilized native form by *in vitro* folding,
reflecting the great potential for manipulation of stability during purification.

Acknowledgements

We thank Dr^s Jens Brange, Daniel Otzen, James Flink, and Benny Wellinder for
valuable discussions and comments. We also thank Jytte Jarsholt and Dorte Worm
for their help in preparing the manuscript.

Abbreviations and Symbols

[ ]	concentration
C_P	heat capacity
ΔC_P	change in heat capacity
DTT	dithiothreitol
E	redox potential of the solution
E_h	redox potential relative to the standard hydrogen electrode
E^{m0}	redox potential at [Ox] = [Red] at pH 0 at unit activity
E^{m7}	redox potential at [Ox] = [Red] at pH 7 at unit activity
E_c	redox potential relative to the calomel electrode
e^-	electron
EDTA	ethylenediaminetetra-acetic acid
F	Faraday's constant
ΔG	change in Gibbs' free energy
$\Delta G°$	change in Gibbs' free energy under standard conditions
ΔG_U	difference in Gibbs' free energy for unfolding
GAD_{65}	glutamic acid decarboxylase EC 4.1.1.15
Gdn	guanidinium
ΔH_u	enthalpy difference for unfolding
ΔH_m	ΔH_u at the melting temperature
HPLC	high-performance liquid chromatography
IGF-1	insulin-like growth factor 1
I_{agg}	aggregated intermediary compound
I_{irr}	stable intermediary compound
K_a	acid constant
K_{eq}	equilibrium constant
MG	molten globule state
MPD	2-methyl-2,4-pentadiol
N	native state
n	number of electrons transferred
Ox	oxidized state

PEG	polyethylene glycol
pI	isoelectric point
pK_a	$-\log K_a$
PLP	pyridoxal 5'-phosphate
PMSF	phenylmethanesulfonyl fluoride
Pt	platinum
R	gas constant
Red	reduced state
RP	reversed-phase
ΔS_m	ΔS_U at the melting temperature
ΔS_U	entropy difference for unfolding
T	absolute temperature (in Kelvin)
T_m	melting temperature
T_s	temperature where $\Delta S_U = 0$
U	unfolded state
UV	ultraviolet

References

[1] Pace, C. N., *Crit Rev Biochem*, 1975, *3*, 1–43.

[2] Privalov, P. L., *Adv Protein Chem*, 1979, *33*, 167–241.

[3] Kauzmann, W., *Adv Protein Chem*, 1959, *14*, 1–63.

[4] Pace, C. N., *J Mol Biol*, 1992, *226*, 29–35.

[5] Burley, S. K. and Petsko, G. A., *Trends Biotechnol*, 1989, *7*, 354–359.

[6] Alber, T., *Annu Rev Biochem*, 1989, *58*, 765–798.

[7] Sali, D., Bycroft, M., Ferscht, A. R., *J Mol Biol*, 1991, *220*, 779–788.

[8] Serrano, L., Kellis, J. T. Jr., Cann, P., Matouschek, A., Ferscht, A. R., *J Mol Biol*, 1992, *224*, 783–804.

[9] Matthews, B. W., *Curr Opin Struct Biol*, 1993, *3*, 589–593.

[10] Stickle, D. F., Presta, L. G., Dill, K. A., Rose, G. D., *J Mol Biol*, 1992, *226*, 1143–1159.

[11] Shirley, B. A., Stanssens, P., Hahn, U., Pace, C. N., *Biochemistry*, 1992, *31*, 725–732.

[12] Dill, K. A., *Biochemistry*, 1990, *29*, 7133–7155.

[13] Burley, S. K., Petsko, G. A., *Science*, 1985, *229*, 23–28.

[14] Creighton, T. E., *Biopolymers*, 1983, *22*, 49–58.

[15] Privalov, P. L., *Annu Rev Biophys Chem*, 1989, *18*, 47–69.

[16] Creighton, T. E., *Curr Opin Struct Biol*, 1991, *1*, 5–16.

[17] Richards, F. M., *Annu Rev Biophys Bioeng*, 1977, *6*, 151–176.

[18] Rose, G. D., *Annu Rev Biophys Biomol Struct*, 1993, *22*, 381–415.

[19] Baker, E. N., Hubbard, R. E., *Prog Biophys Molec Biol*, 1984, *44*, 97–179.

[20] Chothia, C., *J Mol Biol*, 1976, *105*, 1–14.

[21] Matthews, B. W., *Nat Struct Biol*, 1995, *2*, 85–86.

[22] Creighton, T. E., *J Phys Chem*, 1985, *89*, 2452–2459.

[23] Matthews, B. W., *Curr Opin Struct Biol*, 1991, *1*, 17–21.

[24] Jaenicke, R., *Eur J Biochem*, 1991, *202*, 715–728.

[25] Seckler, R., Jaenicke, R., *FASEB J*, 1992, *6*, 2545–2552.

[26] Oobatake, M., Tatsuo, O., *Prog Biophys Mol Biol*, 1993, *59*, 237–284.

[27] Ptitsyn, O. B., Uversky, V. N., *FEBS Lett*, 1994, *341*, 15–18.

[28] Privalov, P. L., *Adv Protein Chem,* 1979, *33,* 167–241.

[29] Jaenicke, R., *Prog Biophys Mol Biol,* 1987, *49,* 117–237.

[30] Creighton, T. E., *Biochem J,* 1990, *270,* 1–16.

[31] Shortle, D., *FASEB J,* 1996, *10,* 27–34.

[32] Neri, D., Billeter, M., Wider, G., Wüthrich, K., *Science,* 1992, *257,* 1559–1563.

[33] Creighton, T. E., in: *Mechnisms of Protein Folding:* Pain, R. H. (Ed.), Oxford: IRL Press, 1994; pp. 1–25.

[34] Brange, J., Andersen, K., Laursen, E. D., Meyn, G., Rasmussen, E., *J Pharm Sci,* 1997, in press.

[35] North, M. J., in: *Proteolytic Enzymes – a practical approach:* Beynon, R. J., Bond, J. S. (Eds.), Oxford: IRL Press, 1989; pp. 105–124.

[36] Battersby, J. E., Hancock, W. S., Canova-Davis, E., Oeswein, J., O'Connor, B., *Int Peptide Protein Res,* 1994, *44,* 215–222.

[37] Schultz, J., *Methods Enzymol,* 1967, *11,* 255–263.

[38] Piszkiewicz, D., Landon, M., Smith, E. L., *Biochem Biophys Res Commun,* 1970, *40,* 1173–1178.

[39] Whitaker, J. R., Feeney, R. E., *Crit Rev Food Sci Nutr,* 1983, *19,* 173–212.

[40] Hayashi, R., Kameda, I., *Agric Biol Chem,* 1980, *44,* 175–181.

[41] Berson, S. A., Yalow, R. S., *Diabetes,* 1966, *15,* 875–879.

[42] Brange, J., Langkjaer, L., Havelund, S., Voelund, A., *Pharm Res,* 1992, *9,* 715–726.

[43] Lewis, U. J., Cheever, E. V., Hopkins, W. C., *Biochim Biophys Acta,* 1970, *214,* 498–508.

[44] Flatmark, T., *Acta Chem Scand,* 1966, *20,* 1487–1496.

[45] Geiger, T., Clarke, S., *J Biol Chem,* 1987, *262,* 785–794.

[46] Clarke, S., Stephenson, R. C., Lowenson, J. D., in: *Stability of Protein Pharmaceuticals:* Ahern, T. J., Manning, M. C., (Eds.), New York: Plenum Press, 1992; Part A, pp. 1–29.

[47] Chazin, W. J., Kossiakoff, A. A., in: *Deamidation and Isoaspartate Formation in Peptides and Proteins:* Aswad, D. W. (Ed.), London: CRC Press, 1995; pp. 194–206.

[48] Slobin, L. I., Carpenter, F. H., *Biochemistry,* 1963, *2,* 22–28.

[49] Brennan, T. V., Clarke, S., in: *Deamidation and Isoaspartate Formation in Peptides and Proteins:* Aswad, D. W. (Ed.), London: CRC Press, 1995, pp. 66–90.

[50] Yüksel, K. U., Gracy, R. W., *Arch Biochem Biophys,* 1986, *248,* 452–459.

[51] Bhatt, N. P., Patel, K., Borchardt, R. T., *Pharm Res,* 1990, *7,* 593–599.

[52] Robinson, A. B., Rudd, C. J., in: *Current Topics in Cellular Regulation:* Horecker, B. L. and Stadtmann, E. R., (Eds.), New York: Academic Press, 1974; pp. 247–295.

[53] Manning, M. C., Patel, K., Borchardt, R. T., *Pharm Res,* 1989, *6,* 903–918.

[54] Wright, H. T., *Protein Eng,* 1991, *4,* 283–294.

[55] Johnson, B. A., Aswad, D. W., in: *Deamidation and Isoaspartate Formation in Peptides and Proteins:* Aswad, D. W. (Ed.), London: CRC Press, 1995; pp. 91–113.

[56] Becker, G. W., Tackitt, P. M., Bromer, W. W., Lefeber, D. S., Riggin, R. M., *Biotech Appl Biochem,* 1988, *10,* 326–337.

[57] Scotchler, J. W., Robinson, A. B., *Anal Biochem,* 1974, *59,* 319–322.

[58] Sen, L. C., Gonzalez-Flores, E., Feeney, R. E., Whitaker, J. R., *J Agric Food Chem,* 1977, *25,* 632–638.

[59] Lee, H. S., Osuga, D. T., Nashef, A. S., Ahmed, A. I., Whitaker, J. R., Feeney, R. E., *J Agric Food Chem,* 1977, *25,* 1153–1158.

[60] Nashef, A. S., Osuga, D. T., Lee, H. S., Ahmed, A. I., Whitaker, J. R., Feeney, R. E., *J Agric Food Chem,* 1977, *25,* 245–251.

[61] Volkin, D. B., Klibanov, A. M., *J Biol Chem,* 1987, *262,* 2945–2950.

[62] Masters, P. M., Friedman, M., *ACS Symp Ser,* 1980, *123,* 165–194.

[63] Hayashi, R., Kameda, I., *Agric Biol Chem,* 1980, *44,* 891–895.

[64] Dedman, M. L., Farmer, T. H., Morris, C. J. O. R., *Biochem J,* 1961, *78,* 348–352.

[65] Tashijan, A. H. J., Ontjes, D. A., Munson, P. L., *Biochemistry,* 1964, *3,* 1175–1182.

[66] Morley, J. S., Tracy, H. J., Gregory, R. A., *Nature,* 1965, *207,* 1356–1359.

[67] Brot, N., Weissbach, H., *Trends Biochem Sci,* 1982, *7,* 137–139.

[68] Cleland, J. L., Powell, M. F., Shire, S. J., *Crit Rev Ther Drug Carrier Sys,* 1993, *10,* 307–377.

[69] Gilbert, H. F., *Adv Enzymol,* 1990, *63,* 69–172.

[70] Wold, F., *Macromolecules: Structure and Function.* London: Prentice-Hall, Inc. 1971; pp. 53–89.

[71] Martin, B. M., Viswanatha, T., *Biochem Biophys Res Commun,* 1975, *63,* 247–253.

[72] Torchinsky, Y. M., in: *Sulfur in Proteins:* Metzler, D. (Ed.), Oxford: Pergamon Press, 1995; pp. 199–217.

[73] Shaked, Z., Szajewski, R. P., Whitesides, G. M., *Biochemistry,* 1980, *19,* 4156–4166.

[74] Soerensen, H. H., Thomsen, J., Bayne, S., Hoejrup, P., Roepstorff, P., *Biomed Environ Mass Spec,* 1990, *19,* 713–720.

[75] Benesch, R. E., Benesch, R., *J Am Chem Soc,* 1958, *80,* 1666–1669.

[76] Florence, T. M., *Biochem J,* 1980, *189,* 507–520.

[77] Torchinsky, Y. M., in: *Sulfurin Proteins:* Metzler, D. (Ed.), Oxford: Pergamon Press, 1981; pp. 57–65.

[78] Misra, H. P., *J Biol Chem,* 1974, *249,* 2151–2155.

[79] Jespersen, A. M., Christensen, T., Klausen, N. K., Nielsen, P. F., Soerensen, H. H., *Eur J Biochem,* 1994, *219,* 365–373.

[80] Tanford, C., *Adv Protein Chem,* 1968, *23,* 218–281.

[81] O'Fágáin, C., Sheehan, H., O'Kennedy, R., Kilty, C., *Process Biochem,* 1988, *XX,* 166–171.

[82] Kunth, I. D. J., in: *The Protein Folding Problem:* Wetlaufer, D. B. (Ed.), AAAS Selected Symposium 89, 1984; pp. 65–81.

[83] Matthiesen, F., Hejnaes, K. R., Skriver, L., *Ann N Y Acad Sci,* 1996, *782,* 413–421.

[84] Hejnaes, K. R., Bayne, S., Noerskov, L., et al., *Protein Eng,* 1992, *5,* 797–806.

[85] Kiefhaber, T., Rudolph, R., Kohler, H.-H., Buchner, J., *Biotechnology,* 1991, *9,* 825–829.

[86] Jaenicke, R., in: *Protein Conformation:* New York: John Wiley and Sons, 1991, pp. 206–217.

[87] Oliveberg, M., Fersht, A. R., *Biochemistry,* 1996, *35,* 2738–2749.

[88] Speed, M. A., Wang, D. I. C., King, J., *Nat Biotech,* 1996, *14,* 1283–1287.

[89] De Young, L. R., Dill, K. A., Fink, A. L., *Biochemistry,* 1993, *32,* 3877–3886.

[90] Moody, A. J., Hejnaes, K. R., Marshall, M. O. et al., *Diabetologia,* 1995, *38,* 14–23.

[91] Chou, P. Y., Fasman, G. D., *Biochemistry,* 1975, *14,* 2536–2541.

[92] Zettlmeissl, G., Rudolph, R., Jaenicke, R., *Biochemistry,* 1979, *18,* 5567–5571.

[93] DeFelippis, M. R., Alter, L. A., Pekar, A. H., Havel, H. A., Brems, D. N., *Biochemistry,* 1993, *32,* 1555–1562.

[94] Sluzky, V., Klibanov, A. M., Langer, R., *Biotech and Bioeng,* 1992, *40,* 895–903.

[95] Mendoza, J. A., Rogers, E., Lorimer, G. H., Horowitz, P. M., *J Biol Chem,* 1991, *266,* 13587–13591.

[96] Bai, Y., Milne, J. S., Mayne, L., Englander, S. W., *Proteins,* 1994, *20,* 4–14.

[97] Cleland, J. L., Wang, I. C., *Biochemistry,* 1990, *29,* 11072–11078.

[98] Tandon, S., Horowitz, P., *J Biol Chem,* 1986, *261,* 15615–15681.

[99] Hartley, D. L., Kane, J. F., *Biochem Soc Trans,* 1988, *16,* 101–102.

[100] Scopes, R. K., in: *Protein Purification – Principles and Practice:* New York: Springer-Verlag, 1987; pp. 41–71.

[101] Hoffmeister, F., *Naun-Schmiedelberg's Archiv Pharmak. Exp. Patholog (Leipzig),* 1988, *24,* 249–260.

[102] Pace, C. N., *Trends Biochem Sci,* 1990, *15,* 14–17.

[103] Becktel, W. J., Schellman, J. A., *Biopolymers,* 1987, *26,* 1859–1877.

[104] Doty, P., Imahori, K., Klemperer, E., *Proc Natl Acad Sci USA,* 1958, *44,* 424–431.

[105] Hippel, P. H., Schleich, T., in: *Structure and Stability of Biological Macromolecules:* Timasheff, S. N., Fasman, G. D. (Eds.). New York: Marcel Dekker Inc. 1969; pp. 417–557.

[106] Tombs, M. P., *J Appl Biochem,* 1985, *7,* 3–24.
[107] Sundby, F., *J Biol Chem,* 1962, *237,* 3406–3411.
[108] Wheelock, E. F., *Science,* 1965, *149,* 310–311.
[109] Hoshino, T., Mikura, Y., Shimidzu, H., Kusumoto, S., Kawai, J., Toguchi, H., *Biochim Biophys Acta,* 1987, *916,* 245–250.
[110] Shami, E. Y., Rothstein, A., Ramjeesingh, M., *Trends Biotechnol,* 1989, *7,* 186–190.
[111] Whitaker, J. R., *ACS Symp Ser,* 1980, *123,* 145–163.
[112] Chalikian, T. V., Gindikin, V. S., Breslauer, K. J., *FASEB J,* 1996, *10,* 164–170.
[113] Porter, W. R., Staack, H., Brand, K., Manning, M. C., *Thromb Res,* 1993, *71,* 265–279.
[114] Janin, J., *Colloids and Surfaces,* 1984, *10,* 1–7.
[115] Makhatadze, G. I., Privalov, P. L., *J Mol Biol,* 1993, *232,* 639–659.
[116] Shellmann, J. A., Lindorfer, M., Hawkes, R., Grutter, M., *Biopolymers,* 1981, *20,* 1989–1999.
[117] Privalov, P. L., Khechinashvili, N. N., *J Mol Biol,* 1974, *86,* 665–684.
[118] Privalov, P. L., Griko, Y. V., Venyaminov, S. Y., Kutyshenko, V. P., *J Mol Biol,* 1986, *190,* 487–498.
[119] Hermans, J. J., Scheraga, H. A., *J Am Chem Soc,* 1961, *83,* 3283–3292.
[120] Volkin, D. B., Middaugh, C. R., in: *Stability of Protein Pharmaceuticals:*
[121] Baldwin, R. L., Eisenberg, D., in: *Protein Engineering:* Oxender, D. L., Fox, C. F. (Eds.), New York: Alan R. Liss, Inc, 1987; pp. 127–148.
[122] Tomazic, S. J., Klibanov, A. M., *J Biol Chem,* 1988, *263,* 3086–3091.
[123] Zale, S. E., Klibanov, A. M., *Biochemistry,* 1986, *25,* 5432–5444.
[124] Cleland, W. W., *Biochemistry,* 1964, *3,* 480–482.
[125] Scott, E. M., Duncan, I. W., Ekstrand, V., *J Biol Chem,* 1963, *238,* 3928-3933.
[126] *CRC Handbook of Chemistry and Physics.* 76th ed. New York: 1996; pp. 8.21–8.33.
[127] Dutton, P. L., *Methods Enzymol,* 1978, *54,* 411–435.
[128] Maeda, Y., Yamada, H., Ueda, T., Imoto, T., *Protein Eng,* 1996, *9,* 461–465.
[129] Hospelhorn, V. D., Cross, B., Jensen, E. V., *J Am Chem Soc,* 19854, *76,* 2827–2829.
[130] Timasheff, S. N., in: *Stability of Protein Pharmaceuticals:* Ahern, T. J., Manning, M. C. (Eds.), New York: Plenum Press, 1992; Part B, pp. 265–285.
[131] Arakawa, T., Prestrelski, S. J., Kenney, W. C., Carpenter, J. F., *Adv Drug Del Rev,* 1993, *10,* 1–28.
[132] Arakawa, T. R., Timasheff, S. N., *Biochemistry,* 1985, *24,* 6756–6762.
[133] Pittz, E. P., Timasheff, S. N., *Biochemistry,* 1978, *17,* 615–623.
[134] London, J., Skrzynia, C., Goldberg, M. E., *Eur J Biochem,* 1974, *47,* 409–415.
[135] Brems, D. N., Havel, H. A., *Proteins,* 1990, *5,* 93–95.
[136] Buchner, J., Rudolph, R., *Biotechnology,* 1991, *9,* 157–162.
[137] Winkler, M. E., Blaber, M., *Biochemistry,* 1986, *25,* 4041–4045.
[138] Zardeneta, G., Horowitz, P. M., *J Biol Chem,* 1992, *267,* 5811–5816.
[139] Kristjansson, M. M., Kinsella, J. E., *Adv Food Nutr Res,* 1991, *35,* 237–316.
[140] Timasheff, S. N., *Annu Rev Biomol Struct,* 1993, *22,* 67–97.
[141] Lougheed, W. D., Albisser, A. M., Martindale, H. M., Chow, J. C., Clement, J. R., *Diabetes,* 1983, *32,* 424–432.
[142] Tandon, S., Horowitz, P. M., *J Biol Chem,* 1989, *264,* 9859–9866.
[143] Tandon, S., Horowitz, P. M., *J Biol Chem,* 1987, *262,* 4486–4491.
[144] Jirgensons, B., *J Biol Chem,* 1967, *242,* 912–918.
[145] Hunt, A. H., Jirgensons, B., *Biochemistry,* 1973, *12,* 4435–4441.
[146] Jirgensons, B., *J Protein Chem,* 1982, *1,* 71–84.
[147] Ryabova, L. A., Desplancq, D., Spirin, A. S., Plückthun, A., *Biochemistry,* 1997, *15,* 79–84.
[148] Timasheff, S. N., *Annu Rev Biophys Biomol Struct,* 1993, *22,* 67–97.
[149] Arakawa, T., Bhat, R., Timasheff, S. N., *Biochemistry,* 1990, *29,* 1914–1923.
[150] Timasheff, S. N., in: *Protein–Solvent Interactions.* Gregory, R. B. (Ed.), New York: Marcel Dekker, Inc., 1994; pp. 445–482.

[151] Beaven, G. H., Gratzer, W. B., Davies, H. G., *Eur J Biochem,* 1969, *121,* 37–42.

[152] Heremans, K., *Annu Rev Biophys Bioeng,* 1982, *11,* 1–21.

[153] Silva, J. L., *Annu Rev Phys Chem,* 1993, *44,* 90–113.

[154] Gross, M., Jaenicke, R., *Eur J Biochem,* 1994, *221,* 617–630.

[155] Goossens, K., Smeller, L., Frank, J., Heremans, K., *Eur J Biochem,* 1996, *236,* 254–262.

[156] Takeda, N., Kato, M., Taniguchi, Y., *Biochemistry,* 1995, *34,* 5980–5987.

[157] Montgomery, D. C., in: *Design and Analysis of Experiments:* New York: John Wiley Sons, 1991; pp. 197–249.

[158] Clonis, Y. D., Stead, C. V., Lowe, C. R., *Biotech Bioeng,* 1987, *30,* 621–627.

[159] Lowe, C. R., Burton, S. J., Burton, N. P., Alderton, W. K., Pitts, J. M., Thomas, J. A., *Trends Biotechnol,* 1992, *10,* 442–448.

3 Production of Transgenic Protein

Gordon Wright and John Noble

3.1 Introduction

As the market penetration of recombinant DNA technologies increases, more and more pharmaceutical companies are looking to this area to strengthen their new product pipelines [1]. Central to the success of this strategy is the development of production techniques that are price competitive with, or demonstrating a higher safety and efficacy than, conventiona) techniques such as surgery, chemotherapy, and existing drug products [2].

One of the most exciting advances within biotechnology in recent years has been transgenic technology, whereby the secretion of human proteins in the mammalian mammary gland could potentially revolutionize commercial protein production. The central strength of the technology is the ability to produce, in high volume, low-cost complex proteins for which current production methods are prohibitively expensive at large scale, or no successful production method is yet available.

The following chapter overviews the technology and examines key issues that must be addressed in process development, facility design, and regulatory affairs as products reach commercial goals.

3.2 Overview of Transgenic Technology

The core technology for the manufacture of proteins in the milk of transgenic animals is illustrated in Fig. 3-1. First, the DNA which encodes the therapeutic protein is linked with a milk gene promoter which ensures that the gene is switched on and the protein is made only in the mammary gland. Then the linked DNA is injected into the pronucleus of a fertilized egg which is placed in a foster mother and develops naturally through pregnancy to full term. The resultant offspring are screened for the presence of the transgene and are mated when mature. The human protein in the milk of the lactating animals is then measured and analysed. The animal which yields the highest levels of the required protein is used to produce a male which will become the founder animal of the production flock. This ensures that genetic varia-

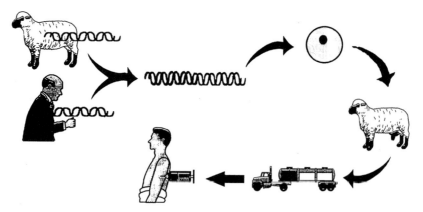

Fig. 3-1. Core technology.

tion is kept to a minimum. Once a founder is established, all further work is carried out using conventional animal breeding.

The levels and bioactivity of the protein in the milk of the transgenic animals are analysed and the purification processes and assays are then designed and developed. The protein is isolated from the milk using a combination of techniques adapted from the dairy industry and conventional pharmaceutical separation techniques such as column chromatography. This approach can be applied to a range of species including sheep, cows, goats, pigs, and rabbits. The choice is determined by issues such as time, cost, and the quantity of product required.

Although the above describes how most transgenic animals are currently produced, the use of newer technologies would allow more precise modifications and be more efficient. For example, in the mouse one can make use of embryonic

Table 3-1. Manufacturing routes for biopharmaceutical products.

Method of Production	Transgenics	Blood fractionation	Bacterial/fungal fermentation	Mammalian cell culture
Secretion levels (g/L)	1 to 30	Product-specific	1 to 3[a]	0.1 to 0.8[a]
Purification yield (%)	50[a]	Product-specific	10[a]	60[a]
Product complexity	Wide-ranging	All blood proteins	No human glycosylation	Wide-ranging
Scale-up	Straightforward	Straightforward	Complex	Very complex
Feedstock	Renewable	Limited supply	Defined	Defined and expensive
Safety issues	Species barrier to viral transfer	Same species viral transfer	Control of GMOs	Control of virus

[a] From reference [2].

stem (ES) cells to target a specific region of the genome, adding or removing DNA sequences. This technology would be extremely beneficial in large animals, but to date no ES cells have been isolated for any of the livestock species. The recent advent of cloning [3] in livestock offers the possibility of many, if not all, of the same advantages. A comparison of transgenics with three well-established production methods: blood fractionation, bacterial fermentation, and mammalian cell culture, is given in Table 3-1.

From Table 3-1 it is clear that transgenic technologies are capable of: (i) very high volume production; (ii) complex molecule production; and (iii) low cost production. The latter point is highlighted further by industry data which demonstrate up to 35 % reduction in production costs through the switch from fermentation to transgenic production routes [2]. It is thus easy to understand why companies such as PPL Therapeutics, Genzyme Transgenics, and Pharming have invested heavily in the technology [4].

3.3 Process Design

A typical transgenic production process is shown in Fig. 3-2. At first sight, although the unit operations involved within the process seem like a standard biopharmaceutical process, closer examination reveals a unique combination of process operations. Progressing through the process we move gradually from an agricultural environment, through processing similar to that employed in dairies, into primary biopharmaceutical production, and then into a classical secondary pharmaceutical environment. The critical issue in process design is to facilitate this progression, through bands of differing technology, without compromising product quality.

Central to this is the specification of relevant standards for each area of process operation. The process can be split into six areas: animal handling, milking, milk handling, primary recovery, polishing, and formulation/filling. A design philosophy

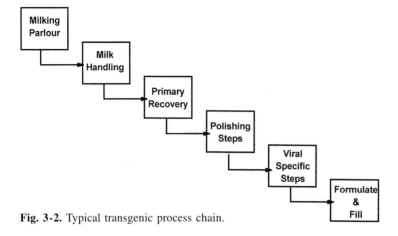

Fig. 3-2. Typical transgenic process chain.

can be developed for each of these areas ensuring that systems are fit for purpose, readily adapted to scale-up, and display a realistic approach to the technologies involved.

3.3.1 Animal Handling

In specifying the animal handling facilities one must have a clear vision of what the end point for a particular product will be. For example, the production of metric tonne quantities of protein in sheep, goats, or cows is likely to require vastly different strategies to the production of gramme quantities in rabbits. This is addressed in more detail in Section 3.4.6.

To provide the necessary quality assurance, each animal within the production group should be effectively tagged, to provide unique identification and allow details of life history and health status to be recorded. Before admission to the milking flock each animal should comply with a minimum defined health status and produce milk consistent with predefined parameters.

3.3.2 Milking

The milking area is a critical point within a transgenic production process as it is the point at which the product is 'harvested' from the production animal. It is thus the first time at which the milk can become contaminated with adventitious agents. It is not practical to milk animals under closed/aseptic conditions. Even in the cleanest dairy environment with pipework sterilization, levels of bacteria up to 10 000 cfu/ml can still be detected [6]. Milking practices, equipment design, and process schedules must be designed to address this problem and ensure end-product quality is not compromised.

Milking equipment is best selected from high-quality dairy systems with additional attention paid to issues of maximizing milk recovery, validation of system cleaning, and specification of control systems. Milk recovery will be critical during start-up when animal numbers are likely to be low. The integration of the parlour control system with the animal tagging system is essential to the maintenance of milk quality, ensuring that milk from risk animals does not enter the process stream.

Typical practice at this stage would be to pool the milk from a single milking for shipping to the milk handling process. Any manipulation of individual milkings must take place within pre-defined acceptance criteria.

3.3.3 Milk Handling

To make best use of downstream processing equipment it is advantageous to be able to hold the milk after collection for subsequent pooling/sub lotting. This allows the purification batch size to be tailored to fit the available equipment and decouples purification scale from animal numbers. Typically, during early production several milkings will be pooled in a single batch while as animal numbers grow milkings may be split into several batches. The key issue at this stage is thus the generation of a stable process stream which can be stored for significant periods. The levels of contamination outlined in Section 3.3.2 could, after only a few hours, lead to product loss. In a conventional dairy process this problem is addressed via pasteurization. This technique may not be acceptable in the production of certain proteins. Historical data indicate that when stored at 4 °C, unpasteurized milk can begin to sour in under 20 h. It is thus unlikely that we would look to hold milk for use in biopharmaceutical production for more than 12 h without stabilization. For robust proteins, freezing and low-temperature storage, −50 °C, may be possible. For labile molecules, techniques such as chemical stabilization should be investigated. If no effective process is available, the holding of milk prior to purification will not be possible and downstream operations will have to be designed accordingly.

Fat removal could also be a key step at this stage of the process as lipid micelles can have a serious effect upon the classical downstream processing operations that will be encountered later in the process. Disruption of the lipid micelles via shear can also lead to souring of the milk and the process stream should be handled with low-shear equipment where possible.

3.3.4 Primary Recovery

The main aim at this stage in the purification is to separate the target protein from the bulk of the high concentration matrix of sugars, lipids, and other proteins which constitute the milk. A further key issue at this stage is that the process stream may well have a significant bioburden. It is thus essential that any purification strategy generates a process stream for which the bioburden can be controlled by, for example, filtration. Of prime importance is the removal of caseins (which form the bulk of the milk protein) from the process stream. One potential approach involves the solubilization of the caseins with the capture of the target protein on a chromatography column [7].

Typically, one would anticipate several purification steps resulting in bulk-purified target protein in a suitable buffer solution which can be filtered through a 0.2 micron filter. For proteins that are sensitive to protease attack the speed and temperature of these operations will be critical to product yield.

Equipment encountered at this stage will be conventional biopharmaceutical recovery systems [8] such as chromatography and centrifuges. Given the special nature of the milk, real system trials are essential, even if milk reserves are scarce. Care has to be taken in educating vendors as to the nature of the process, as few will be familiar with the protein levels encountered in product and waste streams. This is especially true where precipitation is used.

3.3.5 Polishing Steps

Within this phase of the purification process the aim is to move from a bulk-purified product to a final product of the required purity. For human use, purities in excess of 99.9 % may be required [2].

Equipment and operations will be the same as for most modern therapeutic protein production systems [8], including chromatography and tangential flow filtration.

3.3.6 Formulation and Filling

As with the final polishing steps the formulation and filling operation will most likely be identical to classical liquid dosage from products including buffer exchange, concentration, sterile filtration, filling, and freeze-drying.

3.3.7 Viral-Specific Steps

The control of viral/prion contaminants is a critical issue with mammalian-derived products. The removal of these components must be built into the purification strategy. Operations such as precipitation, chromatography, and freeze-drying all have a viral inactivation capacity and as such one can consider the whole process to act as a viral screen. Despite this, for less well-characterized feedstocks, of which milk is one, manufacturing processes will normally incorporate generic viral inactivation or removal steps [9]. Examples of such steps include holding at low pH, pasteurization, dry heat, solvent/detergent, filtration, UV irradiation, and microwave irradiation. Typically one would expect to see an inactivation/removal step as the final stage of polishing. The choice of step and its integration into the process can be critical to product yield, facility design, and process economics and must be assessed in detail.

3.3.8 Utilities

Appropriate choice of utilities is central to the design of safe, robust and cost-effective production processes and this is particularly true of transgenic products where there is such a change in production environment with increasing product quality.

The larger the process scale, the more critical these issues will become. Leaving conventional building services aside [8] specialist utilities in transgenic facilities can be considered as: water, cleaning/sanitization, and sterilization.

3.3.8.1 Water

All biopharmaceutical production processes utilize relatively large quantities of water in cleaning and production operations. To ensure that product integrity is not compromised and cost-effective design solutions are maintained the usage outlined in Table 3-2 is proposed. This clearly shows a gradual increase in water quality with increasing product purity.

Table 3-2. Appropriate water grades within production process.

Duty	Milking	Milk handling	Primary recovery	Polishing	Finishing
Cleaning	Towns water	Towns water	Process water	Process water	Process water
Final rinse	Towns water[a]	Process water	WFI	WFI	WFI
Product contact	N/A	N/A	WFI	WFI	WFI

Process water represents a high-purity, low endotoxin, low bioburden water produced by ion – exchange or reverse osmosis.
[a] Plus sterilizing agent.
N/A, not applicable; WFI, water for injection.

3.3.8.2 Cleaning and Sanitization

The cleaning of biopharmaceutical production facilities is a critical activity and it must be integrated into the design from day one. Typically, one would anticipate a cleaning cycle involving a pre-rinse, clean, acid rinse, post-rinse and final rinse, and could be expected to last up to 45 minutes [10]. In transgenic processes the need to control the scrapie agent can lead to the inclusion of an additional step utilizing the recirculation of 1 M NaOH for 1 h at 20 °C or above. The inclusion of such a step in the cleaning cycle will have significant impact upon the batch schedule and effluent production from any process and must be addressed in detail.

In general the majority of the unit operations upstream of final filling are not suitable for steam sterilization and the bioburden within process equipment, especially in chromatography columns, is best controlled utilizing chemical sanitization and storage. Suitable agents include sodium hydroxide, ethanol/water, and hypochlorite.

3.3.8.3 Sterilization

The final filling equipment will need to be sterilized prior to use. Equipment should be sterilised with clean steam, dry heat or a chemical sterilant. Occasional steam sterilization of upstream vessels and pipework may be advantageous.

3.4 Facility Design

For any biopharmaceutical product the facility design is process-led. The building configuration and services must be specified such that the process can be operated in a manner which does not compromise product quality. The following sections highlight the key design features of the critical building areas.

3.4.1 Overview of Facility Design

An overview of a typical transgenic facility design is given in Fig. 3-3. Here, the facility is presented as a central support core from which the various production modules hang as satellites. As one moves clockwise around the production module

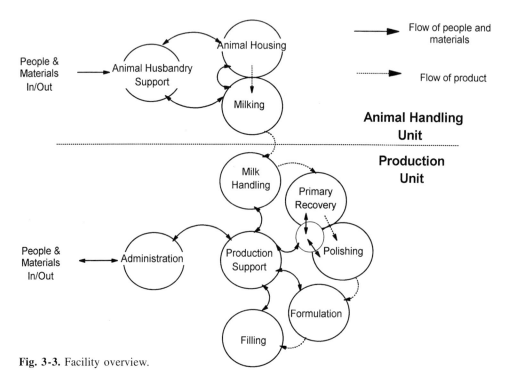

Fig. 3-3. Facility overview.

the product purity and the quality of the facility will increase accordingly. Key points on the design include:

- Complete segregation of personnel in contact with animals from those involved in purification.
- No personnel link between milking and purification only allowance for milk transfer.
- Segregation of primary recovery and polishing operations within a common zone to assist control of bioburden and cross-contamination.
- Segregation of formulation and filling activities from other purification operations to maintain viral/prion free status of final product.

The support core handles the transfer of people from uncontrolled to the controlled production envelope, administration activities, quality control (QC) activities and maintenance activities.

3.4.2 Animal Housing and Milking

It is essential that personnel working in the animal housing unit (AHU) are completely segregated from those producing the purified product. The AHU must be equipped with a dedicated entry and exit facility and enclosed such that the health status of the animals can be maintained. Visitors should be screened for disease and should not be allowed to enter the process building without suitable quarantine/decontamination.

3.4.3 Milk Handling

Milk handling may form a segregated area within the production building with a dedicated entry/exit from the central support core. The transfer of milk from the milking area should, if possible, be via fixed lines. Where volumes or distances preclude this any containers must be suitably decontaminated before entering the production unit. Transfer of the batch to the recovery/purification module should be via fixed lines if possible.

3.4.4 Primary Recovery and Polishing

Primary recovery and polishing are best designed as a suite with a dedicated, common, entry and exit from the central support core. Personnel, equipment, and product can move freely within the zone but there should be a general increase in purity and decrease in bioburden as one moves from recovery through polishing. Transfer to the formulation module should be via fixed lines to maintain the viral-free status of the product.

3.4.5 Formulation and Finishing

Formulation and finishing should be designed as two separate modules each with dedicated entry and exit from the central support core.

3.4.6 Overview of Building Finishes and HVAC

The finish and HVAC conditions within the various areas is heavily dependent upon scale and the nature of the process. At large scale, (> 500 L of milk/batch), it is possible that a closed process could be designed which would limit the need for specialist HVAC and building finishes to the formulation and filling areas. At smaller scales, the building finishes and HVAC in the production area are likely to be more critical to product quality. An overview of finishes and HVAC in the various building areas is given in Table 3-3. Where closed processing is not possible, the potential for contamination via HVAC systems must be addressed. Typically, each building module should have a dedicated system and within milk handling, recovery and polishing 100 % exhaust of the air may be a sensible option.

Table 3-3. Overview of building finishes and HVAC.

Type of production	Animal housing	Milking parlour	Milk handling	Recovery/ polishing	Formulation/ filling
Small mammals	Laboratory	Laboratory	Class D Lab.	Class C Lab.	Class C/ Class A Lab.
Large mammals, small scale	Agricultural	High quality dairy	Class D Lab.	Class C Lab.	Class C/ Class A Lab
Large mammals, intermediate scale	Agricultural	High quality dairy	Hygienic area	Class C clean rooms	Class C/ Class A clean rooms
Large mammals, large scale	Agricultural	High quality dairy	Hygienic area	Hygienic area	Class C/ Class A clean rooms

3.4.7 Site Planning Issues

For large-volume production the location and number of production modules is a key decision. Typically, the adoption of two or more animal sites is recommended to ensure against production loss through disease. Given the potential instability of the raw milk it will be beneficial to have milk handling operations local to these facilities. So long as the milk can be stabilized the use of a single, centralized, unit for purification, formulation and filling is possible.

3.5 cGMP and Regulatory Issues

The following section highlights issues of product safety, efficacy and reliability. It also discusses containment requirements for genetically modified organisms.

3.5.1 Good Manufacturing Practice

Within the pharmaceutical industry there is a high level of legislation to safeguard the safety and efficacy of drug products and there are well-established guidelines covering Good Manufacturing Practice (GMP) for biological products [11,12]. In recent years the emergence of transgenic technology and its move towards commercialization has spurred the production of several regulatory documents detailing critical regulatory issues [13,14]. Key issues arising between transgenic and classical rDNA products are as listed below with suggested actions:

- *The need to redefine concepts of the master and working cell banks.* In many ways the first transgenic animal of the line (Genetic Founder) can be equated to the master cell bank and one or more male offspring (Production Founders) to a working cell bank. However, although this comparison can be useful, care should be taken since any new technology will best be dealt with by developing concepts which are customized to its needs.
- *The, potential variation of milk composition with lactation period, feed composition, individual animals and generations and the need to validate production processes to cater for this.* Experience at PPL is that the composition of milk is less variable than might be suggested in the literature of this area. This is especially true if feeding, maintenance and housing of animals is carefully controlled.
- *Declaration of viral/prion status of animals and control of adventitious contamination.* Animals can be serologically tested periodically for pathogens of specific concern. Unfortunately, control of prion-like diseases can only be controlled by clinical examination, although periodic culling of animals followed by post-mortem examination should be considered.
- *Control of sick animals and their segregation from the production flock.* If an animal is identified which potentially could put the product at risk, it is important that segregation is well controlled and that policies on reintroduction of such animals back into the production herd are well considered.
- *Use of medicines during animal husbandry.* This is a complex area since the number of chemicals that an animal may come into contact with may be very large, either given as medicines or through feeding. A great deal of information

is available in this area from the dairy industry and careful adaption of this information can go a long way to helping solve this problem.
- *Gene stability during breeding process:* Not all transgenic lines ultimately prove to be stable. The number of copies of the transgene can become reduced for a variety of reasons and this normally results in a reduction in the level of the product found in the milk. Decreases in expression level of proteins has also been observed without an apparent loss of transgene copies [15]. The importance of using a stable transgenic line with a constant level of product in their milk cannot be overemphasized.

3.5.2 Containment of Genetically Modified Organisms (GMO)

The manipulation of GMO is carried out under strict government legislation both in Europe and in the USA. For example, in the UK two Acts of Parliament, the contained use [16] and the deliberate release regulations [17,18] outline rules under which work must be undertaken.

Once outside the animal housing the process stream does not constitute a biocontainment hazard.

References

[1] Ernst Young, *European Biotec 95 Gathering Momentum,* Ernst and Young International.
[2] Werner, R. G., *Transgenic Technology a Challenge for Biotechnology,* Biotechnica, 1995.
[3] Wilmut, I., et al, Viable offspring derived from fetal and adult mammalian cells, *Nature, 1997, 385.*
[4] Competition seeks to follow Genzyme Transgenics with a product into Clinic, *Genetic, Engineering News,* November 1, 1996, p 5.
[5] Harvesting Proteins, Biotechnology News, BMB Initiative, September 1996, EMAP Maclaren Ltd, p. 8.
[6] *Sheep, Dairy News,* 1986, Vol. *3,* No. 2, p. 7.
[7] Cole, E. S., *Production and Characterisation of Human Anti-Thrombin III Produced in the milk of Transgenic Goats,* Biotechnica, 1995.
[8] Laydersen, et al., *Bioprocess Engineering, Systems, Equipment Utilities.* John Wiley Sons, ISBN 0-471-03544-0.
[9] ICH, *Quality of biotechnological products viral safety evaluation of biotechnological products derived from cell lines of human or animal origin.* Step 2, Draft, 1/12/95.
[10] Sherwood, D., et al., Experiences with clean in place validation in a Multi product biopharmaceutical manufacturing facility. *Eur J Parenteral Sci,* 1996, *1*(2), 35–41.
[11] FDA Good Manufacturing Practice Regulations for Biologics, Code of Federal Regulations 21, parts 600 to 799, U. S. Government Printing Office, 1990.
[12] The Rules Governing Medicinal Products in The European Community, Vol. 4, Good Manufacturing Practice for Medicinal Products, CEC 1992, ISBN 92-826-31 80-X.
[13] CBER, *Points to consider in the manufacture and testing of therapeutic products for human use derived from transgenic animals.* Food and Drug Administration Docket No. 95D-0131, 1995.

[14] Medicines Control Agency, Use of transgenic animals in the manufacture of biological and medicinal products for human use. Eurodirect Publication No. 3612/93, 1993.

[15] Caver, A., et al., Transgenic livestock as bioreactors: stable expression of human alpha-1-antitrypsin by a flock of sheep. *Biotechnology*, 1993, 2, XX–XX.

[16] The Genetically Modified Organisms (contained use) Regulations, Health and Safety, SI 3217, 1992.

[17] The Genetically Modified Organisms (deliberate release) Regulations, SI 3280, 1992.

[18] The Deliberate Release of Genetically Modified Organisms to the Environment, Directive 90/220/EEC, May 1992.

4 Harvesting Recombinant Protein Inclusion Bodies

Anton P. J. Middelberg and Brian K. O'Neill

4.1 Introduction

The overexpression of recombinant protein in *Escherichia coli* often leads to concentration of that protein as a solid granule called an inclusion body. Numerous proteins are reported to produce inclusion bodies in *E. coli,* including the somatotropins, growth factors, tissue plasminogen activator, and insulin.

Techniques for processing inclusion bodies follow a highly conserved sequence of operations. As discussed in Section 4.4, these typically include cell disruption, recovery of the inclusion bodies, dissolution and renaturation to yield active protein, and high-resolution recovery of the final protein using conventional methods such as chromatography. In this chapter we will focus on the large-scale harvesting of recombinant protein inclusion bodies; a scale-up problem largely neglected in the current literature. In Sections 4.2 and 4.3 we review the properties of inclusion bodies and cellular debris. This knowledge is important in optimizing any collection strategy. In later Sections we then examine processes for dealing with inclusion bodies, before an in-depth examination of the alternatives for harvesting inclusion bodies, namely filtration, centrifugation and newer approaches involving *in situ* dissolution.

4.2 What is an Inclusion Body?

Inclusion bodies (IBs) are solid, micron-scale protein particles contained within the cytoplasm of *E. coli.* They appear as electron-dense bodies under electron microscopy and can be observed under phase-contrast microscopy as refractile bodies. Most eukaryotic proteins form inclusion bodies when overexpressed in *E. coli* [1]. Cytoplasmic inclusions are not, however, unique to recombinant proteins. Many inclusions have been observed naturally in various procaryotic strains, including polyhydroxyalkanoate (PHA), polyglucoside, and polyphosphate. While this chapter focuses on the harvesting of recombinant protein inclusion bodies, the recovery techniques discussed here are generally applicable to other inclusions such as PHA.

In this section we examine the factors influencing inclusion body formation before discussing the composition of inclusion bodies and their size and density. This provides the framework necessary for devising optimal strategies for inclusion body recovery.

4.2.1 Inclusion Body Formation

The likelihood of an inclusion body forming for a given protein expressed in *E. coli* is dependent on many factors, including the nature of the protein, the environmental conditions, and the host strain. However, no common characteristic that causes an inclusion body to form has been identified [2]. Wilkinson and Harrison [3] examined data for 81 proteins at 37 °C in *E. coli* that do and do not form inclusion bodies. They found that charge average and turn-forming residue fraction were correlated with the tendency to form inclusion bodies. Cysteine fraction, proline fraction, hydrophilicity, and the total number of residues were weakly correlated. The results were interpreted in terms of the inclusion body formation model proposed by Mitraki and King [4], shown in Eqn 1,

$$\text{translation} \rightarrow I^f \leftrightarrow I^m \rightarrow \text{native}$$
$$I^f \leftrightarrow I^{f*} \rightarrow IB \tag{1}$$

where I is an intermediate species (partially folded, f, that is capable of forming an aggregated inclusion body, or monomer-forming, m). They argued that a protein with many turn-forming residues would fold more slowly, hence giving a higher concentration of partially folded intermediate and an increased probability of inclusion body formation.

It should be noted, however, that the study did not take into account other important factors which can have a major impact on IB formation, such as host-cell characteristics. Recently it has become clear that the *in vivo* folding pathway greatly influences the likelihood of inclusion body formation. For example, overproduction of folding chaperones such as GroEL and DnaK leads to a reduced propensity for inclusion body formation [5]. It is therefore clear that protein characteristics are only one determinant of inclusion body formation, and arguably a minor one. Differences in host-strain physiology, due to differences in chaperone and protease expression, will have a major impact.

4.2.2 Inclusion Body Composition

The recombinant protein of interest typically comprises > 50 % of the inclusion body [2]. This is the key benefit of expression as an inclusion body. Relatively pure product can be simply obtained by physical fractionation, for example by filtration or centrifugation followed by dissolution and simple ion exchange. Such simple purification can easily give a product with > 90 % purity, compared with 1 % to 25 % purity after initial isolation for a soluble protein [2]. Of course, this advantage may be completely lost if protein refolding is inefficient (see Section 4.4).

Isolated IBs typically contain the protein of interest as well as various contaminants, including non-product polypeptides, nucleic acids, and cell-envelope components (e.g., outer-membrane proteins OmpA and OmpC/F). Hart et al. [6] examined *Vitreoscilla* hemoglobin (VHb) production in *E. coli,* and identified two cytoplasmic aggregates of different morphology. The granules differed in the relative fractions of VHb and pre-β-lactamase, the antibiotic-resistance protein encoded on the same plasmid. The inclusion bodies were also contaminated with the cytoplasmic elongation factor Tu, and by outer-membrane proteins OmpA and OmpF that were attributed to cell-envelope contamination following disruption.

Valax and Georgiou [7] argue that many contaminants believed to be incorporated into IBs may, in fact, adhere to the IB following cell disruption. They found that the expression system and growth conditions have a pronounced effect on inclusion body composition for β-lactamase. Inclusion bodies were purified to apparent homogeneity by differential centrifugation in sucrose gradients, maintaining the integrity and surface characteristics of the IBs. β-lactamase lacking the native signal sequence (i.e., Δ(-20-1)-β-lactamase) formed cytoplasmic IBs that were virtually free of contaminating protein following sucrose-gradient purification. This indicates that IB formation *in vivo* can be a highly specific process. Periplasmic inclusion bodies had a consistently higher level of associated contaminants than cytoplasmic IBs, possibly due to different surface properties leading to enhanced adsorption. This belief was supported by the selective removal of contaminants through detergent washing. Overall, the level of contaminating polypeptides ranged from 5 % to 50 %, while phospholipids constituted 0.5 % to 13 % of the IBs. Nucleic acids were a minor contaminant.

The study by Valax and Georgiou [7] concluded that most contaminants were not incorporated into the inclusion body during synthesis, but rather adsorbed to the IB following cell disruption. This suggests considerable scope for process optimisation by careful inclusion body washing prior to subsequent processing (see Section 4.4).

4.2.3 Size and Density of Inclusion Bodies

Taylor et al. [8] were the first to examine the physical characteristics of inclusion bodies in any detail. They employed a combination of centrifugal disc photosedimentation (CDS) and electrical sensing zone measurement (ESZ). The mean diameters of γ-interferon and calf prochymosin inclusion bodies were 0.81 and 1.28 μm, respectively. Sedimentation studies showed that the density of the IBs increased with the density of the suspending fluid, indicating an accessible voidage within the granules. In deionized water, the IBs had buoyant densities of 1124 kg m^{-3} for γ-interferon and 1034 kg m^{-3} for prochymosin. Voidages of 70 % for γ-interferon and 85 % for prochymosin were determined by matching data from the two sizing methods.

Olbrich [9] determined the size of prochymosin inclusion bodies to be 0.85 μm using ESZ. By matching ESZ and CDS data, an apparent buoyant IB density of 1140–1160 kg m^{-3} was obtained after separating contaminant debris by low-speed centrifugation. In later experiments, Jin [10] measured prochymosin inclusion bodies by PCS and CDS and determined median diameters of 0.98 μm and 0.94 μm, respectively. Samples were again fractionated prior to analysis using low-speed centrifugation. The IB density determined by Olbrich [9] was used for CDS measurements.

The physical characteristics of porcine somatotropin (pST) inclusion bodies have also been examined by CDS. A median IB size of 0.35–0.45 μm was obtained by high-density, fed-batch fermentation using minimal media [11,12]. The reported size was based on an IB density of ~ 1260 kg m^{-3}, determined by cesium chloride gradients (Bresatec Ltd., Adelaide, Australia, personal communication), and compared well with estimates by electron microscopy. Unlike the inclusion bodies examined by Taylor et al. [8], the pST inclusion bodies were highly packed granules with a density approaching that of a crystalline protein precipitate. A range of insulin-like growth factor analogs expressed in *E. coli* has also been examined by CDS [13]. Typical mean inclusion body sizes were 0.3–0.5 μm, based on a density of 1260 kg m^{-3}.

Given the large variation in reported IB size, it is clear that independent determinations must be made for any specific protein. Also, it is often easier to report size as an apparent Stokes settling velocity rather than diameter, particularly if the aim is to recover centrifugally the IBs (see Section 4.6). This avoids the need to directly determine IB density, which can be difficult if the IBs have an accessible voidage.

4.3 Properties of Cellular Debris

The properties of inclusion bodies were discussed in Section 4.2. It is clear that many contaminants adhere to the IB surface, and hence will be collected with the IB unless washing protocols are employed (Section 4.4). However, many more contaminants remain independent of the inclusion bodies in the homogenate following cell disruption. These include soluble components (e.g., nucleic acids, soluble cell proteins, lipid membrane vesicles) and insoluble fragments of the cell wall (e.g., peptidoglycan and associated cell-wall proteins and lipids). The soluble fragments may be readily separated by filtration or centrifugation (see Sections 4.5 and 4.6). However, fractionation of the insoluble IBs from the insoluble debris can be considerably more difficult, particularly if their physical sizes or settling velocities are similar. This fractionation is aided by an understanding of the size of the insoluble cell debris.

A range of sizing techniques is available for characterizing cellular debris, including photon correlation spectroscopy (PCS), CDS, and ESZ. Each has significant disadvantages for debris sizing. PCS is a low-resolution technique that requires extensive sample preparation prior to size analysis. For example, Olbrich [9] centrifuged homogenate samples at 2000 g for 26 minutes, and took the supernatant as the cell debris sample and the pellet as the inclusion body sample. This is necessary as PCS cannot resolve the inclusion body and debris size distributions. A problem with this approach is preferential removal of large debris from the sample being analysed. Jin [10] demonstrated that up to 47 % of the cell debris co-sedimented with the inclusion bodies using Olbrich's [9] fractionation scheme, while 14 % of inclusion body proteins remained in the supernatant (i.e., the debris sample). It is clear that the apparent size distribution is affected by the pretreatment employed. Other methods also have disadvantages. ESZ is prone to orifice blocking, limiting the size range that this technique can analyze and generally making it unsuitable for sizing *E. coli* cell debris. CDS has low sensitivity below 0.2 μm [11] where much of the cell debris is located, and is prone to baseline drift and errors in extinction coefficient [11]. Results from electron microscopy are also prone to error because of sample preparation techniques, including drying and plating, prior to analysis [14].

A newly developed method for cell-debris size analysis [15] overcomes the limitations of traditional methods. It employs cumulative sedimentation analysis (CSA) in a laboratory centrifuge coupled with SDS–PAGE quantitation of sedimentable outer-membrane proteins. The technique provides an accurate assessment of debris size even in the presence of inclusion bodies, without the need for sample pretreatment. However, it is rather laborious for routine quality control, and is more amenable to process optimization during developmental research.

Key results from *E. coli* cell debris-sizing studies are presented in Chapter 6, and in particular Section 6.9 and Table 6-4. CSA indicates a median *E. coli* debris size of 0.5 μm after two homogenizer passes at 55 MPa, decreasing to 0.33 μm after 10 homogenizer passes [15].

Models for the effect of repeated homogenization on debris size are presented in Section 6.9.2, and clearly show that debris size reduces with repeated homogenization. This dependence suggests that there will be a strong interaction between the selected disruption protocol and the ease of IB and debris fractionation. This is discussed further in Section 4.6.

4.4 Process Synthesis

4.4.1 Laboratory-scale Processes

Figure 4-1 shows a summary of strategies employed to obtain active purified protein from inclusion bodies at laboratory scale. The key steps are cell breakage, IB sedimentation, and washing and solubilization [16]. The sedimentation step is often conducted at moderate centrifugal force in a laboratory centrifuge (e.g., 5 min at 10 000 *g*) [16] to effect some differential separation of the inclusion bodies and insoluble cell debris. Following protein renaturation (i.e., refolding), active product may be recovered by conventional methods such as chromatography.

There are many variations to this standard approach, most of which are only minor deviations. For example, an additional cell debris-removal step is often added after solubilization to minimize the carry-over of debris to the chromatographic recovery system. At laboratory scale this usually consists of a high-speed centrifugation or depth-filtration step. Other variations are quite significant, and can include *in situ* solubilization of the inclusion bodies without prior cell disruption. These major variations are discussed further in Section 4.7.

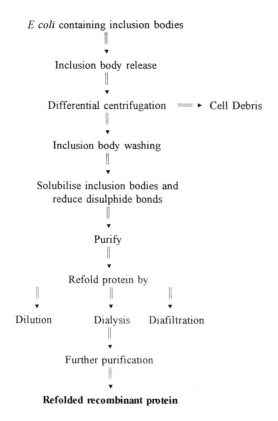

Fig. 4-1. Strategies for processing inclusion bodies at laboratory scale. (From Chaudhuri [17]; reproduced with permission of *the Annals of the New York Academy of Sciences.*)

The area of protein refolding is a field of very active research. Some proteins are particularly difficult to refold with low consequent yields. An excellent review covering methods of dissolution and renaturation is available [18].

4.4.2 Considerations in Synthesizing a Large-scale Process

Laboratory-scale processes for handling inclusion bodies may often be uneconomic if scaled directly. It is preferable to consider the scale of the final process, and then develop a laboratory process that is essentially a scaled-down version of the final anticipated design. This requires a knowledge of the key issues in synthesizing a large-scale process. Two key considerations are the efficiency of large-scale protein refolding, and problems associated with the collection of inclusion bodies.

4.4.2.1 Protein Refolding

Poor refolding performance can destroy the economic viability of a given process, as demonstrated for tissue plasminogen activator (tPA) [19]. Two key factors must be considered in developing a laboratory-scale refolding process amenable to scale-up, namely the effect of protein refolding concentration, and the selected reactor operation mode.

The choice of protein refolding concentration is perhaps the most critical. A simplified generic protein refolding scheme is give by Eqns (2) and (3),

$$D \rightarrow I \leftrightarrow N \tag{2}$$

$$2I \rightarrow A \tag{3}$$

where D is the denatured protein obtained by protein solubilization, I is an aggregating intermediate, N is native protein, and A is a non-native aggregate. Eqn (2) is typically first order, while Eqn (3) is approximately second order. Clearly, with competing first- and second-order equations, the yield of N will be maximized by minimizing the protein refolding concentration (i.e., the concentration of D and hence I). Refolding is therefore normally conducted at extremely low protein concentrations in the laboratory. For process-scale work it may be desirable to increase the refolding concentration. For example, Kiefhaber et al. [20] showed that an increase in the refolding concentration of porcine muscle lactic dehydrogenase from 10^{-3} M to 10^{-1} µM led to a decrease in protein yield of only 25 %. At process scale this may represent an acceptable economic trade-off. For other proteins, an increase in concentration will decrease the yield to unacceptable levels. For example, tPA refolding at less than 2.5 mg L^{-1} was necessary to achieve a reasonable renaturation yield of at least 25 % [19]. This concentration was still too low for an economic process, as 75 % of the process capital cost was associated with inordinately large refolding tanks. Clearly, the effect of protein concentration must therefore be examined at an early stage in process development.

The choice of refolding reactor mode can also impact on process economics. Eqns (2) and (3) represent classical first- and second-order competition. The best conventional reactor mode for this scheme will be one that maintains a low concentration of reactant, which is either a continuous system or a fed-batch system with gradual addition of denatured protein [21]. Batch refolding will be the least successful, but continues to be extensively employed in laboratory studies.

Cost-optimal reactor operating strategies have been developed for continuous protein refolding [21]. Recently, several novel methods have been proposed for increasing refolding yield. Vicik and DeBernadez-Clark [22] conducted a mathematical optimization of human carbonic anhydrase B refolding using diafiltration to reduce the denaturant concentration. The optimal diafiltration protocol and final urea concentration were found to be 0.088 min^{-1} and < 0.4 M, respectively. Low diafiltration rates were required for high yields. Aggregation has also been reduced by conducting refolding in a gel-filtration column [23], although the cost of resin may preclude use at large scale. Reverse micelles have also been investigated [24], as they have the

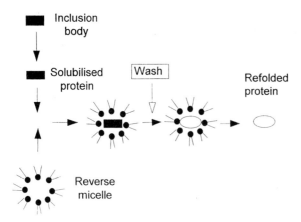

Fig. 4-2. The refolding of proteins in reverse micelles, showing the protective isolation of the protein molecule by the micelle that leads to reduced protein aggregation. (From Chaudhuri [17]; reproduced with permission of *the Annals of the New York Academy of Sciences.*)

potential to isolate individual protein molecules, thus preventing the second-order aggregation reaction as shown in Fig. 4-2 [17]. Complete renaturation of pure, denatured ribonuclease-A has been obtained using this method [24]. General applicability to other proteins must still be demonstrated.

4.4.2.2 Inclusion Body Recovery

The recovery of inclusion bodies inevitably leads to the co-recovery of some contaminants. These may be contaminants incorporated into the inclusion body during synthesis or through adherence to the IB surface during cell release (see Section 4.2.2), or may be insoluble cell debris collected with the inclusion bodies during their recovery by centrifugation or filtration.

There is some evidence that contaminants associated with β-lactamase inclusion bodies may reduce protein refolding yields [25]. The yield of active protein obtained by refolding from IBs was only 20–40 % of that obtained from purified protein under the same conditions. Furthermore, virtually all contaminating protein present initially in the IBs appeared in aggregates that formed following removal of the denaturant. This suggests that the contaminating proteins somehow tended to promote aggregation. Contaminants may also affect refolding through proteolysis of the product. Babbitt et al. [26] have shown that proteases significantly reduced creatine kinase yield during the solubilization and refolding of IBs expressed in *E. coli.* The protease activity was presumably associated with cellular debris that co-sedimented with the IBs during batch centrifugation. Phenylmethanesulfonyl fluoride (a protease inhibitor) had no apparent affect on yield. However, significant yield improvements (ca. 100 ×) could be obtained by washing the IBs with the non-ionic detergent octylglucoside prior to solubilization. Insoluble debris collected during IB recovery may also foul chromatography columns downstream of the refolding operation. This will reduce the effective life of chromatographic resins and consequently increase process cost, and will also complicate process validation. Such debris may be easily removed by filtration or ultracentrifugation at laboratory scale following solubilization. Only depth filtration and ultrafiltration are practical at process scale, although both necessitate the inclusion of an additional process unit with its associated capital, operating, and validation costs.

The three problems identified above, namely reduced refolding yields, proteolysis, and the fouling of chromatographic resins, suggests that inclusion body collection warrants close examination during process development. As demonstrated by Babbitt et al. [26], contaminants may be reduced by selective washing of the IBs. A variety of wash chemical are available [18]. The non-ionic detergent Triton X-100 (0.1 % to 4 %) and low concentrations of denaturant (e.g., 2 M urea) are common choices. However, the use of wash chemicals may also cause problems at process scale. Washing with urea can significantly increase waste disposal costs because of its high biological oxygen demand. Detergents should be chosen with reference to their cost and ease of disposal. Triton X-100 has been employed in the industrial-scale manufacture of porcine somatotropin (pST) [27] and insulin [28]. In the case of pST, Triton X-100 costs represented $2 M per annum, or 33 % of the total annual process chemicals and consumables cost. The cost for insulin manufacture was considerably less [28]. Other effects of adding wash chemicals during inclusion body collection also need to be considered. Removal of the added chemical will be necessary for validation purposes, and in some cases the chemical may have detrimental effects on downstream units. For example, incomplete removal of added detergent may lead to reduced membrane and chromatographic resin life. This is particularly problematic if the process is not robust enough to handle process disturbances that may lead to variable removal of the added chemical, again complicating process validation.

An obvious but difficult method of reducing contaminant loads is to improve the removal of insoluble debris during the IB recovery stage. This is achieved at laboratory scale by low-speed centrifugation that sediments the majority of the dense IBs, while leaving the majority of debris in the supernatant. At process scale this benefit is often forgotten, and the aim is often to maximize the collection of inclusion bodies, and hence debris, without reference to the downstream problems. This represents the perceived difficulty of good fractionation at process scale. For example, Krueger et al. [16] state that 'differential centrifugation may be difficult or impossible on the process or commercial scale'. Wash chemicals are therefore extensively employed. However, recent biochemical engineering progress has been made in differential centrifugation, as discussed in Section 4.6.

For some proteins, refolding yields are neither substantially affected by contaminants nor proteolysis. In such cases, filtration may be considered for IB collection, particularly if the capital cost of a centrifuge is considered prohibitive and an additional filtration module is available after solubilization to reduce insoluble debris contaminants. The collection of inclusion bodies by filtration is therefore considered further in the next section.

4.5 Filtration

Harvesting of inclusion bodies following fermentation and homogenization is tradi-
tionally conducted by centrifugation. This can be a costly exercise when scale-up is
attempted. Pressure-driven membrane processes such as ultrafiltration (UF) and
microfiltration (MF) have been demonstrated to be attractive processes for cell harvesting [29,30] and the recovery of enzymes [31,32]. Such membrane processes may
also be applied to the harvesting of inclusion bodies [33,34], where the presence of
cellular contaminants during refolding does not affect yield.

4.5.1 Modes of Filtration

Two different modes of filtration are possible, namely surface filtration and depth filtration. Filtration for the recovery of inclusion bodies is exclusively by surface filtration at a membrane. The highly fouling nature of cell debris, particularly cell membrane-based components, means that cross-flow membrane filtration is the sole
practical mode.

  Cross-flow filtration refers to a type of surface filtration wherein the main direction of the suspension flow is perpendicular to the flow direction of the recovered
liquid. A wide variety of industrial processes may be classified as cross-flow filtra-
tions. Examples include reverse osmosis, ultrafiltration, and microfiltration. However, an often accepted convention is that cross-flow filtration is normally considered
as a process removing particles whose sizes range from 0.1 to 10 µm. This size range
clearly covers the size of typical inclusion bodies and thus it is a potential technology
for the fractionation of inclusion bodies from soluble proteins as an initial step in the
purification of insoluble cellular proteins.

  In cross-flow filtration, the suspension flows under the applied pressure gradient
along the porous membrane and liquid permeation occurs across the membrane. Consequently, as the liquid permeates the membrane, a fraction of the particles suspended in the feed will be deposited at the surface of the membrane and a solid
cake will form. The thickness of this cake will increase with time as the processing
proceeds. This is accompanied by a resultant decline in the rate of liquid permeation
with time. Such time-dependent behaviour is characteristic of cross-flow filtration.

  The prime advantage of operation in the cross-flow mode is the mininization of
cake formation as a consequence of the scouring action of the flow. To ensure that
the scouring action is effective, the flow conditions should be maintained in the turbulent regime. This may lead to large pressure losses around the retentate loop and
consequent high energy costs. Occasionally, it may be necessary to cool the retentate.

  Membranes are normally cleaned *in situ* by pulsed backwashing. Pulses are
usually of short duration and the filtered product is forced back through the membrane to detach the filter cake from the surface of the membrane. The backwashed
liquor is simply mixed into the retentate flow. The flow's scouring action ensures
that the solids are resuspended. Chemical cleaning may be required to reverse the

effects of long-term flux decline due to adsorption or precipitation leading to gross fouling or pore blockage. There are few available data on the extent of such fouling in the IB harvesting area.

Inclusion body washing or diafiltration is readily accomplished by the addition of water and appropriate buffers. This enables the removal of low-molecular weight contaminating solutes.

4.5.2 Theory

In cross-flow microfiltration or ultrafiltration, the filtration flux will depend primarily upon the transmembrane pressure, the bulk concentration of inclusion bodies, the gel layer concentration, and the mass transfer coefficient. The transmembrane pressure is readily defined as an average pressure drop across the filter according to Eqn (4).

$$P_{tm} = \frac{P_{in} - P_{out}}{2} + P_p \qquad (4)$$

Permeate flux normally declines over operating time as a consequence of the resistance produced by concentration polarization, cake formation, and other causes of membrane fouling. As stated earlier, such a decline is partially reversible by backflushing and/or pulsing of the membrane or in extreme cases by chemical cleaning. Clearly, cross-flow mode will limit the build-up of inclusion body particles and cellular debris on the membrane surface, and at steady state the particle layer is assumed to attain a thickness that is time-invariant but increases with distance from the filter entrance. The flow resistances of the cake and the flowing (polarized) layer of particles plus the intrinsic resistance of the membrane may be assumed to act in series. Hence, the permeate flux may be described by Darcy's law. The transmembrane flux J is then,

$$J = \frac{\Delta P}{\mu(R_m + R_g)} \qquad (5)$$

where R_m is the membrane resistance, R_g is the gel layer resistance, and ΔP is the applied pressure drop. At high solute concentrations, solute precipitates at the membrane's surface forming a thixotropic gel which is permeable to the solvent. If the concentration of the solute is moderate to high, the resistance of this gel layer will increasingly dominate the resistance contribution from the membrane. The permeate flux then becomes independent of the membrane's permeability. As the transmembrane pressure rises, the resistance of the gel layer R_g will increase as solid accumulates at the membrane/liquid interface. This resistance will continue rising until an equilibrium is attained. At this point, nett transport of the solute towards the membrane is balanced by back diffusion of the solute towards the solution bulk as a consequence of the applied pressure gradient. The thickness of the gel layer is thus determined by the convective and diffusive transport mechanisms controlling the concentration of particles near the membrane surface. Applying an elementary mass balance to the particle layer and assuming that steady state conditions prevail yields Eqn (6),

$$\frac{\partial}{\partial x}(uC) + \frac{\partial}{\partial y}(vC) = \frac{\partial}{\partial y}\left(D\frac{\partial C}{\partial y}\right)$$ (6)

where C is the concentration of the solute, u and v represent the axial and radial velocity components of the fluid, and D is the diffusivity of the solute. The diffusivity includes contributions from Brownian motion and diffusion induced by shear. The standard approach to simplifying this balance is to neglect axial convection [35,36]. Integrating the resultant reduced form of Eqn (6) yields the following 'macroscopic' mass balance:

$$JC + D\frac{dC}{dy} = 0$$ (7)

The assumption of complete rejection at the membrane is implicit in this equation's derivation. It may be integrated to suggest that the limiting value of the flux will decrease logarithmically with increasing concentration in the solution bulk,

$$J = k \ln\left(\frac{C_m - C_p}{C_b - C_p}\right)$$ (8)

where the subscripts b, c, p, and m denote the bulk, cake, permeate, and membrane, respectively. The mass transfer coefficient k is primarily a function of the flow geometry of the membrane, fluid properties, and temperature.

4.5.3 Commercial Equipment and Operating Parameters

There are three types of commercial membrane configurations, namely capillary membranes, flat sheet modules, and spiral modules. In the first category, retentate flows inside tubes grouped into bundles and enclosed in cylindrical modules. These provide a large surface area in a moderately sized module, but incur high energy costs to maintain turbulent flow. Capillary types (< 2 mm) are only suitable for low concentration, non-fouling solids. Tubular membranes (2–25 mm diameter) have lower pumping costs and are suitable for high suspended solids loads (to about 20 % dry weight), including solids which may blind such as fermentation broths. Flat-sheet modules resemble plate and frame filters and the retentate and permeate are separated by sheets of the filter membrane. Typical spacings are 1–2 mm and turbulence is promoted by embossed dimples and screens. The narrow passages and presence of screens normally limits suspended solids to a maximum of approximately 10 %. Spiral modules are essentially a wound sandwich of flat-sheet-type membranes. The operating constraints are similar to those of flat sheets.

The cross-flow concept is applicable over a wide spectrum of operations including ultrafiltration, microfiltration, and reverse osmosis. Cross-flow filtration is normally characterized by a particle cut-off size of 0.1 µm and greater. Membrane pore sizes are commonly 0.2, 0.45 and 1 µm with typical average filtration rates for cell harvesting and washing in the range 50–200 L m^{-2} h^{-1}. In general, flux rates are below 100 L m^{-2} h^{-1}, compared with a desired flux rate of approximately 150 L m^{-2} h^{-1} for economic operation [37].

4.5.4 Inclusion Body Recovery by Filtration

Two studies of cross-flow filtration for the separation of inclusion bodies from soluble proteins in recombinant *E. coli* have been reported. The initial study was undertaken by Forman et al. [33]. A recombinant strain of *E. coli* containing a gene encoding a portion of gp41, the transmembrane protein of the AIDS (HTLV-III) virus, was used. Their study examined the effects of key operating variables on soluble protein removal, namely cross-flow rate, transmembrane pressure, initial concentration, and ionic environment. Hydrophilic Durapore™ membranes with mean pore sizes of 0.22 and 0.45 µm provided adequate retention for a feed material containing 5 % solids. The relationship between flux and transmembrane pressure corresponded qualitatively to predictions from the gel-polarization theory. A maximum flux of 8 L m^{-2} h^{-1} was achieved at feed concentrations of 25 and 50 g L^{-1}. As expected the removal of soluble proteins was enhanced at low ionic strengths as such conditions will minimize the tendency for protein aggregation. By contrast, increasing the cross-flow rate did not yield the predicted enhancement of solute passage through the membrane as a consequence of the scouring action of the increased tangential flow. A diafiltration experiment was also performed. A removal of 87 % of the soluble protein was reported (flux = 1.7 L m^{-2} h^{-1} at a constant average transmembrane pressure of 0.5 kPa). Operation in this mode provided an efficient method for continuously performing washing and concentration steps.

Meagher et al. [34] developed a cross-flow filtration process for the purification of rIL-2 inclusion bodies from homogenized *E. coli*. Key elements of the process were a two-step diafiltration, an extraction using 7 M GuHCl, followed by a dilution of the solubilized inclusion bodies into an appropriate buffer. The process provided a threefold increase in the yield of rIL-2 product in the diluted extract when compared with an alternative centrifugation strategy. An additional benefit of the membrane process was the significant improvement in product purity quantified by high-performance liquid chromatography (HPLC). The second diafiltration step using 1.75 M GuHCl appears to have solubilized a significant proportion of the contaminants, allowing their easy removal. This improvement in purity may provide significant benefits and simplifications in subsequent processing steps such as refolding.

Clearly, interest in membrane fractionation of IBs is in its infancy. However, cross-flow filtration does offer significant potential advantages when compared with centrifugation for IB recovery from cell lysate. First, the energy requirements are significantly reduced and initial capital investment may be lower. Second, the ease of

operation in the diafiltration mode simplifies washing and soluble-protein removal steps when compared with the repeated resuspension and sedimentation steps for centrifugation. However, significant improvements in the flux rates and percent transmissions to minimize buffer and membrane capital requirements may be necessary before adoption at commercial scales. Also, the working life of typical membranes is poorly quantified.

4.6 Centrifugation

4.6.1 Modes of Centrifugation

A range of centrifuge designs is available. The most common are the chamber or multichamber bowl centrifuge, the decanter centrifuge, and the disc-stack centrifuge. Each is best suited to a particular application. For example, bowl centrifuges are best suited to streams with a low solids content, while decanter designs are capable of handling high solids throughputs, and are commonly employed in sludge-dewatering applications. Of the available designs, the disc-stack centrifuge is best suited to the recovery of protein inclusion bodies in bioprocessing.

Figure 4-3 shows a cross-section of a disc-stack centrifuge. Material is fed through the center periphery to the outer disc edge. It then flows between the discs which

1 Feed
2 Discharge
3 Photocell
4 Discs
5 Sediment holding space
6 Solids ejection ports
7 Operating-water valve
8 Drain hole
9 Opening chamber
10 Closing chamber
11 Annular piston
12 Timing unit
13 Discharge pump

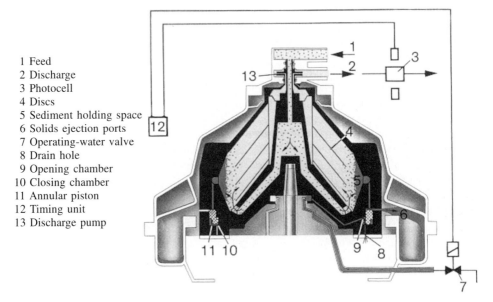

Fig. 4-3. Cross-section of a disc-stack centrifuge. (Reproduced from the Westfalia Separator booklet: *Centrifugal Clarifiers and Decanters for Biotechnology* with the kind permission of the GEA Process Technology Division and Westfalia Separator Australia Pty Ltd.)

rotate at high velocity (ca. 10 000 r.p.m.). Solids are collected on the bottom surface of each disc due to centrifugal force, before being flung outwards to the bowl periphery where a sludge is formed. The sludge may be collected from this region by opening the bowl. Clarified liquid flows from the top of the discs, forming a clarified supernatant stream.

Several designs are available for solids discharge. Split-bowl designs periodically separate the upper and lower halves of the bowl, permitting solids discharge under extreme pressure. Other designs employ nozzles at the bowl periphery that permit continuous or periodic solids discharge. Regardless, essentially continuous solids discharge is possible for extended operational periods, enabling continuous operation without the need for centrifuge disassembly.

4.6.2 Commercial Centrifugation Equipment

A variety of commercial centrifuges are available. The field is, however, dominated by two main manufacturers of disc-stack centrifuges, namely Westfalia Separator AG (Oelde, Germany) and Alfa Laval Separation AB (Tumba, Sweden). Both manufacturers offer contained, sterilizable disc-stack centrifuges suitable for inclusion body recovery, in a range of sizes from pilot scale (e.g., the CSA8 from Westfalia and the BTPX 205 from Alfa Laval) to production scale. Figures 4-4 and 4-5 provide

Fig. 4-4. Westfalia SC35 disc centrifuges. (Reproduced from the Westfalia Separator booklet: *Centrifugal Clarifiers and Decanters for Biotechnology* with the kind permission of the GEA Process Technology Division and Westfalia Separator Australia Pty Ltd.)

Fig. 4-5. The Alfa-Laval BTUX 510 disc-stack centrifuge. (Reproduced from the Alfa Laval booklet: *BTUX 510 Contained Separation System for Commercial Biotech Production,* with the kind permission of Alfa Laval Pty Ltd Australia.)

examples of production-scale centrifuges suitable for the recovery of protein inclusion bodies.

4.6.3 Theory

The aim of modeling centrifuge performance is to predict what fraction of material of a specific settling velocity will be collected. This may be used, with information on the properties of inclusion bodies and cell debris, to optimize centrifuge performance in a given application.

The simplest description of centrifuge performance is provided by the so-called Sigma model. The fraction of particles collected in the centrifuge is given by equation (9).

$$f(d) = v_g(d)\frac{\Sigma}{Q} = \left(\frac{d^2 \Delta \rho g}{18\mu}\right)\frac{\Sigma}{Q} \tag{9}$$

where Q is the centrifuge feedrate, Σ is the equivalent settling area of the centrifuge, v_g is the particle settling velocity, d is particle diameter, $\Delta\rho$ is the density difference between the particle and fluid, μ is fluid viscosity, and g is gravitational acceleration.

The equivalent settling area, Σ, in Eqn (9) is a machine-specific parameter depending on the design of the centrifuge. For disc-stack designs, Σ is given by equation (10),

$$\Sigma = \frac{2\pi}{3g}\omega^2 N \cot\theta (r_1^3 - r_2^3) \tag{10}$$

where ω is the centrifuge angular velocity, N is the number of channels between discs, θ is the inclined angle between the axis and the disc surface, and r is the inner (2) or outer (1) radius of the discs. Empirical corrections to the exponents have been applied to Eqn (10) to take into account flow non-idealities. Sullivan and Erikson [38] defined the 'KQ' empirical correction by substituting $\omega^{1.5}$ and $r^{2.75}$ into equation (10) in place of ω^2 and r^3.

Equation (9) is based on several assumptions regarding centrifuge behavior. The most notable are that flow between discs is ideal and that particles are collected when they reach the centrifuge disc. The assumptions are so restrictive that the predictive behavior of (9) is extremely limited. It invariably overpredicts the fractional collection efficiency, $f(d)$. As a result, collection efficiency is often determined empirically for a given suspension and centrifuge design using Eqn (11),

$$f(d) = 1 - \frac{C_L(d)}{C_o(d)} \tag{11}$$

where C_o is the concentration of particles of size d entering the centrifuge, and C_L is the concentration of particles of size d leaving in the centrifuge supernatant.

A plot of $f(d)$ versus d is termed the centrifuge grade efficiency, and is a critical determinant of the ability to separate solids with similar settling characteristics such as debris and inclusion bodies. Eqn (9) predicts that the grade efficiency curve will be given by Eqn (12),

$$f(d) = \left(\frac{d}{d_c}\right)^s \tag{12}$$

with $s = 2$, where d_c is the critical diameter for the given centrifuge (Eqn (13)).

$$d_c = \sqrt{\frac{18\mu Q}{\Delta\rho g \Sigma}} \tag{13}$$

The critical diameter is the minimum sized particle that will be fully collected in the centrifuge (i.e., $f(d_c) = 1$). Again, this equation will often provide a poor prediction of centrifuge performance as it is based on the assumptions underlying Eqn (9). Empirical grade-efficiency curves are therefore widely used. The simplest allows the exponent s in Eqn (12) to be determined by regression to experimental data. Alternatively, Eqn (14) has been widely employed to model centrifuge grade efficiency,

$$f(d) = 1 - \exp\left[-k\left(\frac{d}{d_c}\right)^n\right] \tag{14}$$

where k and n are parameters determined by regression to experimental data.

4.6.4 Scale-up and Scale-down of Centrifuges

Several researchers have presented similarity criteria for centrifuge scaling, but none has found widespread use. In some respects this is not surprising considering the hydrodynamic complexity involved.

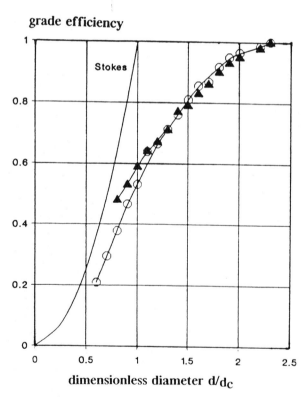

Fig. 4-6. Grade efficiency curves for a Westfalia BSB-7 disc-stack with a reduced sedimentation area, ○, and a full set of discs, ▲, determined by Mannweiler [39]. The Stokes line corresponds to equation (9) (From Keshavarz-Moore et al. [41]; reproduced with permission of *the Annals of the New York Academy of Sciences.*)

Generally, centrifuge scale-up is done using manufacturer's guidelines after determining performance on a pilot-scale machine. Direct scaling according to the equivalent settling area of the centrifuge is possible, provided that a sufficiently large machine was used for pilot-scale tests. Alfa-Laval typically scale results using the KQ correlation, while Westfalia applies the so-called Leistung, or power, factor (another effective area measure). Scale-up will generally be conservative because of the way in which scale-up factors are defined by the manufacturer. Considerable variation between machines with the same nominal equivalent area is also possible through minor variations in caulk thickness (i.e., the distance between parallel discs).

A detailed experimental study of the scale-down of a Westfalia BSB-7 centrifuge has been undertaken using polyvinylacetate emulsion [39,40]. Grade-efficiency curves were determined experimentally and could be well described by Eqn (14). Significant departure from ideality was observed, as shown in Fig. 4-6. Mannweiler and Hoare [40] concluded that the Westfalia BSB-7 centrifuge may be scaled down to 10 % of its total available separation area by removing active discs. If the centrifuge is used to collect the majority of particles, as is normally the case for inclusion-body processing, then accurate prediction of full-scale throughput capacity for the production machine was possible. Such information is extremely important for bioprocessing applications, where there is often limited material available in initial pilot-scale testing.

4.6.5 Inclusion Body Recovery by Centrifugation

Several studies into the collection of prochymosin inclusion bodies have been conducted. The grade-efficiency curves generated using PVA (Section 4.6.4) have been used to simulate the fractionation of prochymosin inclusion bodies and cell debris in a Westfalia SB-7 centrifuge [9]. High inclusion body collection was predicted to result in poor purity (i.e., a high collection of cell debris), while high centrifuge feedrates gave good paste purity but a reduced fractional inclusion body collection (Fig. 4-7). The results were not confirmed experimentally. Experimental trials into the separation of prochymosin IBs and cellular debris using a disc-stack centrifuge followed the simulation studies. Simulated overall collection efficiencies were compared with experimental data [10]. Deviations were observed although general trends were correct. A method for on-line control of inclusion body recovery using turbidity was also developed [10,42]. The ratio of two absorbance measurements (OD^{600nm}/OD^{400nm}) was shown to be a good correlator of the amount of inclusion body material in the centrifuge supernatant. The robustness of the method to changes in inclusion body size and other feed properties has not been defined, and a separate empirical correlating equation will clearly be required for each different feed stream. Nevertheless, the simplicity of the method suggests considerable potential for the on-line control of large-scale processing.

Numerous centrifuge simulation studies have also been undertaken using porcine somatotropin (pST) as a model system. These inclusion bodies are considerably smaller than the prochymosin IBs (Section 4.2.3), thus complicating differential

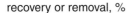

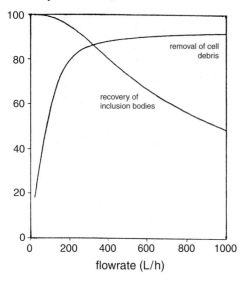

Fig. 4-7. Simulated recovery curves for prochymosin inclusion bodies and cell debris, based on the work of Mannweiler [39]. (From Keshavarz-Moore et al. [41]; reproduced with permission of *the Annals of the New York Academy of Sciences.*)

separation. Approximate IB and debris size distributions were determined for pST using differential sedimentation [43]. The important effect of homogenate viscosity was demonstrated using the ideal Sigma centrifuge model. An experimental study of the recovery of pST inclusion bodies in a Westfalia SB-7 centrifuge was then undertaken, allowing the fractional recovery of inclusion bodies to be defined for a range of feedrates [44–46]. An empirical curve for collection efficiency was determined and used to simulate the classification performance. The simulation results suggested that overall process cost is determined to a large extent by debris size reduction in the homogenizer. Multiple homogenizer passes were shown to be beneficial because of reduced debris size, enabling improved IB purification. The results also suggested that a given inclusion body purity could be attained at lower cost using multiple centrifuge passes. This is simply a consequence of the centrifuge grade-efficiency characteristics.

Considerable experimental work into the recovery of IGF inclusion bodies in a disc-stack centrifuge has recently been completed [13,15]. The impact that centrifugation has on overall product yield for a proteolytically sensitive analog of insulin-like growth factor has been defined. Significant yield improvements were obtained by optimizing the centrifugation protocol. Specifically, multiple centrifuge passes without detergent or chaotrope washing significantly improved protein yield following solubilization. Under standard solubilization conditions, a twofold increase in recoverable protein was achieved by introducing two buffer washing steps, even though this increased the inclusion body loss from 25 % to 42 %. The increased yield resulted from improved cell-debris removal with multiple centrifuge passes, as suggested by the simulation studies discussed above.

Rational optimization of the fractionation process requires accurate debris and IB size distributions, and centrifuge grade-efficiency curves that are determined directly for the type of material being separated. Accurate debris size distributions have been determined using CSA (see Section 4.3).

Wong [15] has also used CSA to determine experimental grade-efficiency curves for cell debris from *E. coli*. The curves for a Veronesi KLE-160 disc-stack centrifuge could be modeled using Eqn (14). By combining the experimentally determined data and experimentally verified models, an extensive simulation study into optimizing the fractionation of IGF inclusion bodies from cell debris was completed [15]. Figure 4-8 shows the effect of centrifuge feedrate on the collection of inclusion bodies and the removal of cell debris for a single centrifuge pass. The benefit of repeated homogenization is clearly demonstrated, as it reduces debris size thus facilitating fractionation. Figure 4-9 shows the simulation results for multiple centrifuge passes. The benefit of repeated homogenization and centrifugation for improved purity are again clear.

It is worth mentioning that the study by Wong [15] is the first to employ experimentally determined grade-efficiency curves for cell debris. Furthermore, debris size distributions were determined using a sizing method that does not require pretreatment of the sample that can compromise the results, and models were validated using experimental data. Consequently, a high degree of confidence can be placed

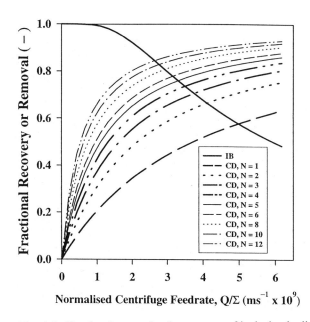

Fig. 4-8. Simulated curves for the recovery of inclusion bodies and the removal of cell debris in a disc-stack centrifuge. IB, inclusion bodies; CD, cell debris; N, number of homogenizer passes. (From Wong [15]; reproduced with the kind permission of Heng-Ho Wong).

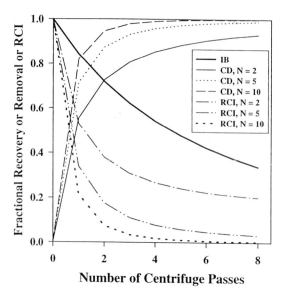

Fig. 4-9. Simulated curves for the recovery of inclusion bodies and the removal of cell debris in a disc-stack centrifuge. IB, inclusion bodies; CD, cell debris; N, number of homogenizer passes; RCI, ratio of cell debris to inclusion bodies, and hence a measure of purity, and is normalized relative to the ratio in the homogenate before centrifugation. (From Wong [15]; reproduced with the kind permission of Heng-Ho Wong).

in the simulation results, which conclusively demonstrate that it is possible to optimize centrifuge operation through careful manipulation of process conditions.

It is worth commenting briefly on the application of chemical washes during centrifugal recovery of inclusion bodies (see Section 4.4.2). The preceding results clearly show that the purity of the inclusion body paste can be controlled by optimizing the process conditions, and in particular the centrifuge feedrate, the number of homogenizer passes, and the number of centrifuge passes. While washing with detergents or chaotropes may be necessary in some cases, their use should only be investigated after the optimal centrifugation procedure has been defined without their use. In this way, it may often be possible to minimize downstream impacts by careful optimization of centrifugation conditions, and it may be possible to minimize or in some cases eliminate the use of wash chemicals.

4.7 Alternatives to Inclusion Body Recovery

Traditional methods of dealing with inclusion bodies described in previous sections are well established. However, they have largely evolved from laboratory-scale approaches based around inclusion body release followed by *in vitro* solubilization. Recently, new approaches with improved potential for scale-up have been developed.

Work has shown that certain inclusion bodies can be partially solubilized in low concentrations of denaturant. Hart and Bailey [47] examined the solubilization of *Vitreoscilla* hemoglobin inclusion bodies in urea. Significant solubilization was achieved with low concentrations (< 3 M urea), suggesting that some inclusion body protein had a partially folded conformation and that a fractional dissolution and refolding process may be advantageous. Chang and Swartz [48] have shown that periplasmic inclusion bodies of insulin-like growth factor (IGF-I) can be partially solubilized and refolded, following cell disruption, in 2–3 M urea with < 10 mM dithiothreitol. Greenwood et al. [49] have also shown that an analog of IGF-I can be recovered in refolded form following dissolution in low concentrations of urea (< 4 M). Relatively low overall yields were obtained in all studies. Nevertheless, the work suggests that it is not necessary to fully denature the protein structure from inclusion bodies using highly concentrated denaturant. It may therefore be possible to develop large-scale processes that do not require denaturant concentrations close to saturation (e.g., 8 M urea), and that have the dissolution and refolding stages integrated into one unit.

In situ dissolution methods for recombinant inclusion bodies have received interest recently. Effectively, these take intact cells containing recombinant inclusion bodies and selectively treat them with chemicals to effect solubilization of the inclusion body. The advantages of this approach are clear. First, culture may be taken directly from the fermenter and treated to release the protein of interest in a soluble form. This deletes several operations from the traditional approach, namely cell disruption and inclusion body collection and washing.

Hart et al. [50] have used this approach to recover IGF-I with an endogenous secretory signal sequence from *E. coli*. The protein was distributed equally between the soluble phase and periplasmic inclusion bodies, so a traditional IB recovery strategy gave low product yields. Treatment of alkaline fermentation broth (pH 10) with a chaotrope (2 M urea) and reductant (10 mM dithiothreitol) gave ~ 90 % of all IGF-I in an isolated supernatant. A two-phase extraction procedure was developed to provide direct extraction of the solubilized IGF-I [50,51]. Urea tended to inhibit two-phase formation and residual solids sedimentation. Highest recovery was obtained with a two-phase systems composed of 5 % sodium sulfate and 14 % PEG-8000. The partition coefficient in this system was approximately 8, indicating significant potential for process-scale application.

This approach of *in situ* solubilization has been extended to cytoplasmic inclusion bodies of an analog of insulin-like growth factor (Long-R^3-IGF-I, R.J. Falconer, Dissertation in preparation, The University of Adelaide). Effective dissolution was obtained by treatment of intact cells with 6 M urea, 3 mM ethylenediaminetetra-acetate, and 20 mM dithiothreitol. The treatment gave comparable solubilized IGF to the traditional approach of mechanical cell disruption followed by *in vitro* solubilization. Low protein release was achieved below pH 9. High cell concentrations led to a significant increase in viscosity, possibly limiting application at large scale. The technique was also extended to give selective release of the IGF protein. Replacement of the dithiothreitol in the treatment mixture with 2-hydroxyethyldisulfide inhibited IB solubilization but still permeabilized the host cells and released soluble contaminant proteins from the cytoplasm. This presumably acted by cross-linking proteins on the IB surface through disulfide bond formation, thus inhibiting urea solubilization. The soluble contaminating proteins were washed away prior to solubilization using the original treatment buffer. The purity of extracted IGF by this method was 50 % of the total protein, constituting a purification factor of greater than 2.5. Total IGF recovery exceeded 80 %.

Methods involving *in situ* dissolution are limited to proteins that are not significantly affected by proteases and contaminants in the refolding mixture. Further development of these selective chemical release methods seems warranted given their potential for industrial-scale application. Other novel methods will undoubtedly emerge as further proteins expressed as inclusion bodies reach commercialization.

Abbreviations and Symbols

C	concentration (kg m^{-3})
D	diffusion coefficent (m^2 s^{-1})
d	particle diameter (m)
d_c	critical particle diameter (m)
f(d)	fractional collection of particles of size d
g	acceleration due to gravity (m s^{-2})
h	distance between centrifuge discs (i.e., caulk height) (m)
J	flux (kg m^{-2} s^{-1})
k	parameter in equation (8)
n	grade-efficiency exponent in equation (14)
P	pressure (Pa)
Q	flowrate (m^3 s^{-1})
R	resistance (m^2 kg^{-1})
s	grade-efficiency exponent in equation (12)
u	axial fluid velocity (m s^{-1})
v	radial fluid velocity (m s^{-1})
v_g	particle Stokes velocity under gravitational force (m s^{-1})

Subscripts

g	gel layer
m	membrane
p	permeate

Greek symbol

ν	fluid kinematic viscosity (m^2 s^{-1})
$\Delta\rho$	density difference between particle and fluid (kg m^{-3})
μ	fluid dynamic viscosity (Pa s^{-1})
Σ	centrifuge equivalent settling area (m^2)
ω	centrifuge-disc angular velocity (rad s^{-1})

Abbreviations

CDS	centrifugal disc photosedimentation
CSA	cumulative sedimentation analysis
ESZ	electrical sensing zone measurement
IB	inclusion body
IGF	insulin-like growth factor
PCS	photon correlation spectroscopy
VHb	*Vitreoscilla* hemoglobin

References

[1] Marston, F. A. O., *Biochem J*, 1986, *240*, 1–12.

[2] Kane, J. F., Hartley, D. L., *Trends Biotechnol*, 1988, *6*, 95–101.

[3] Wilkinson, D. L., Harrison, R. G., *Biotechnology*, 1991, *9*, 443–448.

[4] Mitraki, A., King, J., *Biotechnology*, 1989, *7*, 690–697.

[5] Lee, S. C., Olins, P. O., *J Biol Chem*, 1992, *267*, 2849–2852.

[6] Hart, R. A., Rinas, U., Bailey, J. E., *J Biol Chem*, 1990, *265*, 12728–12733.

[7] Valax, P., Georgiou, G., *Biotechnol Prog*, 1993, *9*, 539–547.

[8] Taylor, G., Hoare, M., Gray, D. R., Marston, F. A. O., *Biotechnology*, 1986, *4*, 553–557.

[9] Olbrich, R., Dissertation, University of London, 1989.

[10] Jin, K., Dissertation, University of London, 1992.

[11] Middelberg, A. P. J., Bogle, I. D. L., Snoswell, M. A., *Biotechnol Prog*, 1990, *6*, 255–261.

[12] Middelberg, A. P. J., O'Neill, B. K., Bogle, I. D. L., Snoswell, M., *Biotechnol Bioeng*, 1991, *38*, 363–370.

[13] Wong, H. H., O'Neill, B. K., Middelberg, A. P. J., *Bioseparation*, 1996, *6*, 185–192.

[14] Bailey, S. M., Blum, P. H., Meagher, M. M., *Biotechnol Prog*, 1995, *11*, 53–539.

[15] Wong, H. H., Dissertation, University of Adelaide, 1997.

[16] Krueger, J. K., Kulke, M. H., Stock, J., *BioPharm*, 1989 *March*, 40–45.

[17] Chaudhuri, J. B., in: *Annals of the New York Academy of Sciences: Recombinant DNA Technology II;* Bajpai, R. K., Prokop, A. (Eds.), New York: New York Academy of Sciences, 1994; Vol. 721, pp. 374–385.

[18] Fischer, B., Sumner, I., Goodenough, P., *Biotechnol Bioeng*, 1993, *41*, 3–13.

[19] Datar, R. V., Cartwright, T., Rosen, C.-G., *Biotechnology*, 1993, *11*, 349–357.

[20] Kiefhaber, T., Rudolfph, R., Kohler, H.-H., Bucher, J., *Biotechnology*, 1991, *9*, 825–829.

[21] Middelberg, A. P. J.,*Chem Eng J*, 1996, *61*, 41–52.

[22] Vicik, S., DeBernadez-Clark, E., in: *Refolding. ACS Symposium Series 470:* Georgiou, G., DeBernadez-Clark, E. (Eds.), Washington: American Chemical Society, 1991; Vol. 470, pp. 180–196.

[23] Batas, B., Chaudhuri, J. B., *Biotechnol Bioeng*, 1996, *50*, 16–23.

[24] Hagen, A. J., Hatton, T. A., Wang, D. I. C., *Biotechnol Bioeng*, 1990, *35*, 955-965.

[25] Valax, P., Georgiou, G., in: *Biocatalyst design for stability and specificity:* Himmel, M. E., Georgiou, G. (Eds.), Washington: American Chemical Society, 1993; Vol. 516, pp. 126–139.

[26] Babbitt, P. C., West, B. L., Buechter, D. D., Kuntz, I. D., Kenyon, G. L., *Biotechnology*, 1990, *8*, 945–949.

[27] Petrides, D., Cooney, C. L., Evans, L. B., Field, R. P., Snoswell, M., *Computers chem Engng*, 1989, *13*, 553–561.

[28] Petrides, D., Sapidou, E., Calandranis, J., *Biotechnol Bioeng*, 1995, *48*, 529–541.

[29] Dostalek, M., Haggstrom, M., *Biotechnol Bieong*, 1982, *24*, 2077–2084.

[30] Haarstrick, A., Rau, U., Wagner, F., *Bioproc Eng*, 1991, *6*, 179–184.

[31] Kroner, K. H., *Biotech Forum,* 1986, *3,* 20–21.

[32] Quirk, A. V., Woodrow, J. R., *Enzyme Microb Technol,* 1984, *6,* 201–206.

[33] Forman, S. M., DeBernardez, E. R., Feldberg, R. S., Swartz, R. W., *J Membr Sci,* 1990, *48,* 263–279.

[34] Meagher, M., Barlett, R. T., Rai, V. R., Khan, F. R., *Biotechnol Bioeng,* 1994, *43,* 969–977.

[35] Trettin, D. R., Doshi, M. R., *Chem Engng Commun,* 1980, *4,* 507–522.

[36] Romero, C. A., Davis, R. H., *Chem Eng Sci,* 1990, *45,* 13–25.

[37] Kroner, K. H., Schütte, H., Hustedt, H., Kula, M.-R., *Proc Biochem,* 1984, *4,* 67–74.

[38] Sullivan, F. E., Erikson, R. A., *Ind Eng Chem,* 1961, *53,* 434–438.

[39] Mannweiler, K., Dissertation, University of London, 1989.

[40] Mannweiler, K., Hoare, M., *Bioproc Eng,* 1992, *8,* 19–25.

[41] Keshavarz-Moore, E., Olbrich, R., Hoare, M., Dunnill, P., in: *Annals of the New York Academy of Sciences: Recombinant DNA Technology I:* Prokop, A., Bajpai, R. K. (Eds.), New York: New York Academy of Sciences, 1991; Vol. 646, pp. 307–314.

[42] Jin, K., Thomas, O. R. T., Dunnill, P., *Biotechnol Bioeng,* 1994, *43,* 455–460.

[43] Middelberg, A. P. J., Bogle, I. D. L., Snoswell, M., in: *Proceedings of Chemeca 89 (Broadbeach, Qld), Australian Chemical Engineering Conference.* Canberra: Institution of Engineers Australia, 1989, pp. 671–678.

[44] Middelberg, A. P. J., O'Neill, B. K., *Aust J Biotech,* 1991, *5,* 87–92.

[45] Middelberg, A. P. J., O'Neill, B. K., Bogle, I. D. L., in: *Proceedings of Chemeca 91 (Newcastle, NSW), Australian Chemical Engineering Conference.* Canberra: Institution of Engineers Australia, 1991, pp. 706–712.

[46] Middelberg, A. P. J., O'Neill, B. K., Bogle, I. D. L., *Trans I Chem E part C,* 1992, *70,* 8–12.

[47] Hart, R. A., Bailey, J. E., *Biotechnol Bioeng,* 1992, *39,* 1112–1120.

[48] Chang, J. Y., Swartz, J. R., in: *Protein Folding: in vivo and in vitro. ACS Symposium Series 526;* Cleland, J. L. (Ed.), Washington: American Chemical Society, 1993; Vol. 526, pp. 178–188.

[49] Greenwood, M., Kotlarksi, N., O'Neill, B. K., Falconer, R., Francis, G., Middelberg, A. P. J., in: *Proceedings of the 1994 IChemE Research Event.* London: Institution of Chemical Engineers, 1994, pp. 250–252.

[50] Hart, R. A., Lester, P. M., Reifsnyder, D. H., Ogez, J. R., Builder, S. E., *Biotechnology,* 1994, *12,* 1113–1117.

[51] Hart, R. A., Ogez, J. R., Builder, S. E., *Bioseparation,* 1995, *5,* 113–121.

5 The Application of Glycobiology for the Generation of Recombinant Glycoprotein Therapeutics

Jan B. L. Damm

5.1 Introduction

This paper discusses the role that glycosylation plays in the development of recombinant glycoprotein therapeutics. It is generally conceived that the opportunities that recombinant DNA technology offers for the generation of new therapeutics can hardly be overestimated. However, this does not mean that there are no pitfalls and potential problems. In this chapter some of the problems that the pharmaceutical industry faces in producing recombinant glycoprotein drugs which are related to the glycosylation of the molecules, and some of the potential solutions and opportunities that glycobiology offers to circumvent these problems and ultimately to develop recombinant drugs with superior characteristics, are addressed.

Recombinant DNA-based biotechnology emerged in the early 1970s as a challenging opportunity for the development of new glycoprotein pharmaceuticals. Although it took nearly a decade before recombinant DNA technology could be put into the day-to-day practice of the pharmaceutical industry, and it took until 1982 before the first biotechnology-based drug, human insulin, was marketed, it was clear from the beginning that biotechnology offered unseen possibilities and held great promise for breakthroughs in the development of new, complex glycoprotein drugs. Human insulin, which is a pure protein drug and not a glycoprotein, was followed by Somatrem for treatment of human growth hormone (hGH) deficiency in children in 1985. Thereafter, many recombinant drugs for treatment of various diseases followed. The approved recombinant therapeutics and vaccines till 1994 are listed in Table 5-1. Figure 5-1 shows the cumulative number of approved recombinant therapeutics and vaccines until 1994, and the US sales. Evidently, after a slow start the number of recombinant DNA therapeutics and vaccines, indicated by the bars in Fig. 5-1, has increased steadily to about 30 recombinant drugs in 1994. The line in Fig. 5-1 indicates U.S. sales of recombinant therapeutics in billion dollars. Thus, in 1990 recombinant therapeutics accounted for $1 billion sales in the United States, while in 1993 this figure already amounted to $3 billion.

It is generally believed that the number of recombinant therapeutics and their sales will continue to grow in the near future. At present, about 150 biotech products have been evaluated in the clinic, and a far larger number is engaged in the research phase. A recent survey among the main American pharmaceutical companies showed that

Table 5-1. Approved recombinant therapeutics and vaccines until 1994.

Product	Indication	Company (Trade name)
1982		
Human insulin	Diabetes	Eli Lilly/Genentech(Humulin)
1985		
Somatrem for injection	hGH deficiency in children	Genentech (Protropin)
1986		
Hepatitis B vaccine MSD	Hepatitis B prevention	Merck (Recombivax HB); Chiron
Interferon alfa-2a	Hairy cell leukemia	Hoffmann-La Roche (Roferon)
Interferon alfa-2b	Hairy cell leukemia	Schering-Plough/Biogen (Intron A)
Muromonab-CD3	Reversal of acute kidney transplant rejection	Ortho Biotech (Orthoclone OKT3)
1987		
Alteplase (tPA)	Acute myocardial infarction	Genentech (Activase)
Somatotropin for injection	hGH deficiency in children	Eli Lilly (Humatrope)
1988		
Interferon alpha-2a	AIDS-related Kaposi's sarcoma	Hoffmann-La Roche (Roferon)
Interferon alpha-2b	AIDS-related Kaposi's sarcoma Genital warts	Schering-Plough/Biogen (IntronA)
1989		
Interferon alfa-n3	Genital warts	Interferon Sciences (Alferon N inj.)
Hepatitis B vaccine	Hepatitis B prevention	SmithKline Beecham (Engerix-B); Biogen
Erythropoietin	anemia associated with chronic renal failure	Amgen (EPOGEN); Johnson&Johnson; Kirin
1990		
Erythropoietin	anemia associated with AIDS/AZT	Amgen (Procrit); Ortho Biotech
PEG-adenosine	ADA-deficient SCID	Enzon; Eastman Kodak
Interferon-gamma-1b	Management of chronic granulomatous disease	Genentech (Actimmune)
Alteplase (tPA)	Acute pulmonary embolism	Genentech (Activase)
Erythropoietin	anemia associated with chronic renal failure	Ortho Biotech (Procrit)
CMV immune globulin	CMV prevention in kidney transplant patients	MedImmune (Cyto-Gam)
1991		
Filgrastim (G-CSF)	Chemotherapy-induced neutropenia	Amgen (Neutrogen)
β-glucocerebrosidase	Type I Gaucher's disease	Genzyme (Ceredase)
Sargramostim (GM-CSF)	Authologous bone marrow transplantation	Hoechst-Roussel (Prokine); Immunex
Sagramostim (GM-CSF)	Neutrophil recovery following bone marrow transplantation	Immunex (Leukine); Hoechst-Roussel
Interferon-alpha-2b	Hepatitis C	Schering-Plough/Biogen (Intron A)

Table 5-1. (continued).

Product	Indication	Company (Trade name)
1992		
Antihemophilic factor	Hemophilia B	Armour (Mononine)
Aidesleukine (interleukin-2)	Renal cell carcinoma	Chiron (Proleukin)
Indium-111 labeled antibody	Detection, staging, and follow-up of colorectal cancer	Cytogen (OncoScint CR103); Knoll
Indium-111 labeled antibody	Detection, staging, and follow-up of ovarian cancer	Cytogen (OncoScint OV103); Knoll
Antihemophilic factor	Hemophilia A	Genetics Inst.; Baxter Healthcare (Recombinate)
Interferon-alpha-2b	Hepatitis B	Schering-Plough/Biogen (Intron A)
1993		
Erythropoietin	Chemotherapy-associated anemia in non-myloid malignancy patients	Amgen (Procrit); Ortho Biotech
Interferon-beta	Relapsing/remitting multiple sclerosis	Chiron; Berlex (Betaseron)
DNAse	Cystic fibrosis	Genentech (Pulmozyme)
Factor VIII	Hemophilia A	Genentech; Miles (Kogenate)
Erythropoietin	Anemia associated with cancer and chemotherapy	Ortho Biotech (Procrit)
1994		
Filgrastim (G-CSF)	Bone marrow transplant	Amgen (Neupogen)
Enzyme (PEG-L-asparaginases)	Refractory childhood acute lymphoblastic leukemia	Enzon (Oncaspar)
Human growth hormone	Short stature caused by hGH deficiency	Genentech (Nutropin)
Glucocerebrosidase	Type I Gaucher's disease	Genzyme (Cerezyme)

Source: *Biotechnology in the U.S.* (Institute for Biotechnology Information, Research Triangle Park, NC, 1995).

more than 30 % of their research projects are biotechnology-based. These data are reflected in the expectations for sales and market share of recombinant therapeutics. According to the Institute of Biotechnology Information of the North Carolina Biotechnology Centre, by the turn of the century the U.S. market for recombinant therapeutics will amount to at least $9 billion. This forecast is based on the still increasing demand for complex and selective therapeutics, such as for the treatment of AIDS, autoimmune diseases, Alzheimer's disease, and cancer.

However, there is also reason for modesty: although impressive, the current U.S. sales of recombinant therapeutics accounts for (only) 5 % of total U.S. sales of therapeutics and undoubtedly investors and speculators had expected more. Why is the list of presently marketed recombinant therapeutics not as long as one might had anticipated some 10 years ago? Potential problems in the production of recombinant drugs include:

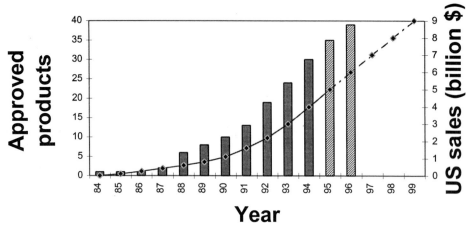

Fig. 5-1. Cumulative number of approved recombinant therapeutics and vaccines until 1994 and their sales in the U.S, 1995 + 1996: preliminary data.

– Instability
– Insolubitily
– Disappointing production rates
– Instability of the selected cell line
– Difficulties in up-scaling of production and purification of the recombinant product
– Undesired biodistribution
– Too long or too short circulatory half-lives
– Complex registration procedures
– Patent issues

In addition to these problems, the fact that many recombinant drugs are glycoproteins rather than proteins is also partly responsible. This means that glycosylation, next to other post-transcriptional modifications of the protein, is a phenomenon which has to be dealt with in the right way in order to obtain suitable products.

However, apart from being a complicating factor which sometimes jeopardizes successful development of recombinant drugs, glycosylation may also create new opportunities to overcome some of the aforementioned problems and in fact it may open the way to development of improved recombinant therapeutics.

5.2 Structure and Function of Glycoproteins

For a proper understanding of the opportunities that glycosylation offers to circumvent some of the aforementioned problems in the generation of recombinant glycoprotein therapeutics, a brief discussion of some general structural and functional features of glycoproteins is necessary. For more details, the reader is referred to Chapter 12.

Glycoproteins are ubiquitous in nature. Table 5-2 shows some examples of glycoproteins and their function. Glycoproteins may function for instance as enzymes, hormones, growth factors, lectins (glycoproteins that bind particular carbohydrate sequences), membrane constituents, serum constituents, or structural components.

The impact of glycosylation on the physico-chemical and functional properties of a (recombinant) glycoprotein drug are best illustrated by an example. Chorionic gonadotropin (CG) and the pituitary hormones lutropin (LH), follitropin (FSH), and thyrotropin (TSH) are members of a glycoprotein hormone family which have different physiological functions. Each glycoprotein hormone is a heterodimer consisting of two non-covalently associated subunits, designated α and β. Within a given animal species the α-subunits of LH, FSH, TSH and CG arise from a single gene [1] and have identical amino acid sequence [2–4]. However, the two homologously localized [5] N-linked oligosaccharide chains differ significantly in structure [6,7]. Although highly homologous (> 80 %) in protein, the β-subunits of LH, FSH, TSH, and CG arise from separate genes and differ in amino acid sequenc[8], number of glycosylation sites (varying from 1 to 6), and type (only N-glycosidic versus. N-

Table 5-2. Occurrence and functions of some representative glycoproteins.

Glycoprotein	Source	Molecular weight (kDa)	Carbohydrate content (%)
Enzymes			
Alkaline phosphatase	Mouse liver	130	18
Carboxy-peptidase Y	Yeast	51	17
Hormones			
Human chorionic gonadotropin	Human urine	38	31
Erythropoietin	Human urine	34	29
Lectins			
	Potato	50	50
	Soybean	120	6
Membrane constituents			
Glycophorin	Human erythrocytes	31	60
Rhodopsin	Bovine retina	40	7
GP-120	HIV virus	120	50
Serum glycoproteins			
IgG	Human serum	150	10
Thyroglobulin	Calf thyroid	670	8
Structural glycoproteins			
Collagen	Rat skin	300	0.4
Other			
Interferon	Human leukocytes	26	20
tPA	Human serum	65	15
Mucins	Mucosal epithelium	$10^3–10^4$	80

and O-glycosidic) and structure of carbohydrate chains (general aspects of N-and O-glycosylation are discussed in Chapter 12; the specific differences in glycosylation between CG, FSH, LH, and TSH have been reviewed previously [5]).

Figure 5-2 depicts the three-dimensional model of human chorionic gonadotropin (hCG), based on its recently determined crystal structure [9]. In humans, hCG is produced by the syncytiotrophoblasts of the placenta, a cell type which lacks secretory granules and releases the hormone constitutively following synthesis [10]. Human chorionic gonadotropin plays an important role during pregnancy and is a typical example of a complex glycoprotein. hCG exhibits LH-like effects and is responsible

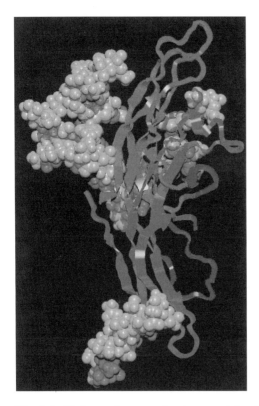

Fig. 5-2. (Colours please see front cover picture.) Three-dimensional model of human chorio-gonadotropin. The model is based on the crystal structure of deglycosylated hCG[9] (PDB code 1hrp). The protein part of the molecule (ribbon) and the four N-linked carbohydrate chains (spheres) are shown on the same scale. The oligosaccharides are attached to Asn52 (top, right) and Asn78 (bottom) of the α-subunit (green), and to Asn13 and 30 (top, left) of the β-subunit (blue). The binding region is indicated in red. It should be noted that the spatial orientation of the carbohydrate chains is arbitrarily set as they are not present in the crystal structure. The carboxy-terminal peptide of the β-subunit (amino acid residues 131–145) is not depicted because its 3D-structure could not be deduced from the crystal [Figure reproduced by courtesy of Prof. Dr. P. D. J. Grootenhuis (Dept. of Computational Medicinal Chemistry, N.V. Organon, Oss)].

for continued existence of the corpus luteum and stimulation of progesterone production during the first 6–8 weeks of gestation. The serum concentration of hCG in pregnancy can reach levels of up to several hundred nM during the first trimester [11]. For any hormone, this is an extraordinarily high concentration, and indeed, extrapolations from *in vitro* studies suggest that this hormone level is much larger than that required for maximal progesterone production by the corpus luteum. The hormone also stimulates progesterone production by the placenta and probably exhibits other biological actions on the fetus. hCG can be detected in maternal serum about four days after fertilization. Following the peak at the end of the first trimester the serum level falls, but is still significant at term. Free subunits can also be detected in the maternal serum. It is noteworthy that the production of the α-subunit appears to be the limiting factor for hCG formation in early pregnancy; in contrast, it is the synthesis of β-subunit that seems to be limiting late in gestation.

The two non-covalently linked, crest-like-shaped protein subunits of hCG are represented by the ribbon in Fig. 5-2. The α- and β-subunits are indicated in green and blue, respectively. The stretches in red indicate the receptor-binding domain of the hormone, involving parts of both the α-and β-subunits. The dimeric glycoprotein hormone carries four N-linked carbohydrate chains or oligosaccharides, indicated by the spheres, which are covalently linked to the protein backbone via the amide function of asparagine residues. The α-subunit contains two N-linked oligosaccharides at asparagine residues 52 and 78. The remaining two N-linked carbohydrate chains are present at asparagine residues 13 and 30 of the β-subunit. In addition, four O-linked oligosaccharides occur, attached to serine residues 121, 127, 132 and 138 of the carboxy-terminal part of the β-subunit. This portion of hCG-β, comprising about 30 amino acid residues, distinguishes hCG-β from the β-subunits of the pituitary glycoprotein hormones. The carboxy-terminal region has been suggested to have occurred via loss of a termination codon when hCG-β evolved from LH-β, thus permitting the 3′-untranslated region to become incorporated into the coding sequence [12]. This unique region of hCG-β may contribute to some of the properties of this glycoprotein, and it has been exploited for the development of highly specific immunoassays [13]. It should be noted, the hCG-β carboxy terminal peptide is not shown in Fig. 5-2 because so far it could not be deduced from the crystallographic data, probably because it is too flexible.

From the model it is apparent that the carbohydrate chains are relatively large entities – often the true proportion of the glyco-part of a glycoprotein is underestimated – and that the oligosaccharides occur on the outside of the molecule. The relatively large proportion of the carbohydrates, together with the location on the periphery of the molecule are the main reasons for the strong influence on the physico-chemical properties of the glycoprotein hormone. For instance, when the carbohydrate chains are (partly) removed, or when N-glycosylation is prevented by mutation of the glycosylation sites in the recombinant glycoprotein, the molecular mass and size of the molecule are obviously much lower (about 30 %). Also, the molecule is less hydrophilic, which is reflected by a dramatic drop in its solubility, leading to the formation of insoluble aggregates. Furthermore, the biochemical and physical stability of the molecule is affected: it is more prone to proteolytic breakdown in the circulation and the α/β-dimer tends to dissociate into (non-active) subunits.

The effect on the physico-chemical properties in turn modifies the molecule's biological function. For hCG, as for the related glycoprotein FSH [14] and many other glycoproteins, it is known that the amount, type, and composition of the oligosaccharides influences the *in vitro* and *in vivo* bioactivity of the hormone. The role of N-linked oligosaccharides in glycoprotein hormone bioactivity has been extensively investigated and reviewed (see [4]). In general, the effects of chemical or enzymatic deglycosylation, the prevention of N-glycosylation by tunicamycin, or deletion of N-linked carbohydrates by mutations at the DNA level, on receptor binding, clearance, and bioactivity have been evaluated. Removal of sialic acid (a negatively charged monosaccharide which usually occurs in the terminal position of the carbohydrate chain, see Chapter 12) from hCG results in increased receptor binding by the hormone [15]. However, (partly) desialylated hCG is quickly removed from the circulation via asialo glycoprotein receptor-mediated clearance in the liver. The efficiency of this clearing mechanism is illustrated by the $t_{\frac{1}{2}}$ of completely desialylated hCG in humans which is in the order of minutes as opposed to the $t_{\frac{1}{2}}$ of native hCG which is ~ 70 h. Mainly due to the influence of the presence of sialic acid – in the terminal non-reducing position of the oligosaccharides – on hCG receptor binding and liver-mediated clearance, desialylation has an opposite effect on the *in vitro* and *in vivo* activities: the *in vitro* activity is slightly enhanced (because sialic acid is not required for receptor binding or bioactivity *per se*), whereas the *in vivo* activity is dramatically reduced. Complete enzymatic deglycosylation of hCG does not alter receptor binding, but abolishes cAMP production as well as biological responses such as spermatogenesis (see [16]). Studies utilizing enzymatic deglycosylation have, however, yielded conflicting results in a number of instances. Chemical deglycosylation by HF solvolysis has also been used [17] to examine the role of the N-linked oligosaccharides in bioactivity [16,18,19]. With the exception of the Asn-linked N-acetylglucosamine residue, complete removal of the oligosaccharides is accomplished with HF solvolysis with little or no detectable damage to the protein subunits. Alternatively, hCG molecules missing one or more of the N-linked carbohydrates have been obtained by site-directed mutagenesis of the N-glycosylation sites [20–22]. Both the studies with HF-deglycosylated hCG and N-deglycosylated hCG muteins have led to the following conclusions:

1. Post-synthesis removal of the carbohydrate moieties does not hinder α/β subunit recombination.
2. N-linked oligosaccharides are not required for receptor binding *in vitro*.
3. Deglycosylated hormones display significantly reduced abilities to stimulate cAMP production in target cells despite unaffected receptor binding.

In essence, it appears that chemical deglycosylation has the effect of converting the hormone from an agonist into a competitive antagonist, since the deglycosylated product retains the ability to bind to the appropriate receptor, but not to induce the biological effects.

Supporting evidence for the role of the oligosaccharides comes from the properties of hCG produced in patients with choriocarcinom [23]. This hCG, having an identical amino acid composition as native hCG but an altered glycosylation pattern, was found to exhibit a threefold increase in affinity for its receptor, but a much lower biological activity when compared with normal hCG.

The various effects of glycosylation on the physico-chemical and biological properties of glycoproteins in general are dealt with in Section 5.2.2.

The effects that glycosylation may exert on the physico-chemical and biological properties of glycoproteins is one of the main reasons for gathering information about the glycosylation when producing glycoprotein drugs by recombinant DNA technology. It should be noted however, that glycosylation does not always influence biological function; there are examples of recombinant glycoproteins with aberrant glycosylation or no glycosylation at all which seem to function perfectly normal *in vitro* [24–26]. On the other hand there are also examples where a non-native glycosylation results in unwanted biological properties. Therefore, to be on the safe side, the glycosylation of the recombinant product should be identical to, or at least closely resemble, the glycosylation of the natural product. More interestingly, manipulation of the glycosylation may give an opportunity to improve the properties of the drug. Especially, when (on basis of sound analytical and biological data) structure–function relationships can be established, the fact that most protein drugs are glycosylated may be exploited to improve the therapeutic properties (see Section 5.3).

The second reason for gathering information about glycosylation is that the recombinant product, including its carbohydrates, must be characterized to verify for instance identity and batch-to-batch reproducibility, also with respect to the carbohydrate moiety. This is important for the manufacturer to check as to whether the product meets required specifications. Moreover, at present a detailed characterization is included in the package of demands of the registration authorities. Questions to answer with respect to glycosylation are:

- type of carbohydrate chains;
- composition and amount;
- Structure;
- reproducibility of glycosylation;
- differences with natural glycoprotein; and
- biological consequences.

To answer these questions considerable effort is put into the analysis of the carbohydrate moiety of new recombinant glycoprotein drugs. For a discussion of the strategies that can be adopted to carry out the characterization of the carbohydrate moiety of recombinant glycoproteins the reader is referred to Chapter 12.

5.2.1 Difficulties in Establishing Carbohydrate Structure–Function Relationships

Unfortunately, the presently available data (for review, see [27]) show that the effects of glycosylation on the functioning of a (recombinant) therapeutic are not entirely predictable. There are several reasons for the difficulty in predicting specific rules for carbohydrate functions.

1. There is an indirect genetic control over the structure of the protein-bound carbo-hydrates. This is due to the mode of biosynthesis of protein-linked carbohydrate chains which involves both co- and post-translational events, resulting in an inherent, yet characteristic, variability. As a consequence, a glycoprotein thera-peutic generally is not a population of identical molecules, but rather is composed of discrete subsets of differently glycosylated molecules, the so-called glycoforms [28], that have different physico-chemical and biochemical properties which in turn may lead to functional diversity. Consequently, any glycoprotein drug that consists of different glycoforms will exhibit a composite activity, reflecting a weighted average of the activity and incidence of each glycoform. Therefore, structure–function relationships for glycoprotein-linked carbohydrate chains are not necessarily apparent at the level of the native glycoprotein (population). Rather, structure–function relationships should be studied on the level of the indi-vidual glycoforms of the glycoprotein therapeutic. However, the preparative iso-lation of individual glycoforms, if possible at all, may require quite some effort.
2. Carbohydrate functions and protein functions are closely related or even inter-twined. Therefore, it frequently is difficult to distinguish the properties imposed by the carbohydrate moiety from the functions that are intrinsic to the protein part of a glycoprotein drug. Obviously this precludes studies towards specific functions of the carbohydrate moiety.
3. Carbohydrate structure–function relationships are not static but dynamic. A spe-cific carbohydrate structure may serve different purposes in different cells (of the same organism) or at different times in the life cycle or state of development of the same cell. So in fact, a particular structure may have different functions depending on its localization or the physiological status of the organism which of course may obscure structure–function relationships.
4. Multivalency seems to be a key-word in carbohydrate function. In many cases per definition it will not be possible to associate a certain biological function with a specific individual carbohydrate structure. Rather, structure–function relationships must be searched for in the context of glycosylation patterns and concentration of carbohydrate chains, both in time (which again refers to the physiological status of the organism) and space (which relates to multivalency as key factor in recog-nition events). For instanc, it is more or less established that during carbohydrate-mediated recognition events only the concerted interaction between a multitude of carbohydrate ligands and their receptors evokes a biological response. Obviously, this phenomenon complicates the study of structure–function relations.

5.2.2 Glycosylation-associated Effects on the Properties of Glycoprotein Drugs

Nonetheless, some general glycosylation-associated effects on the properties of gly-coprotein drugs, acting in- or interdependently of each other, are recognized. As out-lined in Fig. 5-3, two types of effect on the properties of glycoprotein drugs must be distinguished, namely the influence on the physico-chemical properties, and the

Property	Effect
physico-chemical	size
	mass
	solubility
	viscosity
	charge
biological	antigenicity
	stability
	clearance
	intracell. routing
	biodistribution
	receptor binding
	cell-cell contacts

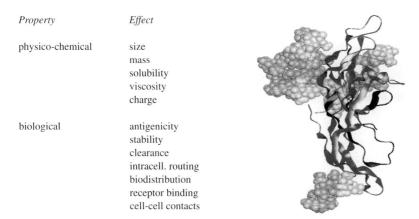

Fig. 5-3. Glycosylation-associated effects on the properties of glycoprotein drugs.

influence on the biological properties. As already discussed for hCG, the effects of the presence of carbohydrate chains on physico-chemical properties of the glycoprotein drug are quite trivial – but are nonetheless sometimes of decisive importance – such as the effects on size, mass, tertiary structure, solubility, viscosity, and charge.

The physico-chemical characteristics may in turn affect – in a more complex way – the biological functioning of the glycoprotein drug, involving antigenicity, stability, plasma half-life, intracellular routing, organ targeting and biodistribution, receptor binding and cell–cell recognition events. Some specific examples of how carbohydrate characteristics influence or even determine the biological properties of a glycoprotein drug include:

1. The influence of the sialic acid/galactose ratio of a glycoprotein therapeutic on its circulatory half-life via determination of the kinetics of liver-mediated clearance [29,30]. The over-riding influence of the presence of sialic acid in terminal position of the oligosaccharides of hCG, masking the penultimate galactose residue and thus preventing rapid hepatic clearance, has already been mentioned. Similar effects are observed for many other glycoprotein (drugs), e.g. recombinant FSH [31] (Puregon[R]) and erythropoietin [32,33].
2. The increased resistance against proteolysis by shielding potential protease cleavage sites. It has been shown for instance, that the carbohydrate chains (in particular the sialic acid residues) of von Willebrand factor protect the protein against amino-terminal proteolytic cleavage and are essential for maintenance of its multimeric structure [34,35]. Similarly, the oligosaccharides of bovine pancreatic ribonuclease protect the molecule against protease degradation [36].
3. The recognition of terminal sialic acids of glycoproteins and glycolipids in the cell membrane by various viruses and bacteria, mediating cellular infection [37–39].
4. Recognition of polylactosamines on erythrocytes [40] and/or platelets [41] by autoimmune antibodies, leading to autoimmune destruction of the cells causing autoimmune hemolytic anemia.

5. Recognition of sialylated, fucosylated lactosaminoglycans on leucocytes by the E-selectin of endothelial cells [42–44], mediating extravasation and also inflammation.

6. The involvement of O-linked carbohydrate chains in mammalian fertilization. Bleil and Wassarman [45] described the essential role of galactose at the terminal non-reducing position of specific O-linked carbohydrate chains of the murine egg zona pellucida glycoprotein ZP3 for the species-specific recognition and primary binding of sperm cells. It is generally conceived that the ZP3-linked oligosaccharides fulfil a similar function in various other species [46–48]. Miller et al. [49] reported the involvement of β1-4 galactosyltransferase in murine sperm–egg recognition. The authors suggested that the galactosyltransferase on the sperm membrane mediates fertilization by binding to the aforementioned O-linked carbohydrate chains located on the zona pellucida glycoprotein ZP3. Hence, the sperm surface galactosyltransferase and the egg coat glycoprotein ZP3 function as complementary adhesion molecules that enable the recognition and primary binding of murine gametes.

An overview of the presently recognized effects of the carbohydrate moiety on the biological functioning of the parent (recombinant) glycoprotein therapeutic can be found in the compendious review paper of Varki [27]. Taken together, the various functions of carbohydrate chains can be summarized in that they either mediate specific recognition events, or that they modulate biological processes. Unfortunately, it is not possible to make general statements about which carbohydrate characteristics in general are best for a(ny) recombinant glycoprotein therapeutic. Rather, for each individual recombinant drug the optimal glycosylation must be unravelled in relation to its desired therapeutic profile.

5.3 Glyco-engineering

To exploit the opportunities of glycosylation and to deal with its restrictions as good as possible, at least two requirements must be fulfilled. The first is the establishment of sound relationships between (protein-linked) carbohydrate structure(s) on the one hand and their influence on the biological functioning of the glycoprotein drug on the other hand. Such a structure–function relation should give an answer to the question '*what* should be made?', or more precisely, 'what should the glycosylation of the recombinant product look like in order to give the (recombinant) product its desired therapeutic properties?' Once it is known what the glycosylation of the ideal product should look like, the second requirement is an answer to the question '*how* can that be accomplished?'.

This is where glyco-engineering comes into the picture. Glyco-engineering may be defined as the ensemble of techniques that allow the controlled production and/or manipulation of the carbohydrate moiety of glycoproteins. Here, the relation between the protein backbone, the selected cell line and the culture conditions on the one hand, and the obtained glycosylation pattern on the other hand is essential.

The importance of glycosylation for production levels of recombinant drugs is well known. Absence or incorrect glycosylation may lead to a dramatic drop in productivity due to decreased or impaired biosynthesis or secretion. In the absence of glycosylation, the newly formed proteins might get stuck in the cell lumen or in the membrane, resulting in reduced secretion or even complete loss of secretion. Furthermore, the earlier discussed complications with respect to (lack of) solubility, stability, biological activity, etc. of the products due to incorrect glycosylation or absence of glycosylation frequently occur.

In addition to obtaining glyco-engineered recombinant therapeutics with increased bioactivity or specificity, controlled modification of the carbohydrate moiety may be used to eliminate, some potential problems, such as antigenicity of recombinant therapeutics that carry potentially immunogenic carbohydrate determinants, to enhance or reduce the clearance rate, or to eliminate unwanted carbohydrate heterogeneity. Reduction of the carbohydrate heterogeneity for instance may make life easier because it facilitates the purification, characterization, quality control, prediction of pharmacokinetic behavior and – if wanted – crystallization of the recombinant therapeutic, which in turn may lead to a shorter development time and registration procedure.

In summary, it can be stated that because the carbohydrate chains influence the physico-chemical and biological properties of the glycoprotein drugs, glycosylation may be exploited to alleviate the mentioned problems and to improve the properties of the recombinant drug by applying glyco-engineering techniques.

Engineering of glycosylation is conceivable at three levels: (i) the DNA-level; (ii) the biosynthesis level; and (iii) the product level (Fig. 5-4).

GLYCO-ENGINEERING

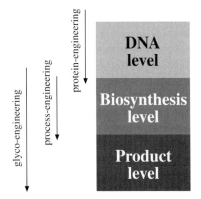

Fig. 5-4. The three levels at which glyco-engineering is possible.

5.3.1 Glyco-engineering at the DNA level

At the DNA level there are again three options to manipulate the glycosylation of the recombinant product, namely via the protein backbone of the recombinant glycoprotein, by choosing the host cell, and by selection of mutants and/or coexpression of glycosyltransferases and glycosidases, (Table 5-3).

Table 5-3. Glyco-engineering at the DNA level.

Carbohydrate structure and protein backbone
– Sites for N- and O-glycosylation, GPI anchors
– Accessibility of glycans to glycosylation enzymes
– Determination of carbohydrate fine structure

Host cell
– Glycosyltransferases
– Glycosidases

Selection of mutants, e.g.

– CHO-tetra	α2-6 ST	sTri/Tetra-antennary	Epo, *in vivo* bioactivity ↑
– CHO-LEC11	α1-3 FT	(s)Le^x/(s)Le^a glycocon jugates	Target ELAM-1/GMP140
– CHO-LEC8	↓ SA/G	terminal GN	Target reticulo-endothelial cells
– CHO-PIR.LEC1	↓ SA/G/GN	terminal mannose	Target reticulo-endothelial cells

GPI, glycosylphosphatidylinositol; CHO, Chinese hamster ovary; α2-6 ST, α2-6 sialyltransferase; α1-3 FT, α1-3 fucosyltransferase; SA, sialic acid (*N*-acetylneuraminic acid); G, galactose; GN, *N*-acetylglucosamine; sTri/Tetra, sialyted tri-/tri' or tetra-antennary oligosaccharides; (s)Le$^{x/a}$, (sialyl)Lewis X or A; Epo, erythropoietin; ELAM, endothelial leucocyte adhesion molecule; GMP, α-granule membrane protein; ↑, increase; ↓, decrease.

5.3.1.1 Carbohydrate Structure and Protein Backbone

First of all, when glyco-engineering at the DNA level is considered it is important to realize that the glycosylation of the protein is affected by the protein itself in at least three ways (this topic is covered in more detail by Cumming [50]). First, and most obviously, the protein presents the potential sites for N-glycosylation, O-glycosylation, and the attachment of glycosylphosphatidylinositol anchors (GPI-anchors, see Chapter 12). For N-glycosylation a consensus amino acid sequence asparagine-X (i.e., any amino acid, except proline) -serine/threonine is required. This implies that by mutating or deleting the consensus sequence, N-glycosylation can be knocked out at specific sites. In contrast, by insertion of consensus sequences additional N-linked oligosaccharides might be attached (it should be noted, however, that presence of a consensus sequence is a necessary, but not sufficient requirement for N-glycosylation). For O-glycosylation and GPI anchors a consensus sequence is not known and hence the situation is less clear.

The second way in which the protein influences its own glycosylation is by determining the accessibility of its attached oligosaccharides to the enzymes involved in glycosylation. By limiting access, or by forcing its oligosaccharide chains into particular conformations, the conversion of the carbohydrate chains into substrates or non-substrates for the (de)glycosylating enzymes, can be determined. In fact, the over-riding influence of the protein backbone on its own glycosylation is exemplified by the establishment of consistent glycosylation patterns for a particular (recombinant) glycoprotein, irrespective of the cell line in which it is produced (there are many exceptions, however, e.g. [51]).

Finally, the specific amino acid sequence may determine the carbohydrate fine structure. This is illustrated for example by the gonadotropic hormone-specific *N*-acetylgalactosamine transferase which is able to attach *N*-acetylgalactosamine residues to the carbohydrate chains linked to luteinizing hormone, but not to the carbohydrate chains of the related follicle stimulating hormone [5], in spite of the fact that both glycoproteins are produced in the same organ and that there are only small overall differences in the protein backbone.

5.3.1.2 Choice of the Host Cell

The second option to manipulate the glycosylation of the recombinant product at the DNA level is by choice of the host cell. It is generally recognized that the glycosylation is more or less host cell-specific due to the repertoire, concentration, and compartmentalization of glycosidases and glycosyltransferases present. Glycosyltransferases add monosaccharides to the growing carbohydrate chain during its biosynthesis, whereas glycosidases are enzymes that remove monosaccharides during the final processing of the oligosaccharides (see [52]). The concerted action of both types of enzymes is needed in the biosynthesis of carbohydrates. Both types of enzymes exhibit a high degree of specificity towards the nucleotide sugar donor, the oligosaccharide acceptor, and the anomeric configuration and type of carbohydrate linkage formed. Since the type of host cell that is chosen for the production of the recombinant drug to a large extent predefines the ensemble of glycosyltransferases and glycosidases present, the right choice of host cell is extremely important for the generation of a 'desired' glycosylation pattern in recombinant glycoproteins.

So far, various types of host cells have been used for the production of recombinant therapeutics. The most important are *Escherichia coli,* yeast, insect, and mammalian cells, (Table 5-4).

E. coli has been the classical host because it is easy to grow and maintain in large quantities, and it gives high protein yields. However, it does not possess the glycosylation machinery of eukaryotes which means that products that normally occur as glycoproteins are produced as proteins. (It is worthy of note that Messner and Sleytr [53] reported the biosynthesis of (bacterial) surface layer glycoproteins in bacteria, but this seems to be an exception.) Because the unglycosylated proteins often are not properly folded, or are insoluble or form aggregates or give problems with respect to *in vivo* activity or antigenicity, prokaryotes like *E. coli* are in many

Table 5-4. Frequently used host cells for the production of recombinant (glyco)protein drugs.

E. coli
– Easy to grow and maintain in large quantities
– No glycosylation machinery → consequences for production, conformation, (in)solubility,
  aggregation, *in vivo* activity, antigenicity?

Yeast cells
– Easy to grow and maintain in large quantities
– Yeast-type glycosylation → Immunogenic compatibility?, bioactivity?

Insect cells
– Relatively easy to grow
– High production rates
– Mammalian-like N- and O-linked carbohydrates

Mammalian cells
– Culture and maintenance not without problems
– Production levels relatively low
– Mammalian glycosylation machinery

cases not suitable for the production of genetically engineered glycoprotein therapeutics.

Yeast cells combine the major advantage of prokaryotes, namely that they are easy to grow and maintain in large quantities, with the main advantage of eukaryotes, namely glycosylation. However, yeast-specific carbohydrate chains differ dramatically from their mammalian counterparts, which may compromize the immunogenic compatibility and/or bioactivity of the glycoprotein drug. Future research will learn whether this problem can be tackled by the use of yeast glycosylation mutants which produce (N-linked) carbohydrate precursors that are also common to mammalian cells, which may then be processed to the desired glycosylation pattern by treatment with the appropriate glycosidases and/or glycosyltransferases.

During the last couple of years the application of baculovirus-transfected insect cells has emerged as a promising possibility for the production of recombinant glycoproteins. Transfected insect cells are relatively easy to grow, allow high production rates of the recombinant glycoprotein and seem to be capable of synthesizing nearly the whole spectrum of N- and O-linked mammalian carbohydrate chains [54,55]. However, there is still debate with respect to the similarity/equivalency of the insect and mammalian glycosylation machinery.

Although the culture and maintenance of large quantities of mammalian cells is not without problems and production rates are frequently lower than in the aforementioned cells, mammalian cells are in general the preferred host for the generation of recombinant glycoprotein therapeutics. The main reason for this is the equivalence of the glycosylation machinery of all types of mammalian cells, including human cells. There may, however, still be considerable differences in the concentration or presence of the glycosidases and glycosyltransferases between the various host cells. Therefore, also in this case the choice of a proper cell line remains a key issue. If the ideal glycosylation pattern for a particular glycoprotein drug is known, or more realistically, if it is known which general characteristics the carbohydrate chains should have or not have, a proper cell line can be selected. Preferably this will be a cell line which is known to express the desired carbohydrate features in its natural products. Frequently used mammalian cell lines to date are BHK and Chinese hamster ovary (CHO) cells and their various mutants.

5.3.1.3 Selection of a Glycosylation Mutant

The third option to manipulate the glycosylation of the recombinant product at the DNA level is by selection of an appropriate glycosylation mutant. At present various (mutant) cell lines have been selected or engineered [56–62], aiming at the synthesis of recombinant glycoprotein therapeutics with increased effectiveness. Mutants can be obtained by screening or by genetic engineering. In the latter case the desired type of glycosylation activity is engineered into the cell line by co-expression of the appropriate glycosyltransferases or glycosidases.

Among the mutant cell lines currently available is the so-called CHO-tetra mutant transfected with the α2-6 sialyltransferase gene for the production of erythropoietin (Epo) with modified carbohydrate chains [32]. For erythropoietin it is known that the biological activity *in vivo* is highly dependent on the structures of its carbohydrate chains [63]. Epo molecules that bear predominantly diantennary carbohydrate chains (carbohydrate chains that contain two branches, see Chapter 12) display a much lower bioactivity than those that bear predominantly tetra-antennary oligosaccharides (carbohydrates that have four branches). By choosing a CHO cell line tansfected with the α2-6 sialytansferase gene, recombinant erythropoietin with fully sialylated, primarily tetra-antennary carbohydrate chains could be produced with significantly increased bioactivity. Probably, the same result may be obtained by transfection of the host with *N*-acetylglucosaminyltransferase-IV and -V, the enzymes which are responsible for the formation of higher branched oligosaccharides, or by supplementing the culture medium with exogenous interleukin-6 (IL-6), which has been shown to shift the *N*-acetylglucosaminyltransferase-III activity in myeloma cell line OPM-1 towards *N*-acetylglucosaminyltransferase-IV and -V activity [64], resulting again in production of glycoproteins with tri-/tri'- and tetra-antennary carbohydrate chains.

Another example is the CHO-LEC11 mutant for the production of glycoconjugates that carry the sialyl-Le x/a determinant [65]. The aim here is to target the drug carrying the sialyl-Le x/a determinant to endothelial cells or lymphocytes that express the cell adhesion molecules ELAM-1 or GMP-140.

A third example of the potential use of mutant cell lines concerns the CHO-LEC8 [66] and CHO-PIR.LEC1 [67] mutants which produce carbohydrates that terminate in *N*-acetylglucosamine and mannose, respectively, in stead of sialic acid (more precisely, *N*-acetylneuraminic acid) or galactose. As already suggested by Stanley [68], this mutant could be used to produce recombinant glycoprotein therapeutics that specifically target the reticuloendothelial cells. It has been shown [69] that administration of exogenous glucocerebrosidase is effective in the treatment of patients with Gaucher's disease, one of the most common diseases of glycolipid metabolism leading to accumulation of glycolipids in various cells and organs, resulting in severe pathological effects. However the drug – glucocerebrosidase – is only effective when it is selectively targeted to the reticuloendothelial cells. This was realized by treatment of the natural enzyme with neuraminidase, β-galactosidase and β-hexosaminidase, respectively, resulting in glucocerbrosidase-carrying carbohydrate chains terminating in mannose residues. The processed glucocerebrosidase molecules are targeted specifically to the reticuloendothelial cells [70] via recognition by the mannose-binding lectin which is present on the reticuloendothelial cells [71]. Obviously, it would be much more straightforward to produce this enzyme in the CHO-LEC8 or CHO-PIR.LEC1 mutant which would directly yield the desired product.

A more detailed description of these and other examples of mutant cell lines that are used or potentially can be used to produce glycoprotein drugs with designed carbohydrate chains can be found in the review by Stanley [68].

5.3.2 Glyco-engineering at the Biosynthesis Level

After selection of a specific cell line for production, glyco-engineering can still be performed at the biosynthesis level, *viz.* during the production of the recombinant product in the roller bottle, fermentor, etc. by adapting the culture conditions in order to manipulate the glycosylation of the recombinant product. Presently, it is recognized that culture conditions, like density, age and growth rate of the cells [72], concentration and type of nutrients [73], presence of growth differentiation factors, cytokines [74], glycosylation inhibitors [75] and glycosylation enzymes [76], pH, CO_2, O_2, and ammonium concentration, all affect glycosylation somehow. Therefore, in theory each of these factors might be exploited to manipulate the glycosylation (for a review of the environmental and bioprocess effects on protein glycosylation the reader is referred to Goochee et al. [74]).

Unfortunately however, at present the relation between the culture conditions and the carbohydrate structures – or more generally, the glycosylation pattern – of the recombinant product is poorly understood. Hence, when applying this strategy a word of warning is in place. Additional and/or irreproducible heterogeneity of the carbohydrate moiety of a recombinant therapeutic as a result of uncontrolled or unstable culture conditions is highly unwanted. Therefore, in general the primary goal is to control and monitor the culture conditions as good as possible in order to obtain a 'reproducible carbohydrate heterogeneity'. However, if in time sound relations between culture conditions and synthesized carbohydrate structures can be established, the second goal will be to engineer the desired glycosylation type into the recombinant product by careful manipulation and control of the culture conditions. So far there are several examples of the feasibility of this approach, like the application of glycosylation inhibitors, the induced expression of silent glycosyltransferase genes by growth factors, cytokines or agents like retinoic acid and butyrate – which affect the differentiation of the cultured cells – and the production at high or low pH, oxygen, CO_2 or glucose levels (for reviews, see [77] and [74].

5.3.3 Glyco-engineering at the Product Level

Finally, glyco-engineering can be performed on the product level. After its biosynthesis and collection, the recombinant product generally has to be purified. During the purification process it may be possible – apart from removing contaminants – to selectively remove, or alternatively selectively concentrate, certain glycoforms. Provided that this can be done under controlled conditions, the purification process may be used to selectively obtain the glycoforms which exhibit the desired product profile.

Another possibility is to modify the carbohydrate chains of the purified product, for instance by treatment with chemicals or glycosidases, in such a way that the desired glycosylation characteristics are obtained. An example of this approach is the preparation of the mannose-terminating glycoforms of glucocerebrosidase to treat Gaucher's disease. In both cases – selective purification and modification of the carbohydrate chains of the purified glycoprotein – the desired glycosylation characteristics are engineered into the finished product afterwards. In general this will be a rather expensive approach, yet in specific cases it might be practically and economically feasible.

5.4 Conclusion and Perspectives

Recombinant glycoprotein therapeutics capture an increasingly important part of the total market for therapeutics. In view of this, possibilities to facilitate and optimize the development of recombinant glycoprotein drugs is the subject of much contemporary research. Figure 5-5 summarizes the main opportunities for the development of improved recombinant glycoprotein therapeutics as discussed in this chapter. The knowledge about the relationships between (protein-linked) carbohydrate structure(s) on the one hand and the protein backbone, selected cell line, and culture conditions on the other hand, and the relation between carbohydrate structure and its biological function or its effect on the biological functioning of the glycoprotein is growing. This fulfils the first requirement for the generation of new, improved glycoprotein drugs. The second requirement is the availability of glyco-engineering techniques and dedicated analytical facilities (Chapter 12). Together, this may ultimately result in rational design of glycosylated recombinant therapeutics with improved characteristics by controlled manipulation of the carbohydrate moiety.

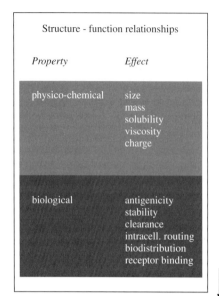

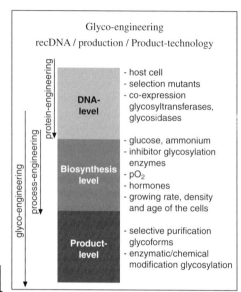

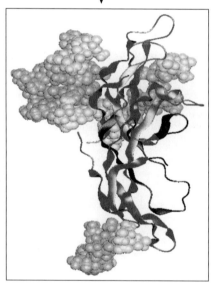

Optimized   glycoprotein

Increase of   Reduction of

reproducibility   heterogeneity
production   antigenicity
solubility   clearing
stability   development time
biological activity
tuning of activity

Fig. 5-5. Possibilities for the development of optimized recombinant glycoprotein drugs on basis of structure–function relationships and by using glyco-engineering techniques.

Acknowledgement

Part of this material is adapted from articles originally published in *Pharmaceutical, Technology* Vol. 7, No. 8, 1995 and *BioPharm*, Vol. 8, No. 9, 1995.

References

[1] Boothby, M., Ruddon, R. W., Anderson, C., McWilliams, D., Boime, I., *J Biol Chem*, 1981, *256*, 5121–5127.
[2] Pierce, J. G., Parsons, T. F. *Ann Rev Biochem*, 1981, *50*, 465–495.
[3] Norman, A. W., Litwack, G., in: *Hormones*, Orlando, x. et al. (Eds.), New York: Academic Press, 1987.
[4] Ryan, R. J., Keutmann, H. T., Charlesworth, M. C., McCormic, D. J., Milius, R. P., Calvo, F. O., Vutyavanich, T., *Rec Prog Horm Res*, 1987, *43*, 383–429.
[5] Baenziger, J. U., Green, E. D., *Biochim Biophys Acta*, 1988, *947*, 287–306.
[6] Green, E. D., Baenziger, J. U., *J Biol Chem*, 1988, *263*, 25–35.
[7] Green, E. D., Baenziger, J. U., *J Biol Chem*, 1988, *263*, 36–44.
[8] Esch, F. S., Mason, A. J., Cooksey, K., Mercado, M., Shimasaki, S., *Proc Natl Acad Sci USA*, 1986, *83*, 6618–6621.
[9] Lapthorn, A. J., Harris, D. C., Littlejohn, A., Lustbader, J. W., Canfield, R. E., Machin, K. J., Morgan, F. J., Isaacs, N. W., *Nature*, 1994, *134*, 455–461.
[10] Hoshina, M., Boothby, M., Boime, I., *J Cell Biol*, 1982, *93*, 190–198.
[11] Cole, L. A., Kroll, T. G., Ruddon, R. W., Hussa, R. O., *J Clin Endocrinol Metab*, 1984, *58*, 1200–1202.
[12] Fiddes, J. C., Talmadge, K., *Rec Prog Horm Res*, 1984, *40*, 43–74.
[13] Birken, S., Canfield, R., Agosto, G., Lewis, J., *Endocrinology*, 1982, *110*, 1555–1563.
[14] De Leeuw, R., Mulders, J. W. M., Voortman, G., Rombout, F., Damm, J. B. L., Kloosterboer, H. J., *Excerpts on Human Reproduction*, 1996, *1*, 13–21.
[15] Kalyan, N. K., Lippes, H. A., Bahl, O. P., *J Biol Chem*, 1982, *257*, 12624–12631.
[16] Sairam, M. R., in: *Hormonal proteins and peptides*, Vol. 11, Li, C. H. (Ed.), New York: Academic Press, 1993, pp. 1–79.
[17] Manunath, P., Sairam, M. R., *J Biol Chem*, 1982, *257*, 7109–7115.
[18] Keutmann, H. T., McIllroy, P. J., Berger, E. R., Ryan, R. J. *Biochemistry*, 1983, *22*, 3067–3072.
[19] Sairam, M. R., Bhargaui, G. N., *Science*, 1985, *229*, 65–67.
[20] Matzuk, M. M., Boime, I., *J Cell Biol*, 1988, *106*, 1049–1059.
[21] Matzuk, M. M., Keene, J. L., Boime, I. *J Biol Chem*, 1989, *264*, 2409–2414.
[22] Matzuk, M. M., Boime, I., *J Biol Chem*, 1988, *263*, 17106–17111.
[23] Mizuochi, T., Nishimura, R., Derappe, C., Taniguchi, T., Hamamoto, T., Mochizuki, M., Kobata, A., *J Biol Chem*, 1983, *258*, 14126–14129.
[24] Moonen, P., Mermod, J. J., Ernst, J. F., Hirschi, M., DeLamarter, J. F., *Proc Natl Acad Sci USA*, 1987, *84*, 4428–4431.
[25] Mackenzie, P. I., *Biochem Biophys Res Commun*, 1990, *166*, 1293–1299.
[26] Lace, D., Olavesen, A. H., Gacesa, P., *Carbohydr Res*, 1990, *208*, 306–311.
[27] Varki, A., *Glycobiology*, 1993, *3*, 97–130.
[28] Rademacher, T. W., Parekh, R. B., Dwek, R. A., *Annu Rev Biochem*, 1988, *57*, 785–838.
[29] Ashwell, G., Harford, J., *Annu Rev Biochem*, 1982, *51*, 531–554.
[30] Smith, P. L., Kaetzel, D., Nilson, J., Baenziger, J. U., *J Biol Chem*, 1990, *265*, 874–881.
[31] De Leeuw, R., Mulders, J. W. M., Voortman, G., Rombout, F., Damm, J. B. L., Kloosterboer, H. J., *Mol Hum Reprod*, 1996, *2*, 361–369.
[32] Takeuchi, M., Inoue, N., Strickland, T. W., Kubota, M., Wada, M., Shimizu, R., Hoshi, S., Kozutsumi, H., Takasaki, S., Kobato, A., *Proc Natl Acad Sci USA*, 1989, *86*, 7819–7822.
[33] Delorme, E., Lorenzini, T., Giffin, J., Martin, F., Jacobsen, F., Boone, T., Elliott, S., *Biochemistry*, 1992, *31*, 9871–9876.
[34] Berkowitz, S. D., Federici, A. B., *Blood*, 1988, *72*, 1790–1796.
[35] Kessler, C. M., Floyd, C. M., Frantz, S. C., Orthner, C., *Thromb Res*, 1990, *57*, 59–76.
[36] Joao, H. C., Scragg, I. G., Dwek, R. A., *FEBS Lett*, 1992, *307*, 343–346.

[37] Varki, A., *Glycobiology,* 1992, *2,* 25–40.

[38] Schauer, R., *Trends Biochem Sci,* 1988, *10,* 357–360.

[39] Weis, W., Brown, J. H., Cusack, S., Paulson, J. C., Skehel, J. J., Wiley, D. C., *Nature,* 1988, *333,* 426–431.

[40] Roelcke, D., Hengge, U., Kirschfink, M., *Vox Sang,* 1990, *59,* 235–239.

[41] Koerner, T. A. W., Weinfeld, H. M., Bullard, L. S. B., Williams, L. C. J., *Blood,* 1989, *74,* 274–284.

[42] Berg, E. L., Robinson, M. K., Mansson, O., Butcher, E. C., Magnani, J. L., *J Biol Chem,* 1991, *266,* 14869–14872.

[43] Larsen, G. R., Sako, D., Ahern, T. J., Shaffer, M., Erban, J., Sajer, S. A., Gibson, R. M., Wagner, D. D., Furie, B. C., Furie, B., *J Biol Chem,* 1992, *267,* 11104–11110.

[44] Phillips, M. L., Nudelman, E., Gaeta, F. C. A., Perez, M., Singhal, A. K., Hakomori, S., Paulson, J. C., *Science,* 1990, *250,* 1130–1132.

[45] Bleil, J. D., Wassarman, P. M., *Proc Natl Acad Sci USA,* 1988, *85,* 6778–6782.

[46] Noguchi, S., Nakano, M., *Eur J Biochem,* 1992, *209,* 883–894.

[47] Wassarman, P. M., *Development,* 1990, *108,* 1–17.

[48] Hokke, C. H., Damm, J. B. L., Penninkhof, B., Aitken, R. H., Kamerling, J. P., Vliegenthart, J. F. G., *Eur J Biochem,* 1994, *221,* 491–512.

[49] Miller, D. J., Macek, M. B., Shur, B. D., *Nature,* 1992, *357,* 589–593.

[50] Cumming, D. A., *Glycobiology,* 1991, *1,* 115–130.

[51] Parekh, R. B., Dwek, R. A., Thomas, J. R., Opdenakker, G., Rademacher, T. W., Wittwer, A. J., Howard, S. C., Nelson, R., Siegel, N. R., Jennings, M. G., *Biochemistry,* 1989, *28,* 7644–7662.

[52] Montreuil, J., Vliegenthart, J. F. G., Schachter, H., in: *Glycoproteins. New Comprehensive Biochemistry,* Vol. 29a, Neuberg, A. and Van Deenen, L. L. M. (Eds.), Amsterdam: Elsevier, 1995.

[53] Messner, P., Sleytr, U. B., *Glycobiology,* 1991, *1,* 545–551.

[54] Luckow, V. A., Summers, M. D., *Bio/Technology,* 1988, *6,* 47–55.

[55] Davidson, D. J., Fraser, M. J., Castellino, F. J., *Biochemistry,* 1990, *29,* 5584–5590.

[56] Lowe, J. B., Kukowska-Latallo, J. F., Nair, R. P., Larsen, R. D., Marks, R. M., Macher, B. A., Kelly, R. J., Ernst, L. K., *J Biol Chem,* 1991, *266,* 17467–17477.

[57] Stanley, P., *Mol Cell Biol,* 1989, *9,* 377–383.

[58] Smith, D. F., Larsen, R. D., Mattox, S., Lowe, J. B., Cummings, R. D., *J Biol Chem,* 1990, *265,* 6225–6234.

[59] Stanley, P., *Glycobiology,* 1991, *1,* 307–314.

[60] Stoll, J., Rosenwald, A., Krag, S. S., *Nato ASI Ser,* 1990, *H40,* 151–166.

[61] Lee, E. U., Roth, J., Paulson, J. C., *J Biol Chem,* 1989, *264,* 13848–13855.

[62] Potvin, B., Kumar, R., Howard, D. R., Stanley, P., *J Biol Chem,* 1990, *265,* 1615–1622.

[63] Takeuchi, M., Kobata, A., *Glycobiology,* 1991, *1,* 337–346.

[64] Nakao, H., Nishikawa, A., Karashuno, T., Nishiura, T., Iida, M., Kanayama Y., Yonezawa, T., Tarui, S., Tanaguchi, N., *Biochem Biophys Res Commun,* 1990, *172,* 1260–1266.

[65] Howard, D. R., Fukuda, M., Fukuda, M. N., Stanley, P., *J Biol Chem,* 1987, *262,* 16830–1687.

[66] Stanley, P., *Trends Genet,* 1987, *3,* 77–81.

[67] Zeng, Y., Lehrman, M. A., *Anal Biochem,* 1991, *193,* 266–271.

[68] Stanley, P., *Glycobiology,* 1992, *2,* 99–107.

[69] Barton, N. W., Furbish, F. S., Murray, G. J., Garfield, M., Brady, R. O., *Proc Natl Acad Sci USA,* 1990, *87,* 1913–1916.

[70] Barton, N. W., Brady, R. O., Dambrosia, J. M., Di Bisceglie, A. M., Doppelt, S. H., Hill, S. C., Mankin, H. J., Murray, G. J., Parker, R. I., Argoff, C. E., *N Engl J Med,* 1991, *324,* 1464–1470.

[71] Stahl, P. D., Rodman, J. S., Miller, M. J., Schlesinger, P. H., *Proc Natl Acad Sci USA,* 1978, *75,* 1399–1403.

[72] Hahn, T. J., Goochee, C. F., *J Biol Chem,* 1992, *267,* 23982–23987.

[73] Hayter, P. M., Curling, E. M. A., Baines, A. J., Jenkins, N., Salmon, I., Strange, P. G., Tong, J. M., Bull, A. T., *Biotechnol Bioeng,* 1992, *39,* 327–335.

[74] Goochee, C. F., Monica, T., *Biotechnology,* 1990, *8,* 421–427.

[75] Elbein, A. D., *FASEB J,* 1991, *5,* 3055–3063.

[76] Srivastava, G., Kaur, K. J., Hindsgaul, O., Palcic, M. M., *J Biol Chem,* 1992, *267,* 22356–22361.

[77] Warren, C. E., *Curr Opin Biotechnol,* 1993, *4,* 596–602.

6 The Release of Intracellular Bioproducts

Anton P. J. Middelberg

6.1 Introduction

Biological molecule recovery is usually done from the aqueous phase using techniques such as chromatography (see Volume 1). These techniques require that the molecule is available for recovery and consequently is not encased in a cell structure. For this reason, one of the first steps in many bioprocesses is the release of protein from the cell cytoplasm to the suspending medium.

Protein excretion technology is developing rapidly for organisms such as *Saccharomyces cerevisiae* and *Escherichia coli* that have a limited natural capacity for transferring biomolecules from the cytoplasm to the suspending medium. However, the approach is often product-specific and is subject to disadvantages, including the need to fuse a signal sequence to the natural protein. Consequently, the usual approach to product release involves a method of cell-wall disruption prior to subsequent downstream processing. Available techniques range from chemical treatment to disorder the wall, to non-specific mechanical methods that physically tear the cell wall apart. This chapter focuses on the release of intracellular bioproducts using such methods.

6.2 Cell Wall Destruction

Destruction of the cell wall is aided by a knowledge of wall structure (Section 6.2.1), allowing effective strategies for cell disruption to be defined (Section 6.2.2). The effectiveness of a particular strategy can then be gauged using a suitable measurement technique for cell disruption (Section 6.2.3).

6.2.1 Cell Wall Structure

Gram-negative bacteria have a relatively simple wall structure. *E. coli* is typical of many Gram-negative bacteria, and is perhaps the best studied. Its wall consists of

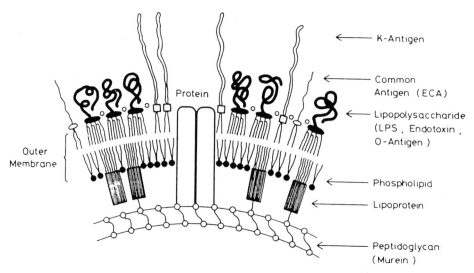

Fig. 6-1. Schematic representation of the Gram-negative bacterial cell wall. (From Rietschel, E. Th., Brade, H., Brade, L., Kawahara, K., Lüderitz, Th., Schade, U., Tacken, A., Zähringer, U. (1986), in: *Biological Properties of Peptidoglycan*. Berlin: Walter de Gruyter Co., 1986; p. 341 Reproduced with the kind permission of Walter de Gruyter Co.)

two basic layers: the outer membrane and a peptidoglycan layer. The outer membrane comprises mainly lipopolysaccharide (LPS), phospholipid (PL), and lipoprotein (LP), arranged as shown in Fig. 6-1. Note that divalent cations such as Ca^{2+} and Mg^{2+} stabilize the outer wall structure by cross-linking the LPS molecules. The outer membrane provides a primary physical barrier retaining molecules such as proteins within the periplasmic space. It is a major barrier to free diffusion and hence the excretion of proteins to the extracellular space.

The peptidoglycan layer serves primarily as a stress-bearing layer within the wall: it provides the cell's mechanical strength. It is considered permeable to protein molecules and salts and therefore is not a major diffusion barrier. Peptidoglycan consists of glycan chains cross-linked by peptide bonds [1]. Recent evidence suggests the *E. coli* wall has essentially one stress-bearing layer of peptidoglycan across 75–80 % of the cell's surface and localized multi-layered patches [2]. A simplified representation of *E. coli* peptidoglycan structure is shown in Fig. 6-2.

In addition to the wall, a thin lipid membrane separates the culture medium and the wall from the cell cytoplasm. This cytoplasmic membrane has little mechanical strength and disrupts when the wall is removed or compromised unless cells are osmotically stabilized. It is a major diffusion barrier separating the cytoplasmic contents from the periplasm and hence the extracellular fluid.

Another Gram-negative host of interest for product release in bioprocessing is *Alcaligenes eutrohpus*. This host has been extensively studied for the production of polyhydroxyalkanoates, which have applications as biodegradable polymers. The wall structure is typical of Gram-negative bacteria, comprising an outer membrane, a peptidoglycan layer, and a cytoplasmic membrane.

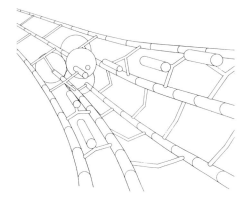

Fig. 6-2. Artist's view of enzyme complex involved in *E. coli* peptidoglycan synthesis. Parallel glycan chains cross-linked by peptide bonds are clearly visible. (From Höltje, J.-V. (1993), in: *Bacterial Growth and Lysis:* de Pedro, M. A., Höltje, J.-V., Löffelhardt, W. (Eds.). New York: Plenum, 1993; p. 425. Reproduced with the permission of Plenum Publishing Corporation.)

Gram-positive bacteria have a wall that lacks the outer membrane, and hence a defined periplasmic space. The peptidoglycan layer is considerably thicker than in Gram-negative bacteria, constituting up to 90 % of the cell-wall mass (compared with 5–20 % in Gram-negative bacteria). Minor components of the wall include the cytoplasmic membrane and acidic polysaccharides (teichoic acids) attached to the outer wall. The increased peptidoglycan content makes Gram-positive bacteria difficult to disrupt. Fortunately, many are natural protein secretors.

Like bacteria, yeast such as *S. cerevisiae* have an inner membrane that separates the cell cytoplasm from the cell wall and the culture medium. However, the similarities end at this point. The wall of *S. cerevisiae* is more complex than that of *E. coli*, and considerably harder to break. It comprises up to 30 % of the cell mass. Recent reviews summarize the current state of knowledge regarding yeast wall structure [3,4]. There appears to be three key layers. Mannan linked with protein (mannoprotein) forms an outer layer that is susceptible to proteolytic attack. It is associated with a central core of alkali-soluble glucan that has an amorphous appearance under electron microscopy, and may confer flexibility to the wall. Finally, an electron-dense layer borders the inner membrane; this is probably alkali-insoluble β1,3 glucan with a high degree of polymerization and some β1,6 links. This layer has a fibrillar appearance under electron microscopy, and has a direct role in maintaining wall rigidity and shape. It is this glucan layer that is believed to provide cell mechanical strength, possibly aided by the amorphous-like central glucan layer, while the external protein layer provides the main permeability barrier for macromolecules. It is worth noting that the structure of the cell wall is highly dynamic, and changes in response to a wide variety of growth and environmental conditions as well as the cell cycle [4]. Furthermore, the boundary between the mannoprotein and glucan layers is not clearly defined. There is evidence that the mannoproteins are interwoven into the glucan matrix, and in some species may penetrate the entire wall [4].

6.2.2 Strategies for Cell Disruption

The release of intracellular bioproducts requires destruction of the barrier separating the product from the extracellular fluid. This process of compramising the cell wall is often called cell disruption. A knowledge of cell wall structure, as given in Section 6.2.1, suggests various selective methods of compromising cell-wall integrity.

The primary diffusion barriers in Gram-negative bacteria such as *E. coli* are the inner and outer membranes. Chemicals that selectively destroy or disorder these membranes, such as detergents and chaotropes, have been used with varying success. Given their specificity, they often disrupt either the inner or outer membrane, but rarely both. Combinations of chemicals are therefore gaining increasing interest. Another approach is to destroy selectively the strength-bearing peptidoglycan in the wall through enzymatic treatment (e.g., lysozyme). Destruction of the peptide cross-links or the glycan chains in peptidoglycan will compromise cell integrity and release intracellular products due to the cell's high osmotic pressure. However, access to the peptidoglycan is limited by the protective outer membrane, so this must often be destabilized prior to enzymatic treatment. The induction of endogenous enzymes that degrade the wall through chemical or physical treatments (e.g., mild temperature rise, mild osmotic shock) is also possible. Certain chemicals (e.g., antibiotics) also inhibit critical growth functions, thus leading to cell lysis with some release of intracellular species. Simple physical processes for cell disruption such as severe thermal treatment and explosive decompression are also possible.

As for Gram-negative bacteria, a range of methods is available to break the yeast cell wall and thus release the intracellular contents. The primary diffusion barriers are the cytoplasmic membrane and the mannoprotein complex in the cell wall. These are difficult to compromise, and chemical treatments generally only permeabilize yeast cell walls. While these treatments facilitate the entry of small molecules or probes into the cell (e.g., for *in situ* assays of enzymatic activity), they are generally unsuitable for large-scale protein release. Selective enzymatic degradation of the yeast wall by successive protease and glucanase attack has, however, been successfully used. This approach uses knowledge of cell wall structure to establish a suitable combination of degrading enzymes.

In general, selective methods of protein release have proved less than satisfactory for the release of intracellular bioproducts. The complexity of the cell wall limits the amount of protein recovered, and these methods are generally used only in specific cases. Combinations of specific treatments must often be established, and results usually vary with the state of the target cell wall. However, increasing research in this area is leading to new combinations of general applicability, particularly for *E. coli* (see Section 6.3).

To overcome the limitations of selective methods, mechanical methods of cell disruption have received widespread use in bioprocessing. Available methods at process scale include high-pressure homogenization and its derivatives, and bead milling. These methods apply brute force to tear apart the wall components non-selectively, and are often capable of effecting complete product release in a contained way and without the need for added chemicals.

6.2.3 Quantifying Cell Disruption

It is clear from the preceding section that a variety of strategies are available for cell disruption, and each affects the cell-wall structure in a unique way. These different methods of release have lead to different methods of quantifying cell disruption, which may be conveniently categorized as direct or indirect.

6.2.3.1 Direct Measurement of Disruption

Conceptually, the simplest method to determine cell disruption is to establish directly how many cells are destroyed during treatment. This can be done by directly counting the number- or volume-fraction of cells destroyed, for example by microscopy. Microscopic observation can be aided by stains that test cell-wall integrity. Nevertheless, it remains tedious even with automated image analysis software. Direct counting may be automated to overcome this. For example, methylenetblue dye exclusion and automatic cell counting using a hemocytometer can be used to quantitate cell disruption [5].

Several researchers have used particle size analyzers to monitor cellular disruption, and also to monitor the size of fragments following disruption (see Section 6.9.1). Elzone particle sizer analyzers (e.g., Coulter counters) are popular, and clearly show the shift to smaller particle size as cells are destroyed. However, they are prone to operational problems such as orifice blocking by cellular debris. Further, accurate quantitation of cellular disruption is difficult due to convolution of the cell and debris peaks. The analytical disc centrifuge has been successfully employed to quantitate *E. coli* disruption accurately, even in the presence of recombinant inclusion bodies [6]. This is a high-resolution technique that determines disruption by ratioing the cell-peak areas before and after disruption. It is not, however, useful for yeast because of excessive overlap of the cell and debris peaks.

Another possibility is to use live cell counts, thus giving an indication of how many cells are inactivated by treatment. Of course, this only gives a measure of cell inactivation and does not guarantee that intracellular contents are actually released.

6.2.3.2 Indirect Methods

Indirect methods estimate the volume fraction of cells destroyed by directly quantitating the release of specific intracellular 'markers', such as the protein of interest. Although these methods provide indirect measures of disruption, they do provide a direct measure of the protein of interest.

The most common approach measures total cell protein in the sample supernatant using dye-binding assays that are readily available (e.g., the Bradford, Lowry, and BCA assays). The fractional release of protein, R_p, is given by Eqn (1),

$$R_p = \frac{C_h - C_o}{C_m - C_o} \qquad (1)$$

where C is the concentration of soluble protein in the supernatant of homogenate (h) or feed (o) samples, and C_m is the maximum possible release of soluble protein, corresponding to complete cellular disruption. For highly concentrated suspensions, it may be necessary to use a dilution technique to correct for the change in liquid-phase volume fraction as cells are destroyed [7].

Direct measurement of protein concentration using the dilution technique may be unsatisfactory when protein denaturation occurs. A mass-balance approach based on total Kjeldahl nitrogen has been developed to overcome this limitation [8]. Although less accurate than the dilution technique, it is particularly useful when protein is denatured.

A range of other specific methods is available for inferring disruption. Several researchers assay for the release of specific enzymes from different cellular locations (e.g., periplasmic, cytoplasmic, cell-wall associated). Monitoring these enzymes gives some indication of the breakdown of cellular structure (e.g., the release of cell-wall-associated enzymes is normally slower, and high levels of release usually indicate an advanced level of cellular breakdown and hence disruption). Direct measurement of product release, for example by HPLC or ELISA, is also possible. In all cases, it is standard to calculate fractional release according to Eqn (1). Correction for changes in the volume fraction, as for dye-binding assays, will generally be necessary.

6.2.3.3 Selecting a Method

No particular method of monitoring cell disruption is ideal for all cases, and choice is often made on the basis of simplicity, cost effectiveness, equipment availability, and personal choice. In general, indirect measurement using dye-binding is commonly employed as it is fast, cheap, and requires no specialized equipment beyond a laboratory centrifuge and a spectrophotometer. Direct and indirect measures generally give similar results when product degradation is minimal.

It is worth mentioning that indirect measures of cellular disruption, such as dye-binding assays, have lowest accuracy at high levels of disruption [9]. The opposite is true of direct methods, which have highest accuracy as disruption approaches 100 %. Direct measurement is therefore preferred when accurate measurements are required, as in modeling and optimization. Also, product degradation is sometimes observed under harsh disruption regimes. Direct determination of cellular disruption will allow the independent processes of cellular disruption and product degradation to be deconvoluted. However, direct disruption measures are also prone to problems. Specifically, cells that appear intact under microscopy may have actually released some intracellular contents. Similarly, cell wall permeability to integrity-testing dyes does not guarantee that cellular contents have actually been released. In all cases, a reasonable degree of caution must be exercised.

6.3 Chemical Disruption

As mentioned in Section 6.2.2, chemical treatments have been most effective at releasing protein from Gram-negative bacteria such as *E. coli*. Chemical treatment of the yeast wall generally only leads to permeabilization without significant protein release, restricting their use to *in situ* enzymatic assays. Methods of permeabilizing the yeast cell wall have been detailed previously [10], so will not be discussed further here.

6.3.1 Antibiotics

Antibiotics are known for their ability to lyse bacteria and prevent further growth, primarily by interfering with the biosynthetic capacity of growing cells [11]. Antibiotics interfere with peptidoglycan synthesis and assembly in Gram-negative bacteria such as *E. coli*, thus permitting enhanced action of endogenous enzymes that degrade peptidoglycan. Although lysis is rapid, the cost of antibiotics is prohibitive at large scale and results depend on the state of the culture. Most antibiotics, for example, do not lyse stationary phase bacteria.

6.3.2 Chelating Agents

Section 6.2.1 shows that the cell wall of Gram-negative bacteria is stabilized by divalent cations such as Mg^{2+} and Ca^{2+} that cross-bridge adjacent LPS patches on the outer membrane. Chelating agents such as ethylenediamine tetra-acetic acid (EDTA) have a unique ability to bind these ions, causing rapid release of significant amounts of LPS [12]. Activity is highest in the presence of Tris buffer because destabilization requires replacement of the cations with large monovalent organic amines [13]. EDTA is primarily of use in releasing periplasmic proteins, or in combination with other treatments such as lysozyme (see Section 6.4), as it has no effect on the cytoplasmic membrane nor the cell-wall peptidoglycan. Periodic EDTA treatment (0.5 to 3.0 mM) has been employed to effect the *in situ* release of soluble periplasmic β-lactamase from immobilized recombinant *E. coli* without significant loss of cell viability [14].

6.3.3 Chaotropic Agents

Chaotropic agents such as guanidium salts and urea disrupt the structure of water, weakening solute–solute interactions. They lead to effective solubilization of integral membrane proteins. The requirement for high chaotrope concentrations generally limits their use at process scale.

Synergistic effects are seen when chaotropes are employed with other chemicals. For example, protein release from *E. coli* is possible with relatively low concentrations of guanidine-HCl (up to 4 M) in the presence of low concentrations (up to 2 %) of the non-ionic detergent Triton X-100 [15].

At very low gu-HCl concentrations (ca. 0.1 M), a direct synergistic effect is seen. Protein release occurs due to molecular alteration of the outer wall and solubilization of the cytoplasmic membrane. However, some resistance to total protein release remains after treatment, limiting the total amount of protein that can be extracted. A subsequent study demonstrated that high protein release (> 75 %) from exponentially growing cells could be obtained using 0.4 M gu-HCl and 0.5 % Triton X-100 [16].

Chaotrope (2 M urea) in the presence of a reducing agent (10 mM dithiothreitol or 50 mM cysteine) has also been used to recover insulin-like growth factor fusion protein expressed as periplasmic inclusive bodies [17]. Under alkaline conditions (pH 10), approximately 90 % protein recovery is achieved. The method was coupled with a direct two-phase extraction to provide initial purification from contaminant protein and insoluble cellular debris.

Chaotrope action can also be enhanced through the synergistic effects of chelating agents. Urea combined with EDTA can release cytoplasmic protein from *E. coli* at the same level as mechanical disruption [18]. Presumably, EDTA disrupts the outer membrane in conjunction with urea providing the chaotrope with access to the cytoplasmic membrane. This approach has been applied to achieve the *in situ* solubilization of recombinant inclusion bodies in *E. coli,* and has been extended to enable selective release of soluble cytoplasmic contaminants prior to *in situ* dissolution of the inclusion body, thus providing crude initial purification (R. J. Falconer, Dissertation in preparation, University of Adelaide). The treatments are effective against exponential and stationary phase *E. coli,* and achieve the same recovery as the traditional mechanical disruption and *in vitro* dissolution and refolding approach.

6.3.4 Detergents

Detergents interact with lipid components in cell membranes, leading to their solubilization. They have been used with varying success to extract proteins from *E. coli* as outer membrane LPS provides partial protection, particularly by restricting detergent access to the cytoplasmic membrane. At low concentration, detergents generally only solubilize membrane proteins from prepared wall samples (e.g., by mechanical disruption) and are particularly useful when the desired protein product is associated with the cell wall. Protein release from whole *E. coli* cells using the anionic detergent sodium dodecyl sulfate (SDS) at high concentration (1 mg SDS per 5–11 OD^{600} absorbance units of cell mass) has been achieved [19]. Fractional protein release was not reported. Complete SDS removal was achieved by adsorption onto zeolite Y. The SDS offered protection against proteases, presumably through selective denaturation. The synergistic use of the non-ionic detergent Triton X-100 and chaotrope has also had some success (see Section 6.3.3).

6.3.5 Alkaline Treatment

Alkaline lysis using hydroxide and hypochlorite is an effective and cheap method for releasing intracellular products and acts by saponifying the cell-wall lipids. The technique is extremely harsh and consequently the product must be resistant to degradation at high pH. This limits the applicability for protein recovery. However, alkaline treatment has received widespread interest for the recovery of polyhydroxyalkanoates (PHA, a biodegradable polymer) from Gram-negative bacteria such as *E. coli* and *A. eutrophus*.

PHA recovery from *A. eutrophus* has been achieved using sodium hypochlorite [20], but with a 50 % reduction in the molecular weight of the polymer. This has been overcome by combining alkaline lysis with chloroform extraction of the polymer, using dispersions of the two chemicals [21]. Under optimal conditions a purity in excess of 97 % and a recovery of 91 % was obtained. Minimal product degradation occurred because of the protective effect of chloroform. PHA recovery from *A. eutrophus* using alkaline treatment in the presence of detergents has also been attempted [22]. Treatment with 1 % Triton (pH 13) followed by a brief hypochlorite treatment gave a purity in excess of 98 %, but with significantly higher product degradation than the hypochlorite-chloroform dispersion method. Hypochlorite extraction of PHA has been extended to recombinant *E. coli* giving high product purity with minimal product degradation and no detectable change in polymer granule size [23,24]. Alkaline treatment prior to mechanical disruption by high-pressure homogenization has also been employed for PHA recovery from *A. eutrophus* (see Section 6.6.3).

6.4 Enzymatic Disruption

The addition of foreign lytic enzymes that selectively degrade the cell wall is a gentle yet powerful method for effecting protein release from a range of organisms. For Gram-negative bacteria, the approach is to use enzymes such as lysozyme that attack the strength-providing peptidoglycan in the wall. Enzymatic access to the peptidoglycan is aided through disruption of the outer membrane, for example using EDTA (see Section 6.3.2). By removing the strength-bearing element, cells that are not osmotically stabilized will lyse and release their cytoplasmic contents. The enzymatic degradation of yeast is somewhat more complex, as a result of the layered design of the cell wall (see Section 6.2.1). Complex enzyme systems comprising proteases that degrade the mannoprotein complex and glucanases that degrade the strength-providing glucan network are required. Treatment with a simple glucanase leads to minimal disruption because of the protective action of the outer mannoprotein complex. Available enzyme systems for yeast have been reviewed [25] and include relatively inexpensive commercial preparations such as Zymolyase 20-T from continuous culture of *Oerskovia xanthineolytica* (Seikagaku America Inc., Rockville, MD) and lytic systems from *Cytophaga* and *Rhizoctonia*. Cell-wall breakdown is aided by thiol reagents that activate endogenous proteases and break disulfide bonds in the cell wall.

Release of cytoplasmic protein from stationary-phase *E. coli* has been achieved using 25–50 µg mL^{-1} lysozyme in combination with 100–800 µg mL^{-1} EDTA [26]. Enhanced release was obtained at 58 °C although the temperature optimum for lysozyme is 35 °C, presumably because of increased outer-membrane disruption at the higher temperature and hence improved lysozyme penetration. Combinations of 5–30 µg mL^{-1} polymyxin and lysozyme also released protein from exponential cells. Polymyxin is a cationic polypeptide antibiotic with an aliphatic chain that disorganizes and penetrates the outer membrane, providing access for the lysozyme to the peptidoglycan.

The disruption of yeast cells by enzymatic treatment has been modelled extensively [27]. Disruption occurs in a step-wise fashion with protease attack of the outer wall preceding glucanase attack. When a sufficient amount of glucan has been solubilized, the inner membrane ruptures, releasing the cytoplasmic contents. Breakdown of subcellular structures follows lysis.

Lysozyme is available relatively cheaply from egg-white preparations, but enzyme cost can still be prohibitive for large-scale applications. This is also the case for commercial systems for yeast digestion. Immobilization of enzymes is a practical way of reducing costs, but is limited by a loss of enzyme activity and inaccessibility of the substrate to enzyme within the pores of solid support matrices. Lysozyme has been immobilized on smooth fibres to improve substrate access [28], but the efficacy of this approach was limited by a rapid loss of enzyme activity. Further work may lead to effective immobilization methods that are generally applicable and suitable for process-scale use.

6.5 Physical Methods of Cell Disruption

Microorganism disruption can also be achieved through physical treatments including supercritical extraction, explosive decompression, and thermal treatment.

Protein may be released from *E. coli* by exposing the outer membrane to elevated temperatures. Periplasmic proteins are released at 50–55 °C [29] while cytoplasmic proteins are released within 10 minutes at 90 °C [30]. Results depend on the strain and in particular its outer membrane characteristics, the rate of heating, and the temperature at which the organisms are stored prior to thermolysis. Thermolysis gives large cell debris, but may also lead to unacceptable rises in viscosity due to DNA denaturation [30]. Mild thermal treatment designed to deactivate *E. coli* prior to homogenization can be detrimental to homogenization efficiency (see Section 6.6.3).

Mild heat treatment can also lyse yeast and release intracellular components through autolysis rather than direct thermal destruction. Treatment at 45–50 °C and pH 5.5 leads to the production of glucanases, proteases, and mannanases endogenous to viable yeast cells. These enzymes act in concert to degrade the wall causing protein release within 24–36 h [31]. This rate of release may be too slow for recombinant products, particularly due to the higher levels of proteases resulting from mild heat treatment, but the technique is routinely employed in preparing yeast hydrolysates.

Explosive decompression has been used to extract product from yeast [32]. Supercritical carbon dioxide is contacted with cell suspension for a specified time and expands rapidly when the pressure is released, causing disruption of the wall. The technique is extremely gentle and extracts off-flavors caused by lipid components, but has a relatively low efficiency.

6.6 High-pressure Homogenization

High-pressure homogenization is a non-specific mechanical method of effecting cell breakage at large scale. It is the most common disruption method for bacteria such as *E. coli*. The usual design employs a positive displacement pump that forces cell suspension at high pressure through a spring-loaded valve. As shown in Fig. 6-3, the suspension accelerates to high velocity in the valve with a concomitant drop in pressure, before impinging on an impact ring and leaving the system at basically atmospheric pressure. The precise mechanism of disruption is unclear. However, it is likely that disruption occurs in response to shear, decompression, impingement, and possibly turbulence and cavitation. Recent evidence suggests that the pressure gradient experienced by the cells is a strong positive correlator of the extent of cell disruption [33,34].

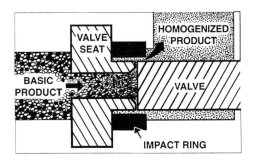

Fig. 6-3. The basic process of cell disruption by high-pressure homogenization. (Reproduced from Pandolfe, W.D., *Cell Disruption by Homogenization,* APV information booklet, with the kind permission of APV.)

6.6.1 Operational Parameters

Key operational parameters in high-pressure homogenization are operating pressure and the number of times that the suspension passes through the homogenizer. The valve design also has a critical effect on disruption efficiency (see Section 6.6.5). Disruption is generally improved by raising the inlet temperature, but at the expense of an increased exit temperature due to heating during homogenization. In general, a temperature rise of approximately 2.4 °C per 10 MPa of operating pressure can be expected. Denaturation of DNA at higher exit temperatures (e.g., > 40 °C) can cause problems in subsequent processing, although DNA can be sheared using a second 'high-shear' homogenizing valve in series with the primary disruption valve. These two-stage machines are no longer widely employed because of higher capital and operating cost, especially as viscosity problems can be reduced through judicious choice of inlet temperature. Cell concentration also has a minor effect on disruption, with decreased disruption at higher feed concentrations [6,35]. The decreased disruption efficiency at high concentration is small compared with the increase in processing time caused by dilution. It was concluded that the optimal *E. coli* feed concentration is the maximum possible that does not lead to practical handling difficulties due to high viscosity.

6.6.2 Commercial Equipment

There are two main suppliers of commercial homogenizer equipment. APV Baker homogenizer division comprises APV Rannie (Copenhagen, Denmark) and APV Gaulin (Massachusetts, USA), and produces homogenizers operating at pressures up to 1500 bar. Niro-Soavi (Parma, Italy) produces machines with operating pressures up to 1000 bar at flowrates up to 9000 L h^{-1}. Higher pressures are possible for custom-manufactured machines (personal correspondence, GEA Australia). Figure 6-4 shows a typical cell-disruption homogenizer. Each manufacturer provides a range of valve designs in various materials, such as ceramic and tungsten carbide. Ceramic valves are particularly recommended when abrasive suspension such as homogenate containing recombinant inclusion bodies must be processed.

Both manufacturers offer a range of laboratory machines available for process testing by the potential client. It is, however, worth noting that no reliable scale-up rules for high-pressure homogenization have yet been developed. The performance of a laboratory-scale machine will typically exceed that of a full-scale production machine as valve radii and lift distances (the distance between the valve and seat when open) will be higher in the production machine, giving lower impact velocities and pressure gradients.

6.6.3 Cell Treatments Before Homogenization

Upstream influences can have a devastating impact on cell disruption during homogenization. For example, batch thermal deactivation of stationary-phase *E. coli* prior to homogenization reduced disruption efficiency at 55 MPa to 25 % from 80 % [36] because cells became smaller and tougher following the thermal treatment. This effect could be negated by charging glucose to the fermenter prior to initiating thermal deactivation.

Fig. 6-4. The Niro Ariete NS 3011 high-pressure homogenizer. (Reproduced with permission from a photograph kindly provided by GEA Process Technology Division.)

Beneficial effects can be achieved through selective treatment of cells prior to homogenization. Brief alkaline treatment of *A. eutrophus* containing PHA at pH 10.5 prior to homogenization increased soluble protein release by 37 % [37]. Pretreatment with the detergent Sarkosyl at 1 % also improved soluble protein release by 38 % following homogenization through synergistic effects (release was minimal prior to homogenization). Enzymatic pretreatment can also aid mechanical disruption. For example, Vogels and Kula [38] have demonstrated that pretreatment of *Bacillus cereus* with lytic enzyme (0.5 mg cellosyl g^{-1} wet cells) prior to homogenization in a Gaulin MC4-TBX homogenizer raises disruption from 40 % to 98 % after a single homogenizer pass at 70 MPa. The homogenization of *E. coli* is also aided by chemical pretreatment. For example, pretreatment of exponential-phase cells with 1.5 M guanidine HCl and 1.5 % Triton X-100 has been shown to significantly improve disruption [39].

6.6.4 Predicting Disruption

High-pressure homogenization data may be described using Eqn (2), which was originally developed for yeast [7]. Soluble protein release, R_P, is a function of the number of homogenizer passes, N, the homogenizer pressure, P, and two empirical constants, a and k.

$$\log\left(\frac{1}{1-R_p}\right) = kNP^a \tag{2}$$

The parameters a and k are dependent on a range of factors, including the type of cell and its growth phase, the growth media composition, the design of the homogenizer and homogenizing valve, and operating parameters such as temperature and to a lesser extent cell concentration. Some studies also demonstrate a dependence of k on P, particularly at high levels of disruption. Table 6-1 gives values for parameters a and k from a range of studies. Despite the large variation in parameter values, Eqn (2) continues to be widely used because of its relative simplicity.

The wall-strength model was developed to address the dependence of cell parameters on feed properties [9,40]. Disruption for multiple passes is calculated by Eqn (3),

$$D = 1 - \int_0^\infty (1 - f_D(S))^N f_S(S) dS \tag{3}$$

where $f_D(S)$ and $f_S(S)$ are the stress and strength distribution functions, respectively.

$$f_S(S) = \frac{1}{\sigma\sqrt{2\pi}} \exp\left(\frac{-(S - \bar{S})^2}{2\sigma^2}\right) \tag{4}$$

$$f_D(S) = \frac{(mP^n)^d}{S^d + (mP^n)^d} \tag{5}$$

Table 6-1. Typical model parameters for equation (2) for common microorganisms from a range of homogenizer studies.

Organism	k (MPa[a])	a	Reference
S. cerevisiae	9.7×10^{-6}	2.9	[1]
E. coli (stationary)	1.5×10^{-3}	1.71	[40]
E. coli (exponential)	0.38	0.64	[40]
E. coli (induced)[b]	4.5×10^{-3}	1.65	[6]
A. eutrophus	4.6×10^{-4}	1.7	[31]

[a] All values based on natural logarithm in equation (2).
[b] Cells containing recombinant protein inclusion bodies.

Table 6-2. Model parameters for the wall-strength model (equations (3) to (5)) for *E. coli* [9] and yeasts [41] for $P > 35$ MPa.

Parameter	E. coli	Brewer's Yeast	Baker's Yeast
d	7.85	10.6	10.6
m	12.6	$12.2Y^{0.18}$	$12.2Y^{0.18}$
n	0.393	0.38	0.38
S	$33X-8.0L+48.8$	51	61 to 64
σ	3.82	4	10 to 14

There is a total of five parameters. Regression to single-pass disruption data for *E. coli* (182 data points, 21 different cultures) gave a good description of data when four parameters were constant (i.e., independent of the feed cell properties). Variation in disruption data was accounted for by a single parameter, $\bar{S}$, termed the mean effective strength of the culture, which is correlated with measurable properties of the feed cells (mean cell length, L, and fractional peptidoglycan cross-linkage, X) as in Table 6-2. The correlation predicts $\bar{S}$ to within 6 %, thus allowing true *a priori* prediction of disruption for a fixed homogenizer system (APV-Gaulin 15M) independently of variations in the feed cells.

The wall-strength model was originally developed for *E. coli*, but subsequent work demonstrated that it is able to describe the disruption of yeast [41]. Parameter values are provided in Table 6-2.

Despite its usefulness in predicting disruption independently of feed cell variation, the wall-strength model remains dependent on assumed functional forms for the strength and stress distributions, and parameters determined by regression. To remove this limitation, fundamental work into defining the true homogenizer-stress and cell-strength distributions is currently being undertaken (A. R. Kleinig, Dissertation in preparation, University of Adelaide).

6.6.5 The Importance of Homogenizer Valve Design

The success of homogenization lies in the valve design. A variety of designs are available from APV Baker, as shown in Fig. 6-5. A comparison of these valves showed that the CD design is more efficient than either the standard or CR design. This is probably due to the efficient conversion of pressure to kinetic energy in the valve inlet, and consequent high impingement velocity at the impact ring. The Niro-Soavi 'Nanovalve' (Fig. 6-6) is designed to have a larger-than-standard valve radius, thus reducing valve lift at a fixed feedrate. It is claimed that this improves disruption above that achievable with a standard valve.

One way of improving disruption efficiency is to reduce the distance between the valve exit and the impact ring. The lateral jet issuing from the valve spreads as it moves toward the impact ring, thus reducing its average velocity. Higher impact velocities are obtained by moving the impact ring closer to the valve. Keshavarz-

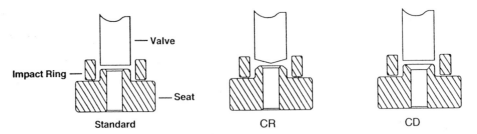

Fig. 6-5. Various homogenizer valve designs. CD, knife-edge configuration; CD, cell disruption configuration. (Reproduced from Pandolfe, W. D., *Cell Disruption by Homogenization*, APV information booklet, with the kind permission of APV.)

Fig. 6-6. The Niro 'Nanovalve' high-pressure homogenizer valve. (Reproduced with permission from a photograph kindly provided by GEA Process Technology Division.)

Moore et al. [42] studied the effect of impact distance for yeast disruption in a high-pressure homogenizer. The group kP^a in Eqn (2) was correlated with stagnation pressure at the impact ring, P_s, which is related to valve gap, h, and impact distance, Y, according to Eqn (6):

$$P_s \propto \frac{1}{Y^2 h^2} \qquad (6)$$

Kleinig et al. [43] examined the effect of impact distance on the disruption of E. coli by high-pressure homogenization. A modified form of Eqn (2) was developed, giving Eqn (7),

$$\log\left(\frac{1}{1-D}\right) = \frac{kNP^a}{(Y^*)^{0.24}} \qquad (7)$$

where Y^* is the dimensionless impact distance (i.e., Y/Y_{std}). Values for parameters k and a in Eqn (7) are given in Table 6-1.

Recent work has suggested that disruption is favored by large pressure gradients [33]. By combining numerical fluid simulations with experimental data, it was shown that disruption is uniquely correlated with the maximum pressure gradient experienced at the valve inlet when the impact ring is removed. Subsequent work has also shown that disruption can be correlated with the pressure gradient experienced at the impact ring after suspension has left the valve [34], and provided an improved correlating equation superseding Eqn (6). Combined, these studies suggest that the 'optimal' homogenizer valve will be one that achieves a high pressure gradient at the valve inlet and a high impingement velocity at the impact ring. Of course, the valve must also have acceptable wear characteristics.

6.7 Bead Milling

Bead mills are relatively simple devices originally developed for the wet grinding of pigments in the paint industry and for the comminution of ceramics and limestone. The basic design is shown in Fig. 6-7. A jacketed grinding chamber is fitted with a rotating shaft through its centre. The shaft is fitted with agitator discs of varying design that rotate at high speed (ca. 4000 r.p.m.) with the shaft. These discs violently agitate glass beads (0.5–1 mm diameter) within the chamber, causing them to collide. Cell suspension is pumped through this violently agitated bead phase, causing cells trapped between colliding beads to be disrupted. Beads are retained in the chamber by a sieve or axial slot at the outlet. Cooling is essential, as virtually all energy input is dissipated as heat. The precise mechanism of disruption in bead milling is not known, although compaction and shearing between the beads and energy transfer from the beads to the cells are believed to be important. Bead milling is normally employed for large organisms such as yeast cells. Smaller cells such as E. coli are more difficult to disrupt by bead milling as they are less easily trapped between the colliding beads.

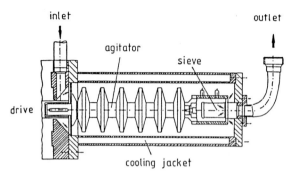

Fig. 6-7. Schematic drawing of the grinding chamber of the Netzsch LME 4 bead mill. (Reprinted from *Chem. Eng. Sci., 47*, Bunge, F., Pietzsch, M., Müller, R., Syldatk, C., Mechanical disruption of *Arthrobacter* sp. DSM 3747 in stirred ball mills for the release of hydantoin-cleaving enzymes, 225–232 [44], with permission from Elsevier Science Ltd, The Boulevard, Langford Lane, Kidlington, OX5 1GB, UK.)

6.7.1 Operational Parameters

A large number of parameters affect the results obtained during bead milling. These include agitator design and velocity, bead loading and size, inlet temperature, suspension flowrate, and cell concentration. Table 6-3 summarizes the effect of bead mill operational parameters. In general, high agitator peripheral velocities (5–10 m s^{-1}), high bead loadings (80–90 % of free grinding chamber volume) and small beads (ca. 0.5 mm) give the highest levels of disruption during bead milling.

Table 6-3. Qualitative observations reported in the literature on the effect of bead milling variables.

Variable	Typical values	Qualitative effects on disruption
Bead size	0.2 to 0.8 mm	Smaller beads are better, but are more difficult to retain, and may tend to be fluidized. In practice, moderate sized beads (e.g., 0.5 mm) are preferred
Agitator peripheral velocity	4 to 12 m s^{-1}	Higher velocities cause higher disruption, but may lead to inefficient energy use if stress probability is limiting (see section 6.6.3)
Cell concentration	5–50 % w/v	Only second-order effects reported
Flow rate	Machine specific	Disruption decreases with increasing flowrate
Bead volume fraction	70–90 %	Disruption increases at higher loadings, but agitator wear and heat generation may be unacceptable. 80–85 % provides best compromise
Temperature	5–40 °C	Slight decrease in disruption as temperature increases. Only second-order effects reported

6.7.2 Commercial Equipment

Two large suppliers of bead mill equipment for bioprocessing applications are Willy A. Bachofen AG Maschinenfabrik (Basel, Switzerland), marketing the Dyno-Mill range, and Netzsch Feinmahltechnik (Germany) who sell the LME range. The Dyno-Mill KDL unit is a laboratory prototype, shown in Fig. 6-8. Larger units (e.g., the KD5 unit with a 5-L working volume) are also available. Netzsch also manufactures bead mills with a range of working volumes (e.g., the LME4 with a 4-L working volume and speeds of 200 to 2500 r.p.m.).

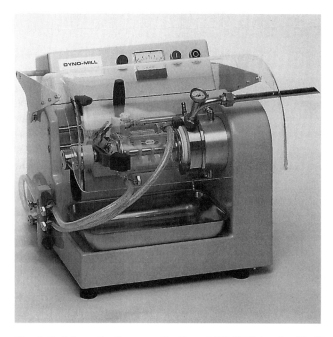

Fig. 6-8. Schematic diagram of a Dyno Mill KDL bead mill. (Reproduced from the Dyno[R]-Mill KDL information booklet with the kind permission of Willy A. Bachofen AG, Basel, Switzerland.)

6.7.3 Predicting Disruption

Limon-Lason et al. [45] showed that the disruption of *S. cerevisiae* in batch bead mills could be described by Eqn (8),

$$\ln\left(\frac{1}{1-D}\right) = kt \tag{8}$$

where, t is the time of disruption and k is a rate constant correlated with agitator velocity. For continuous operation, the bead mill was modelled as a series of j continuous stirred-tank reactors (CSTRs) giving Eqn (9),

$$\frac{1}{1-R_p} = \left(1 + \frac{k\tau}{j}\right)^j \tag{9}$$

where τ is the mean residence time in the mill. Residence time distribution (RTD) studies showed that the Dyno-Mill KD5 mill was equivalent to 5 CSTRs in series, while the Dyno-Mill KDL unit varied between 1.2 and 2.0 CSTRs. For the KDL unit, k was correlated with the agitator peripheral velocity, v_u, according to Eqn (10) independent of the type of agitator design for batch and continuous operation,

$$k = Kv_u \tag{10}$$

where $K = 0.0036$ m^{-1}. Eqn (10) gives k values varying from approximately 0.02 to 0.07 s^{-1} for the KDL unit with typical agitator velocities of 6 to 20 m s^{-1}. For the KD5 mill, k was approximately constant and equal to 0.02 s^{-1} or 0.04 s^{-1} depending on the impeller design but not the agitator velocity.

Melendres et al. [46] studied disruption of *S. cerevisiae* in a Dyno Mill KDL unit and correlated the rate constant, k, with the effective volume between beads, β, and the frequency of bead collisions, f, giving Eqn (11). The parameter η is a constant of proportionality between the bead and agitator velocities. The frequency of bead collisions was related to bead loading, α, bead diameter, d_b, agitator peripheral velocity, v_u, and voidage, ε, using the kinetic theory of gases,

$$k = \beta\eta f = \beta\eta \frac{18\sqrt{2}}{\pi}\left(\frac{v_u(1-\varepsilon)^2\alpha^2}{d_b^4}\right) \tag{11}$$

where the product βf was determined by regression to data, giving Eqn (12):

$$\beta f = 1.05 \times 10^{-6}\, d_b^{3.57} \tag{12}$$

Eqn (11) indicates that the rate constant is proportional to agitator velocity for the KDL unit as also demonstrated by Eqn (10). However, this proportionality fails at low agitator velocities and Melendres et al. [5] concluded that disruption is minimal below some critical agitator velocity equal to 5 m s^{-1} rpm for the KDL unit.

An extensive study on the disruption of *Arthrobacter* sp. DSM 3747 in a Netzsch LME4 bead mill for varying agitator velocities (v_u, 4 to 12 m s^{-1}), bead diameters (d_b, 0.1 to 1.5 mm) and cell concentrations (c, 10 to 55 % w/v) at a bead filling fraction (ψ) of 0.80 has been conducted [44]. Disruption correlated with either specific energy input (Eqn 13) or stress probability (Eqn 14), depending on the operational regime.

$$\ln\left(\frac{1}{1-R_{\mathrm{p}}}\right) = kE = k\left(\frac{2\pi M\omega t}{Vc}\right) \tag{13}$$

$$S_{\mathrm{p}} = \omega t a\frac{d_{\mathrm{p}}}{d_{\mathrm{b}}} \tag{14}$$

In the above equations, M is the torque on the agitator shaft, V is the free grinding volume, c is cell concentration, ω is the rotational speed of the agitator, d_{p} is the cell diameter, and a is the bead filling rate (weight ratio of beads to yeast). Disruption correlated well with specific energy input, E, for small beads ($d_{\mathrm{b}} = 0.355$ mm), moderate to high cell concentrations ($c = 10\%$ to 55%), and low to moderate agitator speeds ($v_{\mathrm{u}} = 4$ to 8 m s^{-1}, as shown in Fig. 6-9. In this regime, an increase in specific energy input lead to a higher stress intensity during bead collision, and this caused higher disruption. However, the correlation was poorer at small (0.110 mm) and large (1.50 mm) bead sizes, and worsened at higher agitator velocities and lower concentrations. Under these conditions, disruption correlated with stress probability as shown in Fig. 6-10. It appears that at low concentrations and high velocities, stress intensity between colliding beads already exceeded cell strength. An increase in specific energy input and hence stress intensity did not increase disruption. The limiting factor in this regime was the need to bring cells into the active volume of the colliding beads. Disruption therefore correlated with stress probability.

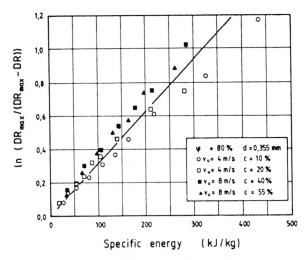

Fig. 6-9. Correlation of disruption with specific energy input to the bead mill for various agitator velocities (v_u) and cell concentrations (c). DR is disintegration rate and is equivalent to R_p defined by equation (1). (Reprinted *from Chem. Eng. Sci., 47,* Bunge, F., Pietzsch, M., Müller, R., Syldatk, C., Mechanical disruption of *Arthrobacter* sp. DSM 3747 in stirred ball mills for the release of hydantoin-cleaving enzymes, 225–232 [44], with permission from Elsevier Science Ltd, The Boulevard, Langford Lane, Kidlington, OX5 1GB, UK.)

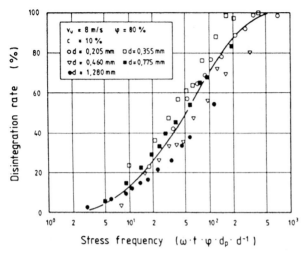

Fig. 6-10. Correlation of bead-mill disruption with stress frequency for various bead sizes (*d*). DR is disintegration rate and is equivalent to R_p, defined by equation (1). Stress frequency is equivalent to stress probability, defined by equation (14). (Reprinted from *Chem. Eng. Sci., 47,* Bunge, F., Pietzsch, M., Müller, R., Syldatk, C., Mechanical disruption of Arthrobacter sp.,DSM 3747 in stirred ball mills for the release of hydantoin-cleaving enzymes, 225–232 [44], with permission from Elsevier Science Ltd, The Boulevard, Langford Lane, Kidlington, OX5 1GB, UK.)

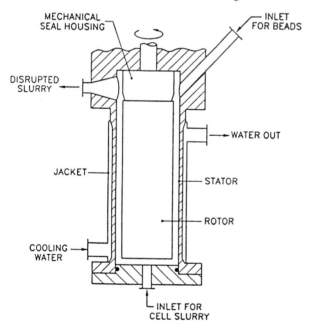

Fig. 6-11. Schematic diagram of the Annu Mill 01. (Reproduced from *Bioseparation, 4,* (1994), 319–328, Disruption of a recombinant yeast for the release of β-galactosidase, Garrido, F., Banerjee, U.C., Chisti, Y., Moo-Young, M. Figure 1. © 1994 Kluwer Academic Publishers [49], with kind permission from Kluwer Academic Publishers.)

The work by Bunge et al. [44] is important from a practical perspective. Studies suggest high disruption is achieved at high agitator velocities, as described in Section 6.7.1. However, this may represent inefficient use of energy if the limiting factor is stress probability. The most efficient use of energy occurs when disruption correlates with stress intensity (i.e., specific energy input) as a sufficient number of cells are transported in zones of active bead collision.

The disruption of *S. cerevisiae* in a novel bead mill, the Annu Mill 01 shown in Fig. 6-11 (Sulzer Brothers Ltd., Winterthur, Switzerland), has been examined [47]. Protein release was described by Eqn (15),

$$\ln\left(\frac{1}{1 - R_p}\right) = kN(1 - a)\frac{V}{Q} \tag{15}$$

where N is the number of passes through the mill and Q is the flowrate. The rate constant k was empirically correlated with the bead filling fraction and flowrate.

6.7.4 The Importance of Agitator Design

Agitator design in bead milling is a key determinant of disruption efficiency. There are two main considerations in selecting an agitator, namely the efficiency of kinetic energy transfer from the agitator to the beads, and the effect that the agitator has on the mill's RTD in continuous operation. Ideally, kinetic energy transfer will be as high as possible to ensure maximum stress intensity between the colliding beads. This should be achieved while maintaining a narrow residence time distribution (i.e., while maximizing the number of CSTRs in series). A narrow RTD ensures that all cells spend approximately the same time in the bead mill, and thus have an equal probability of disruption. Short-circuiting of cells from the feed to the mill exit is thereby avoided. These two effects may often oppose one another, so an optimum trade-off must be identified.

Agitator design has been extensively studied for Netzsch mills [48,49]. For the Netzsch LME 20 mill, a cooled pin agitator (LMJ 15) without stationary counter-pins affixed to the mill wall gave higher *S. cerevisiae* disruption than either a double-disc design (LME 20 'D') or the standard eccentric disc design (LME 20). These designs are shown in Fig. 6-12. Interestingly, the LME 20 'D' design gave a better RTD, but this is clearly offset by improved kinetic energy transfer in the LMJ 15 design. When stationary counter pins are affixed to the mill wall, the RTD of the LMJ 15 design approaches that of an ideal mixer (i.e., a single CSTR) and disruption efficiency is consequently reduced. It is worth noting that although the LMJ 15 design gives higher disruption than the standard agitator, it is also prone to high wear that may increase overall operating costs.

A comprehensive study of 12 different agitator designs has also been undertaken for a Netzsch LME 4 mill [49]. Changing the standard eccentric ring arrangement to a 8-notched, two-cone discs arrangement (Fig. 6-13) increased *S. cerevisiae* disrup-

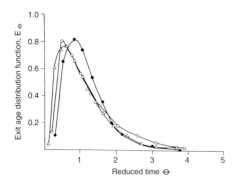

Netzsch LME 20 mill equipped with a double disc stirrer (LME 20 'D'), ●

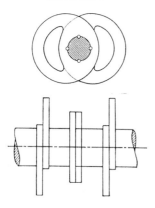

Netzsch LME 20 mill with eccentrically mounted impellers on the drive shaft (LME 20), ○

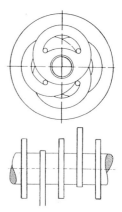

Netzsch LMJ 15 mill with a pin agitator (LMJ 15), △.

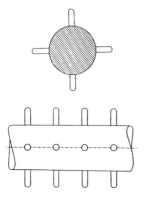

Fig. 6-12. Various agitator designs, showing the effect on RTD. (From Kula, M.-R., Schütte, H. (1987), *Biotech. Prog., 3,* 311–312. Reproduced with permission of the American Institute of Chemical Engineers. Copyright © 1987 AIChE. All rights reserved.)

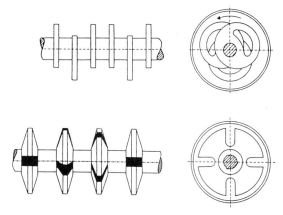

Fig. 6-13. Designs of the eccentric ring agitator (top) and the notched two-cone discs agitator (bottom). (From Schütte et al. [49]; reproduced with permission of the *Annals of the New York Academy of Sciences*.)

tion substantially. The rate constant k in Eqn (8) increased from 0.41 min^{-1} to 1.52 min^{-1}, and this was attributed to improved energy transfer to the beads. The modified agitator also gave higher disruption for *Brevibacterium ammoniagenes*, although optimum disruption for this organism was obtained using a 6-notched disc arrangement with a 45° angular offset between discs. In all cases the rate constant for *B. ammoniagenes* was approximately one order of magnitude below that for *S. cerevisiae*.

The above studies suggest several points that need to be considered when selecting a bead mill and agitator. First, the optimum agitator design is microorganism-specific, necessitating RTD tests and batch tests to assess k for any given application. The aim will be to select a design that gives high segmentation of the mill and a consequent high number of CSTRs while promoting efficient kinetic energy transfer. It will be desirable to operate in the region where disruption correlates with stress intensity to ensure efficient use of energy (see Section 6.7.3). Finally, maximizing disruption through careful agitator selection does not guarantee an optimal solution: the agitator must also have acceptable wear characteristics.

6.8 Other Methods of Mechanical Disruption

Another form of mechanical disruption gaining increased interest is the Microfluidizer (Microfluidics Corp., MA, USA), which operates by impacting two streams of cell suspension at high velocity. Disruption of *E. coli* in a Microfluidizer has been described by Eqn (16),

$$\log\left(\frac{1 - R_o}{1 - R_p}\right) = kN^b P^a \tag{16}$$

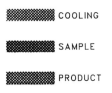

COOLING

SAMPLE

PRODUCT

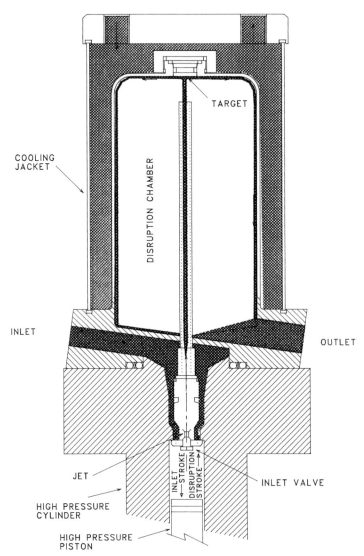

TARGET

COOLING
JACKET

DISRUPTION CHAMBER

INLET

OUTLET

JET

INLET
STROKE

DISRUPTION
STROKE

INLET VALVE

HIGH PRESSURE
CYLINDER

HIGH PRESSURE
PISTON

Fig. 6-14. Schematic diagram of the Constant Systems Ltd. cell disruptor, showing the principle of operation. (From the Constant Systems Ltd. 'Cell Disruption' information booklet. Reproduced with the kind permission of Constant Systems Ltd., Warwick, U. K.)

where R_o is the fractional release of protein prior to mechanical disruption due to enzymatic pretreatment [50]. The value of a varied between 0.6 and 1.77 depending on the properties of the feed cells and k varied from 0.27×10^{-3} to 16×10^{-3} MPa^{-a}. Treatment of *E. coli* with EDTA-lysozyme prior to disruption in a Microfluidizer marginally improved disruption (from 76 % to 88 % for uninduced cells, and from 86 % to 98 % for induced cells) [51]. Disruption using EDTA-lysozyme treatment without Microfluidization was minimal. *S. cerevisiae* has also been disrupted in a Microfluidizer following Zymolyase pretreatment [52]. Four passes through the Microfluidizer at 95 MPa gave almost complete disruption, compared with only 32 % disruption when enzymatic pretreatment was not employed. Only a very low level of disruption (5 %) was achieved using Zymolyase treatment without mechanical disruption, clearly indicating the synergistic effects of enzymatic pretreatment and Microfluidization. Parameters for Eqn (16) were $a = 3.03$, $b = 1.30$, and $k = 2.77 \times 10^{-7}$ MPa^{-a}.

Figure 6-14 shows the general arrangement for a new mechanical disruptor available from Constant Systems Ltd (Warwick, U. K.). High-pressure cell suspension is rapidly accelerated to give a jet with high kinetic energy, that impacts a target before collection at atmospheric pressure. Cooling is provided by a jacket. Disruption presumably occurs in the regions of high pressure gradient (i.e., in forming the jet and impacting the target), as for conventional high-pressure homogenizers. A range of machines from a 'single-shot' laboratory unit (10 mL) to units capable of continuously processing 1.1 L min^{-1} at 1000 bar are available.

6.9 Downstream Impacts of Cell Disruption

From the preceding sections it is clear that upstream units and operating conditions affect the extent of cell disruption by altering cell-wall properties. Similarly, disruption conditions can affect subsequent downstream operations, particularly chromatography. Insoluble cellular debris fouls chromatographic resins and complicates process validation. Further, cell-wall proteases associated with insoluble cellular debris may destroy the product in downstream operations such as dissolution and refolding. This has been demonstrated for analogs of insulin-like growth factor expressed as inclusion bodies in *E. coli* [53]. To minimize these problems it is common to separate cellular debris from the product prior to chromatography, usually by centrifugation or filtration. Optimization of these units is aided by knowledge of the debris size distributions generated during disruption.

6.9.1 Debris Size Analysis

Several studies have addressed the issue of debris size following mechanical disruption. Key results are summarized in Table 6-4 for *E. coli*. All available methods suffer from disadvantages. A common problem is the need for extensive sample pre-

Table 6-4. Summary of results from studies on the effect of mechanical disruption on *E. coli* debris size.

Conditions	Sizing method[a]	Median debris-size (µm)	Reference
Homogenizer, $N = 1$, $P = 55$ MPa	PCS and CDS	0.43	[54]
Homogenizer, $N = 5$, $P = 55$ MPa	PCS and CDS	0.28	[54]
Homogenizer, $N = 3$, $P = 60$ MPa	PCS	0.22	[55]
Microfluidizer, $N = 3$, $P = 60$ MPa	PCS	0.45	[55]
Bead Mill, mean residence time 4 min	PCS	0.53 (bimodal peaks at 0.47 and 0.78 µm)	[55]
Homogenizer, $N = 1$, $P = 62$ MPa	SEM	0.17	[39]
Homogenizer, $N = 1$, $P = 62$ MPa, Pretreated with 4 M guanidine HCl	SEM	75 % of particles < 0.19 µm	[39]
Homogenizer, $N = 2$, $P = 55$ MPa	CSA	0.50	[53]
Homogenizer, $N = 10$, $P = 55$ MPa	CSA	0.33	[53]

[a] Centrifugal disc photosedimentation (CDS); Photon correlations spectroscopy (PCS); Scanning electron microscopy (SEM), or Cumulative sedimentation analysis (CSA).

paration that can destroy the meaning of a size distribution obtained using commercial instruments (e.g., photon correlation spectroscopy). Only one method, cumulative sedimentation analysis (CSA), does not require sample pretreatment. However, this method is rather laborious.

The study by Agerkvist and Enfors [55] for the mechanical disruption of *E. coli* warrants closer inspection as it provides comparative results for different homogenizer systems. The Microfluidizer generated large debris (mode 450 nm), while the homogenizer generated smaller debris (mode 190 nm) after multiple passes. The size distributions from a Dyno-Mill KDL bead mill showed a bimodal distribution with peaks centred at 471 nm and 777 nm after 4 minutes of disruption. Large debris is beneficial where the product is soluble, but detrimental where the product is an insoluble inclusion body. Clearly, selection of a mechanical disruption system needs to be done with reference to the desired debris size, not just the efficiency of product release.

6.9.2 Predicting Debris Size

Particle-size distribution data for *S. cerevisiae* following disruption in an APV-Gaulin LAB6O homogenizer have been determined using electrical sensing zone measurement, and have been fitted to a Boltzmann-type equation,

$$1 - F(d) = \frac{1}{\left[1 + \exp\left(\dfrac{d - d_{50}}{w}\right)\right]} \tag{17}$$

where $F(d)$ is the cumulative undersize mass fraction, d is particle size, d_{50} is the median particle size, and w is a parameter related to the width of the distribution [56]. Parameters d_{50} and w were correlated with experimental data as follows,

$$\ln\left(\frac{1}{d_{50}^*}\right) = \ln\left(\frac{d_{50,o}}{d_{50,o} - d_{50}}\right) = \frac{k_d}{N^{0.4}(P - P_t)} \tag{18}$$

$$w^* = \frac{w_o - w}{w_o} = -2.3 d_{50}^* \quad d_{50}^* < 0.33 \tag{19}$$

$$w^* = \frac{w_o - w}{w_o} = 5.5 d_{50}^* - 2.4 \quad d_{50}^* \geq 0.33 \tag{20}$$

where P is operating pressure, P_t and k_d are parameters (115 bar and 670 bar, respectively), d_{50}^* and w^* are dimensionless forms of the Boltzmann distribution parameters, and subscript o denotes the parameter value before homogenization.

Eqn (17) has been employed to describe debris-size reduction for *E. coli* during high-pressure homogenization in an APV-Gaulin homogenizer [53]. Parameters d_{50} and w for *E. coli* were correlated with the number of homogenizer passes, N, according to Eqns (21) and (22),

$$\ln\left[\frac{1}{d_{50}}\right] = k_1 N^{0.29} \tag{21}$$

$$\ln\left[\frac{1}{w}\right] = k_2 N^{0.1} \tag{22}$$

where k_1 ranged from 0.48 to 0.66 pass$^{-0.29}$ and k_2 ranged from 1.62 to 1.92 pass$^{-0.1}$, depending on the characteristics of the feed cells. The homogenizer operating pressure was fixed at 55 MPa.

In an attempt to provide a rational basis for modelling debris-size reduction, a model for *E. coli* debris-size reduction based on grinding theory has been developed [53]. The model provided an excellent prediction of experimental data, does not require any assumed functional distribution of the debris size, and can be used given information on the initial cell size distribution and the disruption efficiency during homogenisation. The final equation for the cumulative undersize distribution of cell debris after N passes, F_d, is given by Eqn (23),

$$D_N \times F_d(x_i, N) = (D_N - D_{N-1}) \times F_{dc}(x_i, N) + (D_{N-1}) \times F_{dd}(x_i, N) \qquad (23)$$

where x_i is the discretized size (arbitrarily chosen), D_N is the total disruption after N homogenizer passes, F_{dc} is the cumulative undersize distribution of debris generated from whole cells at pass N, and F_{dd} is the cumulative undersize distribution of debris generated by the comminution of existing debris in the homogenizer feed at pass N. F_{dc} and F_{dd} are given by Eqns (24) and (25),

$$F_{dc}(x_i, N) = 1 - [1 - F_c(x_i, N-1)] \exp{(-a_c x_{i-1}^{\alpha'})} \qquad (24)$$

$$F_{dd}(x_i, N) = 1 - [1 - F_d(x_i, N-1)] \exp{(-a_d x_{i-1}^{\alpha'})} \qquad (25)$$

where F_d and F_c are the cumulative undersize distributions of debris and whole cells, respectively, and α', a_c, and a_d are parameters whose values are given in Table 6-5. The total cumulative size distribution of any homogenate is then simply obtained by combining the debris and intact cell distributions according to their relative proportions in the homogenate, giving Eqn (26):

$$F_H(x_i, N) = (1 - D_N) \times F_c(x_i, N) + D_N \times F_d(x_i, N) \qquad (26)$$

Note that the model accounts for the simultaneous processes of debris generation from intact cells and debris size reduction, which are physically quite distinct.

Table 6-5. Parameter values for the *E. coli* debris-size reduction model based on grinding theory developed by Wong [53].

Parameter	Stationary-phase recombinant *E. coli* (not induced)	*E. coli* containing recombinant inclusion bodies (induced)
α	2.4 ± 0.16	2.3 ± 0.13
a_c	1.37 ± 0.11	1.5 ± 0.10
a_d	0.48 ± 0.065	0.85 ± 0.10

Abbreviations and Symbols

a_c	parameter in equations (24) and (25)	$\mu m^{-\alpha'}$
C	concentration	$kg\ m^3$
c	wet biomass concentration	$kg\ L^{-1}$
D	disruption (volume- or number-fraction cells destroyed)	–
d	diameter or model parameter	μm or m or –
d_{50}	median particle size	μm
E	specific energy input	$kJ\ kg^{-1}$
f	frequency of bead collisions	$m^{-3}\ s^{-1}$
F	cumulative undersize mass fraction	–
h	homogenizer valve lift	μm or m
k_d	parameter in equation (18)	bar
k	rate constant	MPa^{-a} or s^{-1}
K	parameter in equation (10)	m^{-1}
L	average cell length	μm
m	wall-strength-model parameter	MPa^{-n}
M	torque on bead mill agitator shaft	$N\ m$
n	wall-strength-model parameter	–
N	number of bead mill or homogenizer passes	–
P	pressure	MPa or bar
P_t	threshold pressure in equation (6-18)	bar
Q	flowrate	$L\ h^{-1}$ or $mL\ min^{-1}$
R_P	fractional release of soluble protein	–
S	effective strength	–
$\bar{S}$	mean effective strength	–
S_P	stress probability	–
t	time	s or min
V	volume	L or m^3
v_u	peripheral agitator velocity	$m\ s^{-1}$
w	Boltzmann distribution parameter	μm
x_i	discretized size interval	μm
X	fractional peptidoglycan crosslinkage	–
Y	distance between valve exit and impact ring	mm
Y^*	dimensionless impact distance (Y/Y_{std})	–

Greek symbols

α'	parameter in equations (24) and (25)	–
α	weight ratio of beads to organisms in bead milling	$kg\ kg^{-1}$
β	effective disruption volume between beads	mm^3
ε	voidage	–
η	proportionality constant in equation (11)	–
σ	wall-strength-model parameter	–
τ	mean bead mill residence time	s or min
ω	angular velocity	$rad\ s^{-1}$

Subscripts

b   beads in bead mill
c   intact (i.e., undisrupted) cells
d   debris (total)
dc   debris generated by cell disruption
dd   debris generated by comminution of debris
h   homogenate
H   homogenate
m   maximum, corresponding to 100 % disruption
N   after N homogenizer passes
o   feed sample (i.e., before disruption)
p   particle (i.e., cell)
s   stagnation
std   standard homogenizer impact ring

Abbreviations

LPS   lipopolysaccharide
PL   phospholipid
LP   lipoprotein
PHA   polyhydroxyalkanoate
SDS   sodium dodecyl sulfate
EDTA ethylenediamine tetra-acetic acid
RTD   residence time distribution
CSTR continuous stirred-tank reactor

References

[1] Höltje, J..-V., Glauner, B., *Inst Pasteur Res Microbiol*, 1990, *141*, 75–103.
[2] Labischinski, H., Hochberg, M., Sidow, T., Maidhof, H., Henze, U., Berger-Bächi, B., Wecke, J., in: *Bacterial Growth and Lysis: Metabolism and Structure of the Bacterial Sacculus:* de Pedro, M. A., Höltje, J.-V., Löffelhardt, W. (Eds.), New York: Plenum Press, 1993; pp. 9–21.
[3] Fleet, G. H., in: *The Yeasts: Yeast Organelles:* Rose, A. H., Harrison, J. S.,(Eds.), London: Academic Press, 1991; Vol. 4, 2nd Ed., pp. 199–277.
[4] Klis, F. M., *Yeast,* 1994, *10*, 851–869.
[5] Melendres, A. V., Unno, H., Shiragami, N., Honda, H., *J Chemical Eng Japan,* 1992, *25*, 354–356.
[6] Middelberg, A. P. J., O'Neill, B. K., Bogle, I. D. L., Snoswell, M., *Biotechnol Bioeng,* 1991, *38*, 363–370.
[7] Hetherington, P. J., Follows, M., Dunnill, P., Lilly, M. D., *Trans Inst Chem Eng,* 1971, *49*, 142–148.
[8] Engler, C. R., Robinson, C. W., *Biotechnol Bioeng,* 1979, *21*, 1861–1869.
[9] Middelberg, A. P. J., Dissertation, University of Adelaide, 1992.
[10] Felix, H., *Anal Biochem,* 1982, *120*, 211–234.
[11] Spratt, B. G., *Philo Trans R Soc Lond B,* 1980, *289*, 273–283.
[12] Marvin, H. J., ter Beest, M. B. A., Witholt, B., *J Bacteriol,* 1989, *171*, 5262–5267.

[13] Nikaido, H., Vaara, M., *Microbiol Rev*, 1985, *49*, 1–32.

[14] Ryan, W., Parulekar, S. J., *Biotechnol Prog*, 1991, *7*, 99–110.

[15] Hettwer, D., Wang, H., *Biotechnol Bioeng*, 1989, *33*, 886–895.

[16] Naglak, T. J., Wang, H. Y., *Biotechnol Bioeng*, 1992, *39*, 732–740.

[17] Hart, R. A., Lester, P. M., Relfsnyder, D. H., Ogez, J. R., Builder, S. E., *Biotechnology*, 1994, *12*, 1113–1117.

[18] Falconer, R. J., O'Neill, B. K., Middelberg, A. P. J., *Biotechnol Bioeng*, 1997, in press.

[19] Eriksson, H., Green, P., *Biotech Tech*, 1992, *6*, 239–244.

[20] Berger, E., Ramsay, B. A., Ramsay, J. A., Chavarie, C., Braunegg, G., *Biotechnol Tech*, 1989, *3*, 227–232.

[21] Hahn, S. K., Chang, Y. K., Kim, B. S., Chang, H. N., *Biotechnol Bioeng*, 1994, *44*, 256–261.

[22] Ramsay, J. A., Berger, E., Ramsay, B. A., Chavarie, C., *Biotechnol Tech*, 1990, *4*, 221–226.

[23] Hahn, S. K., Chang, Y. K., Lee, S. Y., *Appl Environ Microbiol*, 1995, *61*, 34–39.

[24] Middelberg, A. P. J., Lee, S. Y., Martin, J., Williams, D. R. G., Chang, H. N., *Biotechnol lett*, 1995, *17*, 205–210.

[25] Andrews, B. A., Asenjo, J. A., *Trends Biotechnol*, 1987, *5*, 273–277.

[26] Dean, C. R., Ward, O. P., *Biotechnol Tech*, 1992, *6*, 133–138.

[27] Hunter, J. B., Asenjo, J. A., *Biotechnol Bioeng*, 1990, *35*, 31–42.

[28] Lee, C.-K., Ku, M.-C., *Biotechnol Tech*, 1994, *8*, 193–198.

[29] Tsuchido, T., Katsui, N., Takeuchi, A., Takano, M., Shibasaki, I., *Appl Environ Microbiol*, 1985, *50*, 298–303.

[30] Watson, J. S., Cumming, R. H., Street, G., Tuffnell, J. M., in: *Separations for Biotechnology:* Verrall, M. S., Hudson, M. J. (Eds.), London: Ellis Horwood, 1992; pp. 105–109.

[31] Reed, G., Nagodawithana, T. W., *Yeast Technology*. New York: Van Nostrand Reinhold, 1991.

[32] Nakamura, K., Enomota, A., Fukushima, H., Nagai, K., Hakoda, M., *Biosci Biotech Biochem*, 1994, *58*, 1297–1301.

[33] Kleinig, A. R., Middelberg, A. P. J., *Chem Eng Sci* 1996, *51*, 5103–5110.

[34] Kleinig, A. R., Middelberg, A. P. J., *A l Ch E J*, 1997, in press.

[35] Kleinig, A. R., Mansell, C. J., Nguyen, Q. D., Badalyan, A., Middelberg, A. P. J., *Biotech Tech*, 1995, *9*, 759–762.

[36] Collis, M. A., O'Neill, B. K., Middelberg, A. P. J., *Bioseparation*, 1996, *6*, 55–63.

[37] Harrison, S. T. L., Dennis, J. S., Chase, H. A., *Bioseparation*, 1991, *2*, 95–105.

[38] Vogels, G., Kula, M.-R., *Chem Eng Sci*, 1992, *47*, 123–131.

[39] Bailey, S. M., Blum, P. H., Meagher, M. M., *Biotechnol Prog*, 1995, *11*, 533–539.

[40] Middelberg, A. P. J., O'Neill, B. K., Bogle, I. D. L., *Trans Inst Chem Eng*, 1992, *70*, part C, 205–212.

[41] Kleinig, A. R., Middelberg, A. P. J., in: Better living through innovative biochemical engineering: Teo, W. K., Yap, M. G. S., Oh, S. K. W. (Eds.), *APBioChEC 94 Third Asia-Pacific Biochemical Engineering Conference, Singapore, June 1994*, Proceedings, pp. 607–609.

[42] Keshavarz Moore, E., Hoare, M., Dunnill, P., *Enzyme Microb Technol*, 1990, *12*, 764–770.

[43] Kleinig, A. R., O'Neill, B. K., Middelberg, A. P. J., *Biotech Tech*, 1996, *10*, 199–204.

[44] Bunge, F., Pietzsch, M., Müller, R., Slydatk, C., *Chem Eng Sci*, 1992, *47*, 225–232.

[45] Limon-Lason, J., Hoare, M., Orsborn, C. B., Boyle, D. J., Dunnill, P., *Biotechnol Bioeng*, 1979, *21*, 745–774.

[46] Melendres, A. V., Honda, H., Shiragami, N., Unno, H., *Bioseparation*, 1991, *2*, 231–236.

[47] Garrido, F., Banerjee, U. C., Chisti, Y., Moo-Young, M., *Bioseparation*, 1994, *4*, 319–328.

[48] Schütte, H., Kraume-Flügel, R., Kula, M. R., *Ger Chem Eng*, 1986, *9*, 149–156.

[49] Schütte, H., Jürging, B., Papamichael, N., Ott, K., Kula, M.-R., in: *Annals of the New York Academy of Sciences: Enzyme Engineering IX:* Blanch, H. W., Klibanov, A. M. (Eds.), New York: New York Academy of Sciences, 1988; Vol. 542, pp. 121–125.

[50] Sauer, T., Robinson, C. W., Glick, B. R., *Biotechnol Bioeng*, 1989, *33*, 1330–1342.

[51] Lutzer, R. G., Robinson, C. W., Glick, B. R., in *Proceedings of the 6th European Congress on Biotechnology:* Alberghina, A., Frontali, L., Sensi, P. (Eds.), Amsterdam: Elsevier Science BV, 1994; pp. 909–916.
[52] Baldwin, C., Ronbinson, C. W., *Biotechnol Tech,* 1990, *4,* 329–334.
[53] Wong, H. H., Dissertation, University of Adelaide, 1997.
[54] Olbrich, R., Dissertation, University of London, 1989.
[55] Agerkvist, I., Enfors, S.-O., *Biotechnol Bioeng,* 1990, *36,* 1083–1089.
[56] Siddiqi, S. F., Titchener-Hooker, N. J., Shamlou, P. A., *Biotechnol Bioeng,* 1996, *50,* 145–150.

7 Microcarriers in Cell Culture Production

Björn Lundgren and Gerald Blüml

7.1 Introduction

Animal cell culture is currently operated at 10 000-L scale in suspension cultures and 4000-L scale in microcarrier cultures of anchorage-dependent cells (Fig. 7-1). Both are large unit operations. As the size of such operations increases, as does investment in resources and personnel. Cell culture conditions are also more critical. Failures become far more costly.

The pressure in biotechnology production today is for greater speed, lower costs, and more flexibility. Ideally, a production unit should be compact (requires less investment), and modular (for use in different production schemes) (Fig. 7-2). Designs that allow the same plant to be used for different cell types (bacteria, yeast, insect cells, plus suspension and anchorage-dependent animal cells) are thus preferable. Animal cell culture is constantly being developed to increase unit productivity and thus make the production costs of animal cell products more competitive.

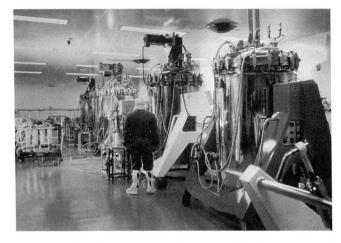

Fig. 7-1. Polio virus vaccine production in 1000-L fermenters. (Courtesy of Institute Merieux, France.)

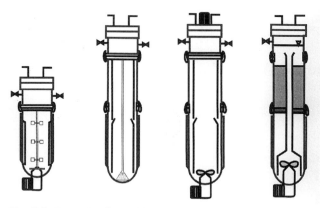

Fig. 7-2. Example of a multipurpose reactor design.

Because animal cells have a relatively low productivity, large amounts of culture supernatant are needed per clinical dose of final product. Extremely large volume cultures will be needed to produce kilogram quantities of a therapeutic monoclonal antibody, for example. The productivity of large-scale cell culture can be increased either by scaling-up to larger volumes with cell densities of $2-3 \times 10^6$ mL^{-1}, or by intensifying the process in smaller volumes but with higher cell densities (up to 2×10^8 cells mL^{-1}. When intensifying cell densities, more frequent media changes are needed and eventually perfusion has to be applied for which many competing technologies are available.

Microcarrier culture of anchorage-dependent or entrapped cells lowers the volume to cell density ratio and thus belongs to the second of these alternatives. This technique has many advantages for the commercial manufacturer. It operates in batch or perfusion modes and is well suited to efficient process development and smooth scale-up (Figs. 7-3 and 7-4). In addition, the reactors can be modified to grow other organisms. This chapter reviews the principles and methods of using microcarriers in cell culture production.

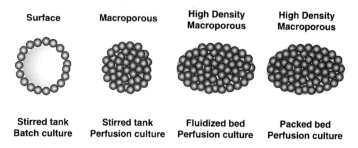

Fig. 7-3. Different microcarrier alternatives and applications.

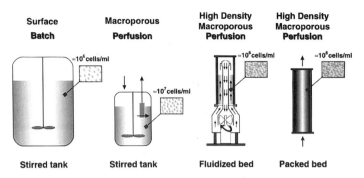

Fig. 7-4. Different microcarrier culture set-ups.

7.2 Production Considerations

7.2.1 Production Economy

Calculating total production economy is complex. Many aspects need to be considered; the organism to be cultured, amount of product needed, medium volumes, equipment, staff, downstream costs, etc. The major part of the Cost Of Goods Sold is, however, fixed costs [1]. Investments in equipment and personnel thus determine whether a project will be profitable or not.

Ideally, investments should be suitable for other projects if the one planned is not successful. Sometimes the choice of technology is restricted. For example, batch/fed-batch processes must be used when producing cytopathic viruses. To maximize the benefits of perfusion, immobilization is also needed to prevent the washout of cells. The cost ratios for perfusion, continuous-flow, and batch techniques to produce monoclonal antibodies are $1:2:3.5$ [2].

7.2.2 Consumable Cost Comparison

The three major consumables are the culture surface (for anchorage-dependent cells), serum (or protein additives), and medium. The medium cost can normally not be controlled, other than by negotiating price with different suppliers.

7.2.2.1 Culture Surface

To make a true cost comparison of cell culture surfaces, grow a specific cell line on different supports and calculate the cost of each support /yield of 10^6 cells. Start by calculating the price per m^2 of material. Take care when comparing porous materials.

Some determinations may also measure surfaces inside pores that are not accessible for cells. The true cell culture surface will thus be smaller than that measured. Remember that the larger the diameter of cell, the smaller the accessible area. Large cells thus give much lower densities in macroporous carriers.

As large a surface area as possible is desirable as it allows a higher number of cell doublings. With a large surface area and by optimizing the inoculation density, it is possible to maximize the cellular multiplication in each culture step. This reduces the number of scale-up steps and equipment necessary to reach final production volume (cell number).

Comparing cost of surface area/yield of 10^6 cells, reveals the following. Microcarriers and the Cell Cube (Costar) have the same order of magnitude and the lowest cost, roller bottles (1400 cm^2) are five times more expensive, and hollow fiber reactors 200 times more expensive in generating the same number of cells. (The cell density used for the hollow fiber is that given by the supplier.)

7.2.2.2 Serum and Additives

The most expensive additive to the medium is serum (proteins). Reducing serum in the medium by 1 % lowers costs by approximately 3 US $ L^{-1}.

Comparing the costs of a suspension culture with a fluidized bed run reveals the following. (It is assumed that serum concentration can be reduced by 1 % due to the higher cell density in the fluidized bed – this is a modest assumption.) Bed volume is 5 L. This is normally perfused with 10 volumes of media/bed volume/ day = 50 L per day. The saving is then $3 \times 50 = 150$ US $ per day. If run for 30 days, the total saving is 4500 US $. Subtracting the cost of the carriers gives a net saving of 3740 US $ per run. A 5-L fluidized bed reactor with perfusion thus produces the same amount of cells and products in 1 month as a 1500-L batch reactor does in one week at a consumable cost saving of 3740 US $. The investment cost is also much lower for the smaller system.

Table 7-1 compares the medium utilization (mL spent medium per mg product) of a human hybridoma cell line in a continuous stirred tank reactor (CSTR) with a fluidized bed reactor.

Table 7-1. Medium utilization in continuous stirred tank and fluidized bed reactors.

	CSTR	FBR
Culture volume	20 L	20 L (10 L of carrier)
Dilution–perfusion rate (L h^{-1})	0.0125	0.105
Medium L per day	6	50
Product (mg) per day	50	400
mL spent medium per mg product	120	125

7.2.3 Important Developments

Continuous development of manufacturing technology is needed if animal cell culture processes are to be competitive. Better cell culture, higher productivity, and simpler media requirements are all important [1]. Yields should increase in relation to space used. Processes should be robust with high success rates and minimum down-time and the number of operations should be as few as possible. Plants should be multipurpose to fully utilize capacity. Remember that fixed costs make up most of the Cost Of Goods Sold, and that R&D normally focuses on work that affects variable costs!

7.3 Microcarrier Background

7.3.1 Adhesion (Cell–Cell, Cell–Surface)

Cell adhesion is a multi-step process (Fig. 7-5). For an adherent cell to attach quickly to a surface, it needs to be round (have a disrupted cytoskeleton) and have exchanged its cell surface receptors for newly produced ones. It is therefore necessary to have trypsinized cells in a single cell suspension (trypsin also acts as a growth factor). Work has shown that when the surfaces on which cells grow are degraded, the cells remain flat and retain their cell–cell contacts. These cells did not attach for several hours when seeded onto a cell culture surface. When they did attach, the distribution over the surface was very uneven due to clumping. This resulted in an inhomogeneous cell distribution on the carriers and a slower growth rate.

Normal and transformed cells seem to adhere in slightly different ways (see Section 7.4.1). *In vitro,* adherent cells bind to extracellular matrix components; type I and IV collagen, fibronectin, vitronectin, laminin, chondronectin, thrombospondin, heparan sulfate and chondroitin sulfate.

The 'normal' *in vitro* adhesion process is considered to occur via integrin receptors binding to primarily fibronectin mol. wt. 220 KDa) and vitronectin (65 KDa). These proteins are abundant in serum-containing media (10% FCS contains 2–3

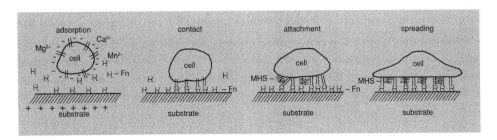

Fig. 7-5. The adhesion process.

μg fibronectin mL^{-1}) and quickly bind to the surface of the cell culture material. The cells in their turn bind to these proteins via the integrin receptors, with Ca^{2+}and Mg^{2+} ions as cofactors. A minimum of 15 ng cm^{-2} of adsorbed fibronectin is required for BHK cell attachment. Fibronectin binds easily to gelatin, which is why gelatin gels are used for the affinity purification of fibronectin. It is also the reason why gelatin is a good cell culture substrate and used to manufacture some microcarriers.

Vitronectin is considered to be more potent and active at even lower concentrations than fibronectin. Vitronectin is also called serum spreading factor and has been shown necessary for the spreading out phase of cells. These proteins are also produced by a number of cell lines. Other integrin receptors present on the cell surface are the collagen and laminin receptors. These proteins are not usually present in serum, but are sometimes added to coat the cell culture surface. Other molecules have been reported to be involved in lymphocyte adhesion but will not be discussed here.

The integrin receptors, together with cadherin receptors, are also responsible for cell–cell binding and the formation of tight cell junctions and sheets. At a later stage of attachment, the cells themselves produce multivalent heparan sulfate, which reinforces the binding. Finally, the cells assemble the cytoskeleton and spread out. Anchorage-dependent cells cannot proliferate without being spread out, so it is essential to enhance all steps involved in adhesion.

Cells can adhere to a wide variety of materials; glass, various plastics, metals (stainless steel 316, titanium used in implants), dextran, cellulose, poly-lysine, collagen, gelatin, and numerous extracellular matrix proteins (see above). Borosilicate glass is normally negatively charged. Attachment to it can be increased by treating the surface with NaOH or by washing with 1 mM Mg acetate. Plastics used in cell culture include polystyrene, polyethylene, polycarbonate, Perspex, PVC, Teflon, cellophane, and cellulose acetate. These organic materials are made wettable by oxidizing, strong acids, high voltage, UV light, or high-energy electron bombardment rendering them negatively charged. One major drawback of plastics is that they do not normally withstand autoclaving. Poly-d(I)-lysine is an artificial molecule that can also induce attachment when coated onto surfaces. It appears that it is the charge density and not the type of charge that is important for attachment [3]. Cells that grow attached on plastic surfaces should readily attach and grow on carriers.

7.3.2 Immobilization Principles

Immobilization was first described as early as 1923 by Carrell in the paper 'Immobilisation of animal cells' [4]. Today, biocatalysts (cells) are defined as immobilized when they are restricted in their motility by chemical or physical methods while their catalytic activity is conserved. Different immobilization techniques include binding biocatalysts to each other (aggregates) or on carriers, and physical entrapment in a polymeric matrix or through membrane separation (Fig. 7-6).

Immobilization allows heterogeneous catalysis, in contrast to catalysis in which substrate and biocatalyst (cells or enzymes) are homogeneously distributed. An eco-

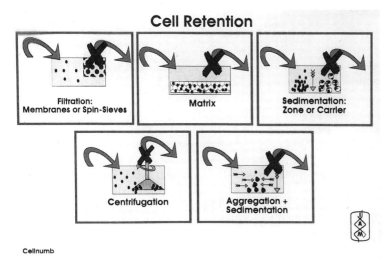

Cellnumb

Fig. 7-6. Different immobilization principles.

nomically viable separation of suspended biocatalysts from product is hardly possible. However, the immobilization technique does enable preliminary separation of product and biocatalyst, even during upstream processes. Immobilization also creates and maintains a high density of cells in a small volume (Table 7-2). Another major advantage is that the medium feed rate is not dependent on the growth rate of the cells. The higher throughput of medium guarantees higher volumetric productivity. In fact, some cell lines cannot grow or produce the desired product without immobilization.

Butler [5] calculated the surface-to-volume ratios of different immobilization systems. Roller bottles had a ratio of 1.25, packed beds (with spherical carriers) 10, and artificial capillaries 30. Microcarriers (25 g L^{-1}) had 150, by far the best surface to volume ratio.

Table 7-2. Cell densities in different immobilization systems.

System	Cell density (m L^{-1})
Suspension	10^6
Cell retention	10^7
Cell immobilization	10^8
Ascites	10^8
Tissue	2×10^9

7.3.3 Materials

Materials are important because of their chemical, physical, and geometric effect on the carrier. For example, they influence toxicity, hydrophilicity, hydrophobicity, microporosity, mechanical stability, diffusion of oxygen or medium components, permeability, specific gravity, and shape (form, size, thickness, etc.).

A wide variety of materials have been utilized to produce microcarriers: plastics, (polystyrene, polyethylene, polyester, polypropylene), glass, acrylamide, silica, silicone rubber, cellulose, dextran, collagen (Fig. 7-7) (gelatin), and glycoseaminoglycans. These materials can be formed into different shapes, with spherical the most common, though fibers, flat discs, woven discs, and cubes are also found.

Carriers are usually positively or negatively charged, though non-charged carriers are also available. These are normally coated with collagen or gelatin or have fibronectin or fibronectin peptides coupled onto the surface. Glycoseaminoglycan microcarriers are slightly negatively charged. Cells bind directly to the collagen, fibronectin, or fibronectin peptides and to the glycoseaminoglycan microcarriers (see Section 7.3.1). Unfortunately, protein-coated and glycoseaminoglycan carriers cannot be autoclaved as the protein structures are destroyed, but gelatin (denatured collagen) may be autoclaved. Gelatin also has a very high affinity for fibronectin, which is why it is so suitable as a cell culture substrate.

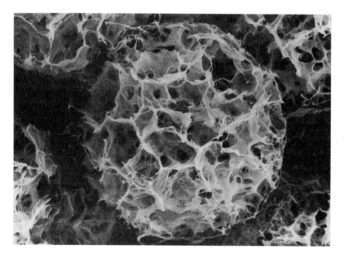

Fig. 7-7. Collagen-based macroporous microcarrier (Verax).

7.3.4 Size Shape and Diffusion Limits

The diameter of the different carriers varies from 10 μm to 5 mm, the smaller carriers being best suited for stirred tanks. The higher sedimentation rates of larger carriers make them suitable for fluidized and packed beds. The smaller the carriers, the larger the surface in the settled bed volume because of the smaller void volume between them. A diameter of 100–300 μm is the ideal size for smooth microcarriers, as a very narrow size distribution is most important for good mixing in the reactor.

The emulsion and droplet techniques produce round carriers, though an exception is the DEAE cellulose carrier, which has a cylindrical shape (Fig. 7-8). Macroporous carriers are on average bigger because their pores may be up to 400 μm wide. However, a large pore size must be balanced against the disadvantages of bigger particles such as diffusion limits and higher shear stress on the outer surface.

Mass transfer in the immobilized cell aggregate is a significant problem in immobilized cultures. The poor solubility of oxygen in the medium at 37 °C and the high consumption rate of the cells make it a marker for limitations in the cell aggregate.

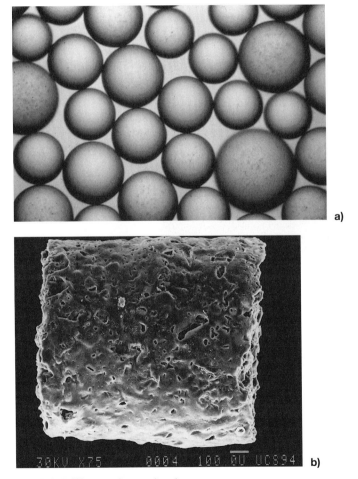

a)

b)

Fig. 7-8. Different microcarrier shapes.
(a) Spherical (Biosilon Nunc). (b) Cylindrical, side view (Immobasil ASL).

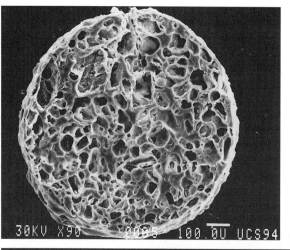

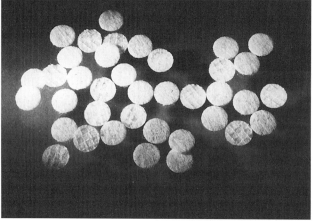

Fig. 7-8. (continued).
(c) Cylindrical, front view. (d) Woven discs (Fibracell, Bibby Sterilin).

Keller [6] reports that in cell cultures with high densities (up to $2 \times 10^{14}/$ m^{-3} in the cell layer), an oxygen consumption rate of 5×10^{-17} mol per cell and second, and a medium volume which is tenfold the cell mass, consumes the oxygen within 3 minutes.

The single oxygen molecule has to overcome three barriers before it reaches the cell in the middle of the carrier; firstly transport from the gas phase into the medium, then transfer from the medium to the cell mass, and finally diffusion and consumption through the cell layers. The OTR (oxygen transfer rate) can be increased by increasing the volume-specific surface or using pure oxygen instead of air. The cells themselves should not be exposed directly to oxygen because of its toxic effect. However, the protection afforded by macroporous structures allows the use of pure oxygen.

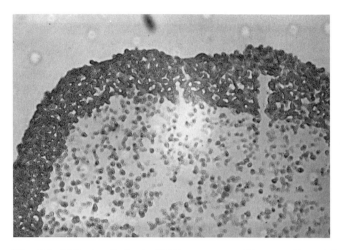

Fig. 7-9. Diffusion limits in CHO aggregates (Hematoxylin staining, size of aggregate 3 mm; (note necrotic center).

Many different methods of oxygenation are described or used. Bubble-free aeration [7], gassing with large bubbles [8] or using microbubbles [9,10] are used in high-density immobilized cultures (see Section 1.6.5). Depending on the size of the carrier, the oxygen concentration on its outside must be increased. Carrier particles up to 500 μm in diameter can be supported with an oxygen tension of about 10 % (Fig. 7-9); if the particles are larger than 900 μm, limitations occur if the oxygen tension is below 35 % [6]. Griffiths [11] found sufficient oxygen penetration into cell layers up to 500 μm thickness.

7.3.5 Specific Density and Sedimentation Velocity

Smooth microcarriers in stirred tank reactors have a specific density just above the medium between 1.02 and 1.04 g cm^{-3}, while Materials or material mixtures used to produce macroporous microcarriers have specific densities between 1.04–2.5 g cm^{-3}. The sedimentation velocity is a better parameter for the suitability of a microcarrier in a specific reactor type. This is because not only the specific density but also size and shape influence the sedimentation velocity. Velocities lower than 30 cm/min^{-1} do not create enough circulation and mixing for efficient nutrient supply throughout the carrier bed [6]. Adherent cells also tend to form bridges between microcarriers in fluidized beds. Higher sedimentation rates (150–250 cm/min^{-1}) will prevent the bridges from forming.

7.3.6 Rigidity and Shear Force

The rigidity of microcarriers is important in long-term cultures. The materials used should withstand the organic acids and proteases found in culture supernatants. Abrasive carriers made of brittle materials such as glass or ceramics could harm cells, valves, and bearings and cause problems filtering culture supernatants. In turbulent fluids, particle/particle collisions and particle/stirrer collisions are highly energetic [12]. In contrast to stirred tanks, the shear forces in fluidized beds are homogeneously distributed and impeller/carrier collisions are not possible. Shear forces in fluidized beds correlate with particle sedimentation velocity and reactor type. Keller [6] measured shear tensions of ~ 0.3–0.5 N cm^{-2}; which are far below the damaging shear tension for kidney cells (10 N cm^{-2}) [13]. Spier et al. [14] transferred shear forces into wind velocities to demonstrate the force affecting the cells. The linear velocity should not be more than 0.3 km h^{-1} during attachment of anchorage-dependent cells while a velocity of 95 km h^{-1} is necessary for cell detachment.

7.3.7 Porosity

Carriers can either be solid or microporous. Microporous carriers allow the cells to take up and to secrete material also on the basolateral side of the cell. Molecules up to mol. wt. 100 KDa can penetrate these carriers (Fig. 7-10). Note that when

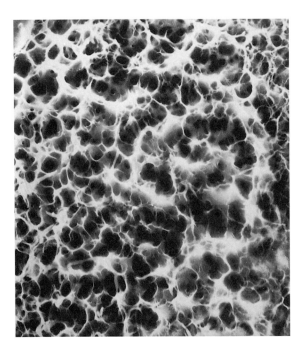

Fig. 7-10. Microporous microcarrier (Cytodex Amersham Pharmacia Biotech). Molecular weight cut-off = 100 000 Da.

the microcarriers are entirely confluent, there can be a different environment inside the beads than on the outside! The latest development in microcarrier technology is macroporous carriers that allow cells to enter the carriers, the average pore size of the different types is being between 30 and 400 µm (Fig. 7-11a,b). As the mean cell diameter of single cells in suspension is about 10 µm, this allows cells easy access into the carriers. Macroporous carriers are also suitable for immobilizing non-adherent cell types (Fig. 7-12); in this case, the cells are forced into the matrix and entrapped. Macroporous carriers provide higher cell densities and are therefore normally used in perfusion culture.

The porosity of macoporous carriers is defined as the percentage volume of pores compared with the total carrier volume and is normally between 60 and 99 %. The advantages and disadvantages of microporous and macroporous carriers are covered

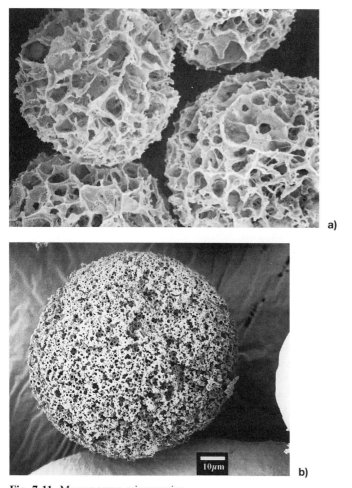

a)

b)

Fig. 7-11. Macroporous microcarrier.
(a) Cytopore (Amersham Pharmacia Biotech). (b) Microporous microcarrier (SoloHill).

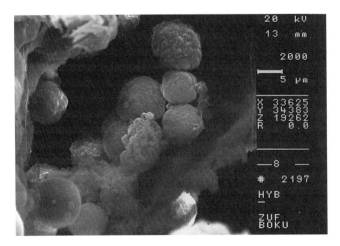

Fig. 7-12. Scanning electron micrograph (SEM) of macroporous carrier with entrapped hybridoma cells. (Cytoline 2 Amersham Pharmacia Biotech)

later in this chapter. In spite of the large number of microcarrier designs and types, very few are still available commercially and even fewer fulfil industrial standards for large-scale manufacturing processes.

7.3.8 Cell Observations

Transparency of a microcarrier is important for simple cell observations in a light microscope (Fig. 7-13a,b). In vaccine production especially, it is important to see the morphology of cells directly on the carrier to identfy the correct moment at which to infect the cells or to harvest the virus. Unfortunately, due to the size, three-dimensional structure, and component materials of a number of microcarriers, the cells cannot be observed clearly with a light microscope.

Because of the long preparation time and the effect of dehydration on cell shape and morphology the scanning electron microscope is not suitable for such observations (Fig. 7-14). However, confocal laser scanning microscopy [15] is an excellent tool for making cells visible in the pores of macroporous beads. With this technique, it is possible to make optical sections through the carrier and create a three-dimensional reconstruction. A viability stain with two fluorescent dyes (FDA fluorescein diacetate for living cells and ethidium bromide for dead cells) allows the viability of cells in three-dimensional structures to be estimated [16,17] (Fig. 7-15). Other staining methods such as MTT [18] also make cells visible in macroporous structures.

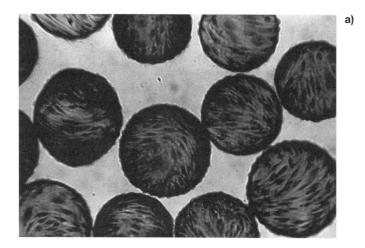

a)

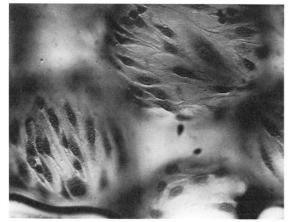

b)

Fig. 7-13. Transparent microcarriers with cells (Cytodex). (a) Hematoxylin staining, magnification 50×. (b) Hematoxylin staining, magnification 250×.

Fig. 7-14. SEM of dehydrated pig kidney on microcarriers. (Courtesy G. Charlier, INVR, Brussels, Belgium).

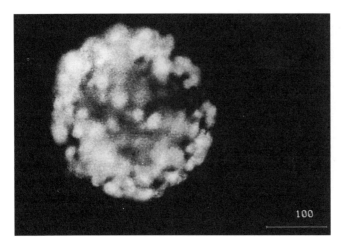

Fig. 7-15. Fluorescein di-acetate viability stained CHO cells in a macroporous microcarrier (Cytopore).

7.4 Microcarrier Technology

7.4.1 Microcarrier History

Microcarrier technology began with a paper by Professor Van Wezel in 1967: 'Growth of cell-strains and primary cells on microcarriers in homogeneous culture' [19], where he described the use of DEAE–Sephadex A-50, a positively charged ion exchanger, to grow human fibroblast-like cells. What he achieved was a homogeneous unit process. Microcarrier cultures then comprised cultivation of anchorage-dependent animal cells on small spherical particles kept suspended in culture medium. In 1969, Van Hemert et al. published the paper 'Homogeneous cultivation of animal cells for the production of virus and virus products' [20], describing the use of human diploid cells grown on microcarriers.

However, the original microcarriers had too high a charge of DEAE and were toxic to cells. Levine et al. [21] suggested lowering the level of DEAE substitution, and thus Cytodex microcarriers – launched in 1981 – were the result of collaboration between Van Wezel and Pharmacia Biotech AB. Early on, Van Wezel found that transformed cells detach from 'smooth' microcarriers and grow in the culture medium as aggregates. Many workers contributed to the research and development of other 'smooth' microcarriers.

The next major development was macroporous gelatine microcarriers developed by Nilsson et al. [22] (Fig. 7-16), which allowed growth inside the beads, thereby increasing cell density and protecting the cells. The final development was presented in an article by Young and Dean [23] describing the use of microcarriers for animal cells in fluidized beds. This step allowed the immobilization of both anchorage and

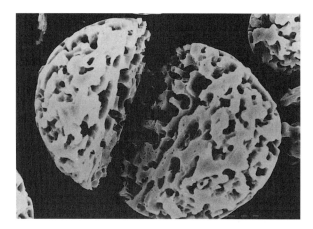

Fig. 7-16. Macroporous gelatin microcarrier (Cultisphere).

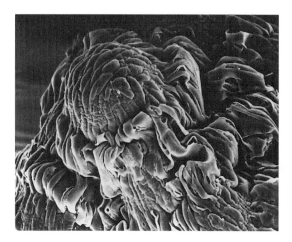

Fig. 7-17. 125 μm microcarriers used to stimulate cell aggregation (SoloHill).

suspension cells in high cell density production systems. Recently, microcarriers with a smaller diameter, 125 μm, have been used to induce and stabilize cell aggregation [24] (Fig. 7-17).

7.4.2 Advantages of Microcarriers

Microcarriers have many advantages: they are essential when surfaces are needed for anchorage dependent cells, and are also inexpensive (in terms of price per m²). Microcarrier technology results in a homogeneous culture system that is truly scaleable. Because of their large surface area-to-volume ratio, they occupy less space in storage, production, and waste handling. The surface also allows cells to secrete and deposit an extracellular matrix. This helps to introduce certain growth factors to cells. The spherical microcarriers have short diffusion paths, which facilitates nutrient supply in general, while the extracellular matrix provides cells with support to build their cytoskeleton, and to organize organelles intracellularly. Sometimes, however, this leads to increased productivity of functional product.

Microporous carriers (see Fig. 7-10) allow the cells to create a micro-environment inside the beads and also facilitate polarization and differentiation of cells. As the cells are immobilized on the microcarriers, it is easier to retain them in culture during perfusion, a process which separates cells from products. At high cell concentrations

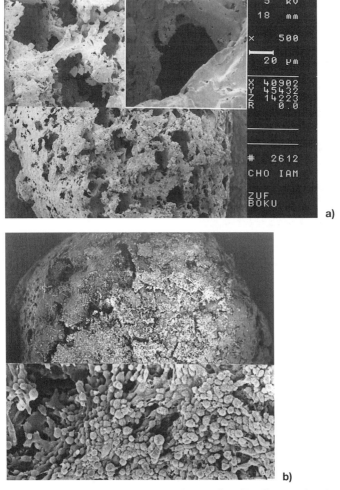

Fig. 7-18. (a) Empty macroporous microcarrier (Cytoline 1). (b) Macroporous microcarrier with high cell density of CHO cells.

and perfusion rates, the residence time of the product at 37 °C is very short, and it can quickly be separated from the cells and cooled down.

Macroporous microcarriers have additional advantages. They allow cells to grow in three dimensions at high densities (Fig. 7-18), which stabilizes the cell population and decreases the need for external growth factors. This makes it easier to use low-serum, serum-free, and even protein-free media which, of course, cuts costs (see Section 7.2.2). The high cell density confers more stability and improves the longevity of the culture, making macroporous microcarriers suitable for long-term cultures. The structure protects the cells from shear forces generated by the stirrer, spin filter and air/oxygen sparging [25], which facilitates oxygen supply. Macroporous microcarriers can be used both for suspension (entrapment) and anchorage-dependent cells. Various cell culture technologies can be applied, including stirred, fluidized, and packed bed reactors. The process advantage of macroporous carriers is that the high perfusion rates maintain a homogeneous environment which not only ensures a sufficient nutrient supply but also removes toxic metabolites. Vournakis and Runstadler [26] note that a regular distribution of oxygen and other nutrients in the pores is secured through a 'micropump'. The convection stream through the fluidized bed creates a pressure drop over the carrier surface that causes medium to flow in and out of the carrier.

Changing media from fetal calf serum-containing to protein-free is sometimes only possible with immobilized suspension cells or has a shorter adaptation and is less time consuming than with non-retained cells [27]. The growth rate of adherent cells in macroporous carriers is reduced by 30–50 % compared with that in T-flasks and is independent of carrier concentration and stirrer speed [28].

7.4.3 Disadvantages of Microcarriers

Some carriers have to be washed and prepared. Scale-up using cells harvested from microcarriers is more complex than suspension expansion and, it is even more difficult to harvest cells from macroporous carriers. Cell enumeration and harvesting are more difficult due to the higher cell density, the latter condition also making it more difficult effectively to infect all cells simultaneously, when especially using non-lytic viruses. Finally, large carriers with small pores may restrict the diffusion of nutrients to some cells.

7.4.4 Types

Tables 7-3 and 7-4 list the major commercially available microcarriers. The lists are not complete and all carriers for research have been omitted.

Table 7-3. Surface microcarriers.

Name	Material	Diameter (μm)	Reactors
Cytodex 1 (Amersham Pharmacia Biotech)	DEAE–Dextran	147–248	STR
Cytodex 2 (Amersham Pharmacia Biotech)	THMAP–Dextran	135–200	STR
Cytodex 3 (Cellex)	Gelatin–Dextran	141–211	STR
Biosilon (NUNC)	Polystyrene	160–300	STR
Bioglass (Solo Hill)	Glass, collagen coated plastic	90–500	STR
FACT III (Solo Hill)	Modified collagen	90–500	
DE 52/53 Whatman	DEAE–Cellulose		STR

Table 7-4. Macroporous microcarriers.

Name	Material	Diameter (μm)	Pore size (μm)	Porosity (%)	Reactors
Cultispher-G, S, GL (Percell Biolytica)	Gelatin	170–500	~50	50	STR
Informatrix (Biomat. Corp.)	Collagen-glycose-aminoglycan	500	40	99	STR
Microsphere (Cellex)	Collagen	500–600	20–40	75	FB, PB
Siran (Schott Glaswerke)	Glass	300–5000	10–400	60	FB, PB
Microporous MC (Solo Hill Labs Inc.)	Polystyrol	250–3000	20–150	90	STR
Cytopore 1, 2 (Amersham Pharmacia Biotech)	Cellulose	180–210	30	95	STR
Cytoline 1, 2 (Amersham Pharmacia Biotech)	Polyethylene	2000–2500	10–400	65	FB, PB, STR
ImmobaSil ASL	Silicone rubber	1000	50–150	> 40	STR

FB, fluidized bed; PB, packed bed; STR, stirred tank reactor.

7.4.5 Scale-up Considerations

Scale-up starts with the creation or choice of cell line! The characteristics of the cell line will then greatly influence the scale-up possibilities and technology choices. When choosing the cell line, it is essential to bear in mind the production technology available and the final scale of production – this is especially important if hardware investments have already been made. For example, the scale and technology used affects the stability of expression as well as the adhesive properties required by the cell. The higher the degree of transformation, the more difficult it is to get the cells to attach and spread onto surfaces. It may thus prove difficult to find a suitable 'industrial' microcarrier if technology screening is left too late after designing the cell line.

When evaluating different microcarriers to be used in a production process, the consequences of the choice at the final process scale must be considered. Normally, it is easier to handle a microcarrier that is autoclavable. It should also be possible to handle the material in the open, e.g. to subdivide lots more easily. If mistakes are made during process start-up, it helps if the microcarriers can be resterilized without loss of material. It is more difficult to sterilize the carriers at larger scale; as an autoclave is no longer viable due to the large volumes involved. Normally, the microcarriers are then sterilized inside the bioreactor.

The type of microcarrier is intimately linked to the bioreactor design. Large, high-density microcarriers always have to be prepared inside the bioreactor. The size of the carriers will influence their transfer through valves and tubing during sterilization and harvesting. The size and density will also influence how easy it is to keep carriers in suspension (i.e. impeller and reactor design, stirrer speed, etc.) and how quickly they sediment to allow large volume media changes (fed batch). These characteristics also affect how easy it is to retain the microcarriers in a perfusion system (settling zone, spin filters).

It must also be considered if the microcarrier generates truly homogeneous cultures, or causes heterogeneity. Homogeneous cultures are normally easier to scale-up as they facilitate oxygen and nutrient supply, and make it easier to remove waste products.

The choice of solid versus porous microcarriers will impact on how easy it is to wash the cells during media changes from serum to serum-free media, and to wash and harvest cells from the carriers. Solid carriers are normally more easy to handle in these respects. Macroporous carriers are the most difficult at harvest. Gelatin or collagen microcarriers are the easiest to harvest, as it is possible enzymatically to degrade the microcarriers and recover the cells.

The simplicity of a low-density system should be weighed against the demands and complexity of high productivity, high cell density cultures. Another aspect to consider is inactivation of the carriers after production. This is especially important when producing hazardous agents. Here, it is once again advantageous if the carriers can be autoclaved prior to disposal. Positively charged carriers will adhere to non-siliconized glass and stainless steel. This can be overcome during cleaning by washing with high-pressure water, automatically via spray balls or manually by using high-pressure washes. This 'stickiness' can be reduced by changing pH to decrease the charge, and increasing the ionic strength by adding salt to the washing solution. The material used to produce the carriers will influence waste disposal, as occasionally the material is degraded naturally.

One further aspect to consider is whether the microcarrier is made of a material that is used in downstream chromatography processing; this will facilitate process validation if it is the case. Finally, it is important to determine whether the microcarrier is an established product that is already registered with regulatory authorities for similar production processes. If so, it should also facilitate in the registration of a new process.

7.4.6 Choice of Supplier

Important aspects to consider when evaluating suppliers are their production capacity, i.e. the batch size available. This affects purchasing and testing during full-scale operation. It is important to evaluate what function and quality testing is performed by the supplier, and if a certificate of analysis is available for each batch. This helps to minimize batch-to-batch variation and ensure process consistency. Normally, at least three different lots should be evaluated before scale-up. The availability of product stability and leakage studies is also important. Regulatory issues to consider are if the company works according to ISO 9000, cGMP, and if it is possible to audit the supplier. Ask if they have Regulatory Support Files or Drug Master Files available for the product (Table 7-5). Support in form of application work, product information, trouble-shooting, and validation support is important. The financial stability of the manufacturer is important to ensure secure supply for the lifetime of the process. Finally, the price of the microcarriers could be an issue; however, as stated previously, the main part of Costs Of Goods Sold is fixed costs!

Table 7-5. Regulatory Support File and Drug Master File.

Regulatory Support File	Drug Master File
+ Information directly to customer	– Information directly to FDA
+ Updated automatically when new information available, information accessible to customer	– Updated annually, information not accessible to customer
+ World-wide support	+ Well accepted system
– No recognized approval	– No recognized approval
– No manufacturing information	+ Contains manufacturing information

7.4.7 The 'Ideal' Microcarrier

There is no such thing as the ideal microcarrier (Fig. 7-19); there are always compromises. However, the desirable features can nevertheless be described as:

- autoclavable for easy handling
- autoclavable attachment functions (for use in serum-protein-free medium)
- available with a Drug Master or Regulatory Support File
- available in large batches for industrial customers
- both non- and specific attachment functions for a large number of cell types (rapid attachment, spreading, and proliferation)
- certificate of analysis for each batch
- good long-term stability (minimal leakage)

Fig 7-19. The 'ideal' microcarrier.

- high batch-to-batch consistency
- high surface-to-volume ratio (large multiplication steps)
- material of non-biological origin (minimize viral risks), inactivation procedures possible and available for material of biological origin
- macroporous for shear force protection
- no nutrient limitation at center of matrix (macroporous)
- non-toxic, non-immunogenic matrix (transplantation purposes, products used as pharmaceuticals)
- possibility to count cells
- porous to allow for cell polarization/differentiation
- thoroughly quality controlled and meet the standards for GMP
- transferable between vessels (ease of scale-up)
- transparent for ocular inspection
- uniform size (cell growth homogeneity)

7.5 Microcarrier Culture Equipment

7.5.1 Unit Process Systems

A unit process system contains all cells used in production within the same vessel (Fig. 7-20). This means that all cells are grown under 'identical' conditions! Examples of truly homogeneous unit process systems for carriers are microcarrier cultures

Fig. 7-20. Surface-to-volume ratio comparison of a unit process system (macroporous micro-carriers) and a multiple unit system (rollers).

in stirred tanks and fluidized bed cultures with sufficient oxygen supply to not limit the scaleability. The Verax system is not truly homogeneous as it creates an oxygen gradient throughout the bed as the medium passes through. This makes it difficult to scale up. Other examples of unit process systems are the glass bead reactor run as a packed bed and the Fibracel woven plastic discs, also run as a packed bed. In both cases, the carriers with cells are stationary, which results in channeling of media in the reactor and which causes necrotic zones to arise inside the carrier bed.

7.5.2 Small-scale Equipment

Small-scale equipment for microcarrier cultures often comprises plastic bacterial Petri-dishes for stationary cultures, roller bottles for mixed cultures, and glass spinner flasks for stirred cultures. Borosilicate glass is normally used in spinners.

The spinners with the best performance for microcarriers are the Techne bottles equipped with bulb stirrers. The spinner flasks are placed on magnetic stirrers in incubators or warm rooms. Normally there are no controls attached. However, the Superspinner supplied by B. Braun Biotech contains a microporous polypropylene membrane for bubble free aeration in the medium. The aeration tubing is connected to a pump which delivers oxygen and CO_2 incubator (Fig. 7-21). Spinners are avail-

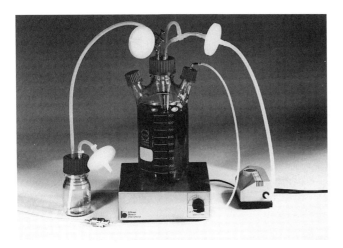

Fig. 7-21. B. Braun Superspinner.

able up to 20-L size, but it is easier to handle 4 × 5-L flasks in incubators and sterile work benches. This volume is sufficient to generate inoculate for 1- to 200-L cultures.

For process development, a number of 1- to 10-L fermenters (at least two) are used (Fig. 7-22) which are usually made of borosilicate glass. However, to prevent cells and positively charged carriers attaching, the glass needs to be siliconized. The silicone solvent must be washed away thoroughly before use.)

The temperature of the fermenter is controlled by circulating water between the double jacket and a waterbath, or by a heating jacket. There is also complete pH, dO_2, and stirrer control. Often, the controllers are also equipped with pumps for adding alkali and/or medium perfusion. The fermenter bottom should be rounded to give better mixing, and a marine impeller used to minimize shear forces. A spin filter is needed to run perfusion and sparge gas into the culture; in smaller reactors, this can be mounted on the stirrer axis. The mesh size is normally between 60–120 μm.

Fig. 7-22. B. Braun laboratory fermenter.

7.5.3 Large-scale Equipment – Stirred Tanks (Low Density)

Larger stirred tanks are normally made of electropolished pharmaceutical grade stainless steel. Before using a stainless steel vessel for the first time; it should be washed with a mixture of 10 % nitric acid, 3.5 % hydroflouric acid, and 86.5 % water. Both specialized sampling devices and spin filters are commercially available for larger reactors. A conical-shaped reactor offers a number of benefits (Fig. 7-23). As the probes are placed low in the reactor, it is possible to vary the volume in the reactor over a greater range and thus retain control. In addition, there is a large head-space-to-surface area ratio for gassing.

Fig. 7-23. Example of a conically shaped stirred tank reactor.

The large stirred tank reactor is the most widely used reactor in the fermentation industry and has already been scaled up to 10 000 L. Its major advantages are its conventional design, proven performance, homogeneous results, and good potential for volumetric scale-up. Its disadvantages are that it is a low-intensity system with high shear forces at large scale and oxygen supply limitations.

7.5.4 Packed Beds

There are two different approaches to this technology. The first is to pack a cage with plastic woven discs (Fig. 7-24) or other microcarriers and then place it inside a stir-

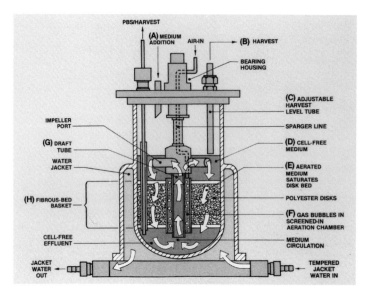

Fig. 7-24. Packed cage reactor (New Brunswick, Fibracel).

red tank. Medium is circulated through the cage using a marine impeller. Gas is supplied via sparging. Reactors of this type are available from New Brunswick (USA) and Meredos (Germany). The second approach is to pack glass beads into a column. Siran beads of 3–5 mm diameter are frequently used [29] (Fig. 7-25), though 5 mm-diameter beads normally give higher cell yields[30]. An airlift-driven system usually gives better oxygenation.

Fig. 7-25. Glass macroporous microcarrier for packed beds (Siran).

Low surface shear, high unit cell density and productivity, and no particle/particle abrasion are some of the major advantages of packed beds. However, poor oxygen transfer, channel blockage and difficulties in recovering biomass from the bed have all limited their use.

7.5.5 Fluidized Beds

The Verax system and the Cytopilot are examples of two different designs of fluidized beds.

7.5.5.1 Fluidized Bed With External Circulation

Verax introduced this technology for animal cells [28]. The fluidized bed is equipped with an external recirculation loop connected with a gas exchanger (hollow fiber cartridge), heating elements, pO_2, pH and temperature sensors, and a recirculation pump (Fig. 7-26). Certain aspects of the external circulation loop can be problematic, such as shear stress created by the pump, oxygenator fouling, and sterilization procedures in large scale. A gas exchanger transfers oxygen to the culture medium in the external loop. A certain oxygen tension at the entrance of the bed is achieved. The supernatant is increasingly depleted of oxygen on its way up the fluidized bed from the bottom to the top of the reactor (Fig. 7-26). The requirement of the oxygen tension from the biological system therefore limits reactor height and scalability. This problem of an oxygen gradient along the reactor, height can be overcome by integrating a membrane module directly into the fluidized bed as described in [31] (see also Fig. 7-29b). Reactors of this type are available from B. Braun, Melsungen Germany) with a settled carrier volume ranging from 20 mL up to 5 L.

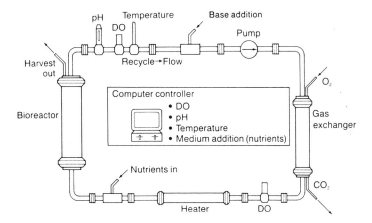

Fig. 7-26. Schematic drawing of the Verax system.

7.5.5.2 Cytopilot – Fluidized Bed With Internal Circulation

The fluidized bed with internal recirculation (Cytopilot) was developed at the Institute for Applied Microbiology in Vienna, Austria, in co-operation with the company Vogelbusch GmbH, Vienna, Austria [32]. Cytopilot comprises a lower and an upper cylindrical chamber (Fig. 7-27). The lower chamber has a bottom adapted to the special flow conditions and is equipped with the following: a heating circuit via a double jacket, sampling and discharge facilities, a magnetic stirrer (rotating in both directions), and probe nozzles for pH and dO_2. The liquid agitated by the magnetic stirrer is conveyed via the distributor plate to the microcarriers in the upper chamber of the vessel. The hydrodynamic pressure lifts the settled microcarriers to form a fluidized bed with a clear boundary between the top of the fluidized microcarriers

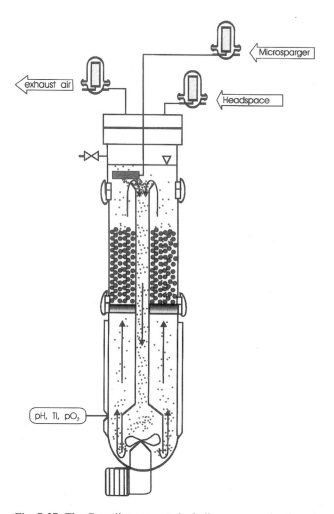

Fig. 7-27. The Cytopilot concept, including oxygenation by microsparging.

and the uppermost part of the medium volume. The bed expands or contracts as a function of the stirrer speed. The medium then flows through a sieve to the internal recirculation loop and back to the stirrer in the lower chamber of the vessel. Micro-bubbles of oxygen are sparged homogeneously into the downflow in the draft tube and then uniformly distributed by the impeller (Fig. 7-27). This system provides oxygen gas hold up via the dispersed oxygen bubbles (continuous transfer of dO_2 from gas to liquid). It both minimizes dO_2 gradients in the system and greatly increases the theoretical height of the fluidized bed.

7.5.5.3 Fluidization and Fluidization Velocity

If a carrier bed laying on a distributor plate is streamed through with medium, it will have a pressure drop (Δp). This pressure drop reduces the pressure (p) of the carrier bed on the plate. If Δp equals p, the bed will expand (Fig. 7-28). This is the fluidiza-tion point. The maximum fluidization speed depends on the sedimentation rate of the microcarriers. When the upward flow rate is higher than the sedimentation rate of the microcarriers, they will be carried to the top, which is known as flush out. Flush out velocities are between 10-to 100-fold higher than the velocity at the fluidization point. The efficiency of mixing in a fluidized bed is much higher than in many other systems. This efficiency increases with the particle size of the carriers.

Because there are no channels between carriers for cells to block, and because it allows cells to be retrieved from the reactor, the fluidized bed represents an improve-ment over the packed bed reactor. Its scale-up potential and very good mass transfer are additional benefits. Drawbacks include particle/particle abrasion and shear stress affecting growth on the bead surface (This affects only 10 % of the available area of macroporous beads.)

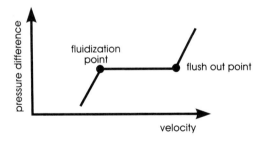

Fig. 7-28. Fluidization principles.

7.6 Culture Conditions

7.6.1 Media and Components

It is important to note that a certain amount of medium will only generate a finite amount of cells of a particular cell line. This is independent of the culture technology chosen! Basic media formulations were originally DME and DMEM with added F10/12 or medium 199 enrichments. RPMI 1640 was often the base for hybridomas.

Today, there is a multitude of different media developed for specific cell types, many of them serum-free. A number of suppliers provide collapsible plastic bags with ready-made media to be hooked-up to the bioreactor. The most recent development is liquid concentrates that can be automatically mixed with water on-line, sterilized, and fed into the bioreactor, (this system is available from Gibco). The media composition during growth and during maintenance of the culture can be quite different. Often serum is reduced or even completely removed at later stages of high cell density cultures. For some cell lines, it is even possible to run protein-free media at later stages of the culture.

The buffer system frequently used in cell culture media, carbon dioxide/sodium bicarbonate, is a weak buffer system. At times, it is beneficial to change to a better buffering system. HEPES (10–20 mM) is, therefore, used as an alternative (33], especially in serum-free media where the buffering capacity of the serum needs replacing.

Nutrients that are often depleted quickly include glutamine and, when growing human diploid cells, cysteine. Amino acid analysis will help determine utilization rates for different cell types. Glucose is used in a wasteful manner by cells, so it is desirable to start at concentrations below 2 g L^{-1} and then add more after two to three days. An alternative is to switch to other sugars as sources of carbon and energy, for example, fructose or galactose.

Additives
Common additives to low-serum or serum-free media are insulin (IGF-1)(5 mg L^{-1}, transferrin (5–35 mg L^{-1}), ethanolamine (20 µM), and selenium (5 µg L^{-1}). Mixtures of these supplements are now commercially available. Media for adherent cells should always contain Ca^{2+} and Mg^{2+} ions as these act as cofactors for adhesion. Sometimes, sodium carboxymethyl cellulose (0.1 %) is added to prevent mechanical damage to cells. Pluronic F68 (0.1 %) is used to reduce foaming and to protect cells from bubble shear forces in sparged cultures, especially when low- or serum-free media are used. Cyclodextrin or dextrans are also used in serum-free media as albumin replacements.

7.6.2 pH

pH is very important during inoculation. Cell attachment to carriers with an electro-static surface is highly dependent on the right pH (e.g., a pH of 7.4 is recommended for Vero cells and Cytodex 1 microcarriers). However, the pH of the medium has lit-tle effect on cell attachment to coated microcarriers [34]. An alkaline pH prevents/ prolongs attachment, and higher pHs kill the cells. The lower setting for pH is nor-mally 7.0, but below 6.8, it becomes growth inhibitory. An incubator set point of 5 % CO_2 is normally used, together with sodium bicarbonate to stabilize pH. Autoclava-ble probes are normally used in fermenter systems and an upper and lower pH set point chosen. If alkali is needed to compensate for an acidic pH, addition of 5.5 % $NaHCO_3$ is preferable to 0.2 M NaOH. NaOH can, however, be used in very well-mixed systems where it is not delivered directly onto cells. Note that silicone tubing is gas-permeable and can cause changes in pH during media transport.

7.6.3 Dissolved Oxygen

The solubility of oxygen in aqueous solutions is very low (7.6 µg mL^{-1}). The mean oxygen utilization rate of cells has been determined as 6 µg per 10^6 cells h^{-1}. Oxygen supply depends on the oxygen transfer rate (OTR = Kla (C*-C). OTR is one of the main limiting factors when scaling-up cell culture technology.

Oxygen supply is often via surface aeration, and can be increased further by using medium perfusion, increased oxygen pressure, membrane diffusion, or by direct sparging of air or oxygen into the culture medium (Fig. 7-29). OTR will increase in vessels having a large height-to-diameter ratio because of the higher hydrody-namic pressure at the base of the vessel. Membrane diffusion is inconvenient because a lot of tubing is needed, which also makes it expensive. Medium perfusion and oxy-

a)

Fig. 7-29. Examples of dO$_2$ supply devices. (a) Microsparger.

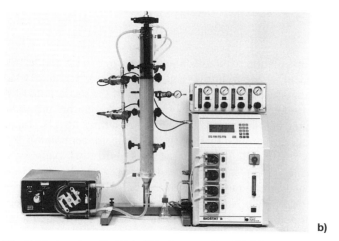

b)

Fig. 7-29. (continued). (b) bubble-free aeration (Fluidized bed system of KFA Julich).

genation in a separate vessel have been particularly effective in microcarrier cultures. Sterilizeable 0.22-μm non-wettable filters are used to supply gases continuously to cultures.

7.6.4 Redox Potential

The Redox potential represents the charge of the medium. It is a balance of oxidative and reducing chemicals, pO_2 concentration, and pH. An optimum level for many cells is +75 mV, which equals a pO_2 concentration of 8–10 % (approximately 50 % of air saturation). The Redox potential falls under logarithmic growth and is at its lowest 24 h before the onset of stationary phase [35,36].

7.6.5 Stirring

Avoid stirrers with moving parts as these will damage both cells and microcarriers. Top-driven reactors are best if the carriers/cells are not physically separated from the stirrer. The geometry and speed of the impeller greatly influence OTR. With certain vessel designs and a marine impeller, 150 r.p.m. has been achieved without being detrimental to the cells.

7.6.6 Control and Feeding Strategies

For good final product quality, it is important to have as steady state conditions as possible. For example, variations in sugar concentration will invariably affect final product glycosylations. In simple cell culture set-ups, carrier samples are taken, cells are counted, and morphology is examined (photographs taken) to document growth. As cells grow, stirrer speed is increased to compensate for the increased weight of the carriers (empty carriers 1.03 g mL^{-1} cells 1.015–1.070 g mL^{-1} depending on cell type). This keeps them suspended and increases gas transfer. After about three days in culture, the medium turns acidic and needs to be changed.

Media supply is determined empirically in relation to growth and productivity, or by amino acid, glucose, or lactic acid analysis. In bioreactors, there is continuous control of pH, dO$_2$, and stirrer speed, usually via a programmable controller. pH is normally initially controlled by adding CO$_2$ to lower pH. Nitrogen will wash out CO$_2$ and increase pH. Adding alkali via sodium hydrogen carbonate or NaOH solutions will increase pH during very high lactic acid and CO$_2$ production.

dO$_2$ is controlled by adding air or, at high cell concentrations, pure oxygen. Increasing stirrer speed can be coupled to the increased demand for oxygen supply and increases OTR. To use oxygen in sparging, the cells need to be protected from the gas bubbles to avoid toxic effects.

Media utilization is normally determined by taking samples to measure glucose concentration. Sampling is either done manually or automatically via a flow injection analysis biosensor system connected on-line to the bioreactor [37,38] (Fig. 7-30).

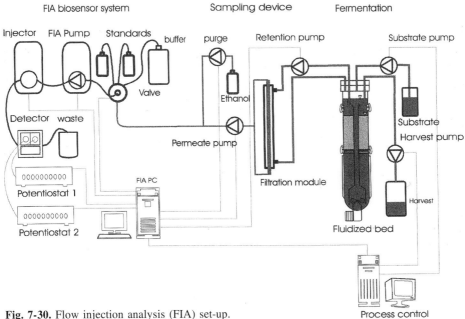

Fig. 7-30. Flow injection analysis (FIA) set-up.

These results are then used to determine the rate of manual media change or to regulate automatically the perfusion rate via pump speed (Fig. 7-30). In addition to glucose, glutamine and even metabolites like lactate can be used to measure media utilization. Ammonia can be measured on-line with the FIA biosensor system. Oxygen consumption rate can also be used to determine cell growth and be used as a parameter for feeding strategies when optimized and correlated with the metabolic rates of the substrates. By switching off the supply and looking at the linear consumption of oxygen over time, the number of cells can be calculated.

7.7 Microcarriers in Practice

7.7.1 Preparation of Carriers

It is a major advantage if the carriers are supplied dry, as the exact amount needed can be weighed and then prepared *in situ* by autoclaving. If a mistake is made during preparation, the carriers can be re-sterilized. Pre-sterilized microcarriers must always be handled under sterile conditions and can normally not be re-sterilized by autoclaving.

At large scale, it is advantageous to prepare the carriers directly inside the reactor because of the problem of heat transfer into vessels standing in autoclaves. The correct amount of carrier is added to an already cleaned and sterilized reactor. Then a volume of distilled water large enough for the microcarriers to swell correctly (hydrated microporous microcarriers) and be covered after settling is added to the fermenter. (Note that for soft microcarriers, the swelling factor in water is much larger than in PBS!) The reactor is then closed and the vessel re-sterilized while the carriers are stirred.

The culture medium is made up from powder, allowing for the water volume remaining inside the reactor. It is then sterile-filtered, added to the reactor and all parameters equilibrated. The reactor is then ready for inoculation.

This method works well with serum-containing media but caution is required when working with serum-free media, where additional washing, sedimentation, and decanting of washing buffer may be necessary prior to sterilization.

Carriers for packed or fluidized beds are also normally prepared inside the reactor. At small scale, the entire reactor can be sterilized by autoclaving, while at larger scale, carriers are sterilized *in situ*.

7.7.2 Microcarrier Concentrations

With packed beds, the airlift or cage is filled with beads or discs. In homogeneous stirred tanks, the concentration of surface microcarriers in batch cultures is normally 2–3 g L^{-1}. However, this can be increased up to 15–20 g L^{-1} (surface area of

60 000–80 000 cm^2 L^{-1} when perfusing the culture. In fluidized beds, 10–50 % of the column volume is filled with carriers. Fluidization is normally better at higher concentrations. In packed beds, the cage or column is normally filled with microcarriers, the amount being related to the size of the cage or column.

7.7.3 Inoculum

It is essential to equilibrate and stabilize all culture parameters before adding the inoculum. Temperature and pH are especially important.

When inoculating, actively growing cells in logarithmic growth should be used and cells in stationary phase avoided. If the cells are stationary, i.e., in G^0, they must be returned to the cell cycle by addition of growth factors. This is normally a 20- to 24-h process, which leads to an initial lag phase in growth. Stationary phase cells can cause the loss of one day process time at each subculture step. The cells used for inoculation should be single cells (Fig. 7-31) and absolutely not aggregates, which will otherwise cause heterogeneity.

Surface carriers are normally inoculated with 0.5–2 × 10^5 cells mL^{-1} or 0.5–2 × 10^4 cells cm^{-2}. Macroporous carriers are normally inoculated with about the same cell concentrations. Inoculate the cells in one-third to one-half of the final volume. Stir immediately at the lowest speed to keep the microcarriers in homogeneous suspension and, after the attachment phase, add media to the final volume. If the vessel is very well suited for microcarriers, it may be possible to start directly with the final volume.

In some processes, large volumes of cells are generated on microcarriers, harvested, and frozen. These cells are then thawed and used immediately as inoculum. The process is then scaled up two more steps before final production, which makes production more flexible.

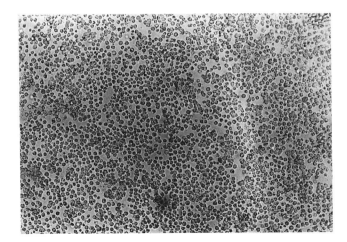

Fig. 7-31. Single-cell suspension.

Fluidized beds are inoculated with $1-2 \times 10^6$ cells mL^{-1} of packed bed carrier volume. The inoculation density is highly dependent on both cell line characteristics and media composition. The stirring rate is also related to the shear sensitivity of the cell line, with normal stirring rates of between $20-100$ r.p.m. Stirring speed is initially low and increased as the cell concentration increases. During inoculation, the fludized bed is run as a packed bed, and after ~ 5 h the bed can be fluidized.

Consider from the beginning how to create the large quantities of cells needed to inoculate a production-scale fluidized bed. Imagine 1 kg of product is needed and that the anchorage dependent cell line produces 100 mg L^{-1} carrier per day. A fluidized bed with 100 L of carriers has to be operated over a period of 100 days. Thus, 2×10^{11} cells would be needed to inoculate these 100 L of carriers (2×10^6 per mL of carrier). This would require 2000 roller bottles, each of 850 cm^2. The problem could be overcome by:

- Preparing the inoculum as aggregates in a stirred tank;
- scaling-up on the same carriers via trypsinization;
- scaling-up on different carriers (smooth carriers to macroporous); and
- Carrier-to-carrier transfer.

For suspension cells, the easiest way is always to create the inoculum in stirred tanks before transfer to macroporous carriers. Figure 7-32 shows that the attachment phase after inoculation is finished after 2 h on macroporous carriers in fluidized bed applications. Fluidization can thus be started very soon after introducing the cells into the reactor.

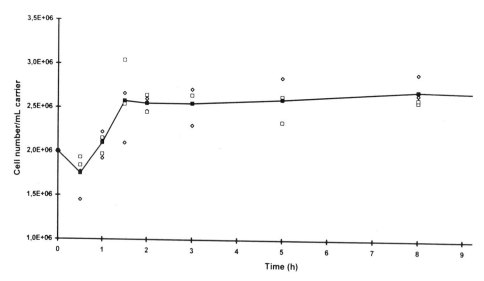

Fig. 7-32. Cell attachment kinetics in fluidized bed culture.

7.7.4 Cell Quantification

Cell numbers can either be determined directly or indirectly.

Direct methods
These include cell enumeration by counting whole cells (attached cells trypsinized) or crystal violet-stained nuclei [39]. Others are determining total cell mass via protein or dry weight measurements. Cell viability is more difficult to quantitate, but is often done via trypan blue (0.4 %) or erythrosine B (0.4 %) staining. Another alternative is the MTT colorimetric method [18]. Some cell culture systems cannot be sampled, and therefore numbers cannot be determined directly, nor morphology studied via hematoxylin staining/fixation.

Indirect methods
These measure metabolic activities such as glucose or oxygen consumption, lactic or pyruvic acid production, CO_2 production, and increase in product concentration. They are quite useful during logarithmic growth, but can be misleading later. Another possibility is to measure enzyme concentrations in the culture. One example is lactate dehydrogenase (LDH), which is less dependent on the different phases of growth. A relatively new technique is to measure capacitance (Aber instruments), the advantage of this technology being that it measures cell mass even within macroporous carriers, and so monitors cell growth continuously within the bioreactor.

All these cell counts provide a good picture of cell growth if the same detection method is always used. It is nearly impossible to compare different culture runs (performed with different cell lines) measured with different cell-counting methods. This is demonstrated by Capiaumont et al. [40], who found up to a factor of 8 difference in between the various cell-counting methods (Table 7-6).

Table 7-6. Cell contentration determination techniques.

Parameters	Cell count at the end of culture (10^6 per ml)
Glucose	2.5
Lactate	4
Alkaline phosphatase	1.14
β-Glucuronidase	0.9
Alanine	0.9
Mean glucose/lactate	3.2
Counted nuclei	3

7.7.5 Scale-up

Scale-up generally means a lengthy development period to ensure that all parameters are under firm control. It is best, therefore, to attempt working with a limited number of cell lines. Process simulation is also an essential part of scale-up and determines that the cells cope with scale-up without alterations and with maintained productivity. The viability and productivity of the cells has to extend beyond the planned time of the production process.

An essential part of scale-up is the cellular multiplication in each scale-up step. If it is possible to inoculate at a low cell density, i.e., 10^4 mL^{-1} and the final yield is 10^6 mL^{-1}, the multiplication factor is 100×. The larger the multiplication factor in each step of scale-up to final production volume, the fewer steps are needed for its achievement. It is important to maximize this as it minimizes the number of steps/operations needed! This also affects the investments required. If the above applies, it is possible to inoculate a 100-L fermenter from a 1-L spinner. In the case of colonization (see Section 7.7.5.2), a 1-L spinner could inoculate a 1000-L fermenter.

Scaling-up by volume normally involves moving from glass to stainless steel vessels, from mobile to static system, and from autoclavable to *in situ* sterilization. Additional equipment is also needed at larger scale, such as seed vessels, medium hold tanks, and sophisticated control systems.

Scale-up of microcarrier cultures can be done by increasing the size of the vessel or by increasing the microcarrier concentration. Production units up to 4000 L have been achieved by increasing size. Factors that influence this strategy include reactor configuration and the power supplied by stirring. The height-to-diameter ratio is one important factor. For surface aeration, the surface area-to-height ratio should be 1:1. Variables that affect impeller function include shape, ratio of impeller to vessel diameter, and impeller tip speed. Larger impellers at lower speeds generate less shear forces, with marine impellers having been found to be more effective for cells.

Pneumatic energy supplied via air bubbles or hydraulic energy in perfusion can both be scaled-up without increasing power input.

Desirable features of a stirred microcarrier reactor include no baffles, curved bottom for better mixing, double jacket for heating and cooling, top-driven stirrer, and a smooth surface finish (electropolished). Scale-up by increasing the microcarrier concentration requires perfusion, which makes it necessary to have a separation device to keep the carriers in the reactor. A settling zone or a spin filter are both suitable for this task.

The limiting factor for higher cell concentrations is usually oxygen supply. It is difficult to use direct sparging with normal microcarriers in stirred tanks, as the carriers may accumulate and float in the foam created. It can however be achieved if large bubbles are used or by sparging inside a filter compartment. Other alternatives that increase oxygen supply include increasing surface aeration, perfusion via an external loop and oxygenation device/vessel [41].

It is easier to increase oxygen supply via sparging in packed or fluidized beds as the cells are protected inside the macroporous carriers. Scale-up of fluidized beds is

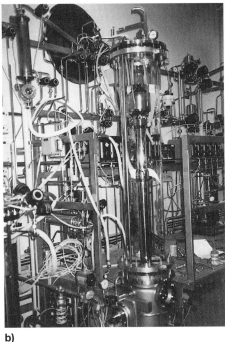

a)   b)

Fig. 7-33. Scale-up of fluidized bed technology. (a) Cytopilot Mini (100–500 ml bed volume). (b) 100 L Cytopilot fluidized bed reactor (up to 40 L bed volume).

linear, i.e., the diameter-to-height ratio is the same, which keeps the fluidization velocity constant. As oxygen is supplied via direct sparging with microbubbles (fluidized bed with internal loop), there is a continuous gas to liquid transfer throughout the bed, which makes the technology easily scaleable (Fig. 7.33a,b). Airlifts have been scaled-up in this way to 10 000-L scale!

7.7.5.1 Harvesting

The aim of this step should be to have an inoculum consisting of a single cell suspension of highly viable cells that were in logarithmic growth prior to harvesting. To achieve this, it is necessary to develop and follow a strict harvesting protocol to apply after the carriers have sedimented and the supernatant been decanted.

The main task is to break the cell–surface and cell–cell interactions (the cells must be harvested prior to confluence) and to round up the cells. As cell binding is dependent on divalent ions, the carriers must be washed with citric acid or EDTA (0.2 %, w/v) containing buffers (phosphate-buffered saline, PBS) to help detach the cells. If the media contains serum, α-1-anti-trypsin, a trypsin-blocking protein abundant in serum, has also to be removed by washing. Microporous carriers require more extensive washing compared with solid carriers, normally two to five washings (this should be optimized depending on cell/microcarrier/medium), with a volume of washing solution equal to the sedimented microcarrier volume.

As trypsin has a pH optimum close to pH 8, it may be necessary to increase the buffer capacity of the PBS used for harvesting to maintain a pH of 7.4 throughout the procedure. Note that the substrate concentration for trypsin increases drastically for sedimented beads compared with ordinary flask cultures as the cell density becomes very high! The same ratio of units of trypsin/number of cells should be maintained, if possible. Normally, this means increasing trypsin concentration ten-fold (i.e., 0.2 % w/v, but this may depend on the supplier) compared with harvesting flasks. To speed up harvesting, both the washing solution and the trypsin solution should be pre-warmed to 37 °C. As trypsin activity is dependent on Ca^{2+} ions, it is advantageous to separate the EDTA washes and the final trypsinization. Different suppliers should be screened for a suitable trypsin for a particular cell line; 'crystalline' trypsin should not be used. It is normally better to use a trypsin contaminated with other proteases, as this is usually more effective!

Some shear force may have to be applied to quickly detach the cells. In spinners, stirring speed can be increased during this step or flasks shaken. At larger scale, the bioreactor can be emptied into a special harvesting reactor developed by Van Wezel (available from B. Braun) (Fig. 7-34). This vessel is divided into two compartments by a stainless steel 60–120 μm mesh filter. The upper compartment contains a Vibromixer, a reciprocating plate with 0.1–0.3 mm holes moving at a frequency of 50 Hz. The microcarriers are collected on top of the mesh. Washing is done by adding washing buffer and draining it through the mesh. Pre-warmed trypsin is added so that it just covers the microcarriers. This is left for some minutes (depending on the cell line), after which the Vibromixer is used for a short period to help detach cells. After detaching, the cells separate from the used carriers by draining through the mesh. Additional washing is done to improve yield. The filters work most efficiently with hard carriers (plastic, glass) that do not readily block the mesh. It is also pos-

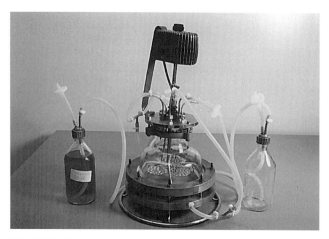

Fig. 7-34. Vibromixer harvesting vessel.

sible to detach only the cells and to transfer the whole mixture of used beads and cells to the new reactor.

The trypsin has to be inactivated before the cells can be used as an inoculate. Add either serum to the harvest or aprotinin, soybean trypsin inhibitor, if serum-free cultures are required. If serum is used, it can be added at the same time as the cells.

7.7.5.2 Colonization

Cultures of cells that do not attach well to carriers or that detach easily during mitosis, can be scaled-up just by diluting the microcarrier culture with fresh microcarriers and more media [42]. A 1:1000 dilution has been achieved with Chinese hamster ovary (CHO) cells (Fig. 7-35). Microcarriers have been used for protease-free trans-

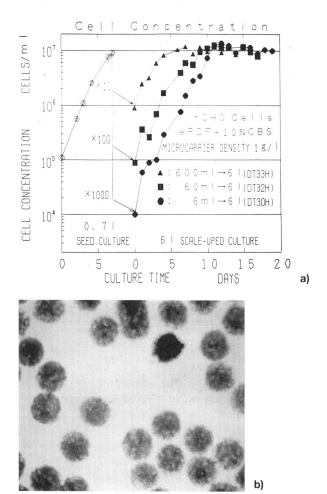

Fig. 7-35. Carrier to carrier transfer (Cytopore). (a) Kinetics. (b) Different phases of colonization.

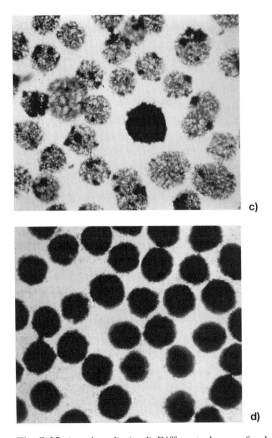

Fig. 7-35. (continued). (c, d) Different phases of colonization.

fer of cells by simply placing them into blood vessels or onto cell culture surfaces and allowing the cells to migrate onto the surface. The migration method normally works better for transformed cell lines such as CHO, hybridomas, etc.

7.7.5.3 Suspension

Cells that grow in suspension can first be scaled-up by increasing the volume of the suspension cultures. During final production, the cells can be immobilized at high densities in carriers and the product continuously harvested via perfusion.

7.7.5.4 Documentation

The most critical variable in cell culture production is the cell line itself. Therefore, it is essential to have thorough documentation, process control, and very strict protocols, for all steps from thawing the cells, through small-scale culture, and up to final production. This naturally applies also to evaluation studies, scale-up, and process development/trouble-shooting. Without such documentation, the reproducibility and quality of the production process will be very poor and the chances of success minimal.

During each phase and step, document:

1. Cells; type, viability, plating efficiency (if attached), doubling time, split ratio at passaging, saturation density (cells cm^{-2}, cells mL^{-1}, yield after harvesting, passage number. Check and keep track of morphology (photographic documentation).
2. Preparation of microcarriers; type, weight/volume, batch, hydration/washing solution and procedures, sterilization, medium equilibration, culture parameter equilibration (pH, dO_2, temperature).
3. Culture vessel; type, preparation, media volume, amount of microcarriers used, stirrer speed, air lift or fluidization rate.
4. Medium; batch, composition, preparation, additives and their batch numbers, buffer system. Serum testing, type, treatment before addition.
5. Gas supply; gases added, batches, set points, flowrates, aids for supply, stirrer, sparging, membrane, filters used.
6. Culture program; inoculation density, preparation of inoculum, inoculation volume, stirring during inoculation, maintenance program, set points, sampling.
7. Results; cell culture results of each step. Parameter control.
8. Cell harvesting and recovery: washing, chelator treatment, enzyme treatment, viability, yield, enzyme inactivation.
9. Scale-up; separation of carriers, colonization, suspension.
10. Productivity: media change, virus infection, multiplicity of infection, product concentration.

To help document your work, detailed microcarrier cell culture record sheets and problem-solving check lists are available from the authors.

7.8 Optimizing Culture Conditions

The quickest way to determine the key variables in a culture is to set up factorial design experiments. Feed a limited number of experimental results into mathematical models and evaluate the variables that require attention and further optimization. The fewer the cell lines, the more in-depth the optimization that can be done.

However, each recombinant construct has different growth properties and requirements. To fully optimize media utilization, analyze the amino acids and determine which factors in the medium are limiting. These factors can be added selectively, or the medium composition changed to better suit the cell type. If there is no time or resources for this analysis, use a surplus of medium to maintain steady-state conditions and to maximize growth rate.

7.9 Trouble-shooting

7.9.1 Stirred Microcarrier Cultures

Some problems may arise when working with stirred microcarrier cultures for the first time. The following list summarizes typical areas of difficulty and the most likely solutions. These points also form a useful check list when culturing new types of cells.

1. Medium turns acidic when microcarriers added.
   – Check that the microcarriers have been properly prepared and hydrated.
2. Medium turns alkaline when microcarriers added.
   – Gas the culture vessel and equilibrate with 95 % air, 5 % CO_2.
3. Microcarriers lost on surface of culture vessel.
   – Check that the vessel has been properly siliconized.
4. Poor attachment of cells and slow initial growth.
   – Ensure that the culture vessel is non-toxic and well washed after siliconization.
   – Dilute culture in PBS remaining after sterilization and rinse microcarriers in growth medium.
   – Modify initial culture conditions, increase length of static attachment period, reduce initial culture volume, or increase the size of the inoculum.
   – Check the condition of the inoculum and ensure it has been harvested at the optimum time with an optimized procedure.
   – Eliminate vibration transmitted from the stirring unit.
   – Change to a more enriched medium for the initial culture phase.
   – Check the quality of the serum supplement.
   – If serum-free medium is used, increasing attachment protein concentration may be necessary (fibronectin, vitronectin, laminin).
   – Check for contamination by mycoplasma.
5. Microcarriers with no cells attached.
   – Modify initial culture conditions, increase length of static attachment period, reduce initial culture volume.
   – Improve circulation of the microcarriers to keep beads in suspension during stirring.
   – Check the condition of the inoculum, especially if it is a single-cell suspension.
   – Check that the inoculation density is correct (number of cells/bead).
6. Aggregation of cells and microcarriers.
   – Modify initial culture conditions, reduce the time that the culture remains static.
   – Increase stirring speed during growth phase, improve circulation of microcarriers.
   – Reduce the concentration of serum supplement as the culture approaches confluence.
   – Reduce the concentration of Ca^{2+} and Mg^{2+} in the medium.
   – Prevent collagen production by the cells by adding proline analogs to the culture medium.

7. Rounded morphology of cells and poor flattening during growth phase.
   - Replenish the medium.
   - Check the pH and osmolality of the culture medium.
   - Reduce the concentration of antibiotics if low concentrations of serum are used.
   - Check for contamination by mycoplasma.
8. Rounding of cells when culture medium is changed.
   - Check temperature, pH, and osmolality of replemshment medium.
   - Reduce the serum concentration.
9. Cessation of growth during culture cycle.
   - Replenish the medium or change to a different medium.
   - Check that pH is optimal for growth.
   - Re-gas the culture vessel or improve supply of gas.
   - Reduce stirring speed.
   - Check for contamination by mycoplasma.
10. Difficulties in controlling pH.
    - Check that the buffer system is appropriate.
    - Improve the supply of gas to the culture vessel, lower the concentration of CO_2 in the headspace, or increase the supply of oxygen.
    - Improve the supply of glutamine, supplement the medium with biotin, or use an alternative carbon source, e.g. galactose.
11. Difficulties in maintaining confluent monolayers.
    - Check that pH and osmolality are optimal.
    - Reduce the concentration of serum supplement.
    - Improve the schedule for medium replenishment.
    - Reduce the concentration of antibiotics.
    - When culturing cell lines that produce proteases in a serum-free medium, it may be necessary to add protease inhibitors to prevent the cells from detaching (CHO cells have been shown to secrete proteases!).
12. Broken microcarriers.
    - Ensure that dry microcarriers are handled carefully.
    - Check the design of the culture vessel/impeller and ensure in case of a bottom-drive agitation system that the double mechanical seal is designed properly, e.g., no existing graps for microcarriers.
13. Difficulty in harvesting cells from microcarriers.
    - Ensure that the carriers have been washed extensively together with mixing.
    - Check that approximately the same amount of protease (U per cell) is used as when harvesting from flasks.
    - Check that the trypsin has not been thawed for too long (loss of activity).
    - Check that sufficient shear force is used in addition to the trypsinization.
14. Flotation of microcarriers in foam due to sparging.
    - Reduce the serum/protein concentration as much as possible.
    - Add pluronic F68 to decrease foaming.
    - Add polymers to increase viscosity.
    - Aerate via silicone tubing (Diesel, bubble-free aeration), via spin filter (New Brunswick, Celligen), via external loop (vessel, hollow fiber).

7.9.2 Fluidized Bed Trouble-shooting

1. Culture medium too acidic.
   – Expel CO_2 by using a sparger that creates large bubbles.
   – Add sodium hydroxide to titrate the pH, observe the osmolarity.
   – Try to increase buffer capacity.
   – Optimize the culture medium and oxygen support to avoid production of lactic acid.
2. Bridging of microcarriers.
   – Use higher circulation rates. This shortens contact time between the microcarriers. Bed expansion should be between 150 and 200 %.
3. No attachment.
   – Initial cell density too low.
   – pH level and dO_2 concentration were incorrect during inoculation.
   – Contact time between cells and the microcarriers was too short. More time is needed in packed bed mode.
   – Microcarriers were not washed properly.
   – Microcarriers were not equilibrated.
   – Cell inoculum was in stationary phase and not in the exponential growth phase.
4. Oxygen supply too low
   – Use microsparging technique with a sparger that creates small bubbles (pore size around 0.5 µm).
   – Use pure oxygen for gasing. The cells inside the pores of the macroporous microcarriers are protected against the toxicity of oxygen.
   – increase the circulation rate.

7.10 Applications

Well over 500 publications reflect the many applications of microcarrier technology and the great number of different cell lines cultured (Table 7-7). Today, its main industrial use is to produce vaccines, natural and recombinant proteins and, increasingly, monoclonal antibodies. An interesting minor application is its use in artificial organs (livers).

The number of applications run using microcarrier technology may be affected by the choice of producer cell and its glycosylation pattern. If more natural adherent cell lines are chosen, this will greatly increase the applications of the technology.

Table 7-7. List of cells grown on microcarriers.

Tissue	Cell line
Adrenal	Mouse cortex tumour – Y-1
Amnion	Human – WISH
Amniotic cells	Human amniotic fluid
Bone marrow	Human – Detroit 6
	Human – Detroit 38
Carcinoma	Human nasal – RPMI 2650
	Human larynx – HEp 2
	Human oral – KB
	Human cervical – HeLa
	Human colon
	Human thuroid
Conjunctiva	Human – Chang D
Cornea	Rabbit – SIRC
Endothelium	Rabbit coronary endothelium
	Human coronary endothelium
	Mouse brain capillary endothelium
	Bovine pulmonary artery endothelium
Epithelium	Human – NITC 2544
Fibroblast	Human foreskin – FS-4
	Human foreskin Detroit 532
	Human – SV40 – transformed WI-38
	Mouse – SC-1, 3T3, 3T6, L-cells, L-929, A9
	Mouse – transformed
	Mouse – embryo
	Chicken – embryo
	Human – embryo
	Rat – embryo
	Rabbit – embryo
	Human – Xeroderma pigmentosum
	Muntjac – adult skin
Fibrosarcoma	Human – HT 1080
	Mouse
Fish	Rainbow trout gonad – RTG
	Fat head minnow – FHM
	Carp epithelioma – EPC
Glial	Rat
Glial tumor	Rat – C6
Glioma	Human
Heart	Human atrial appendage – Girardi heart
Hepatoma	Rat – HTC, Morris MH_1C_1

Table 7-7. (continued).

Tissue	Cell line
Insect	*Drosophila* *Spodoptera* *Trichoplusia*
Kidney	Human embryo Human embryo – Flow 4000, L-132 Bovine embryo – MDBK Monkey – primary Dog – primary, MDCK, transformed Rabbit – primary, NZ White, LLC-RK RK-13 Rat – NRK, transformed Pig – PK-15, IBR Syrian hamster – HaK, BHK, transformed Potroo – Pt-k-1 Rhesus monkey – LLC-MK$_2$ African Green monkey – Vero, CV-1 BSC-1, BGM, GL-V3
Leukemia	Human monocytic – J111
Liver	Human primary hepatocytes Rat primary hepatocytes Chimpanzee Human – Chang liver
Lung	Chinese hamster – Don Chimpanzee embryo – CR-1 Human embryo – L-132, MRC-5, MRC-9, WI-38, IMR-90, Flow 2002, HEL 299 Cat embryo Bat – Tb 1 Lu Mink – Mv 1 Lu
Lymphoid	Human – lymphoblastoid Human – lymphocytes
Macrophage	Mouse – peritoneal, peripheral blood Rat – peritoneal Human – peripheral blood Mouse – P388D1
Melanoma	Human Mouse
Muscle	Chicken myoblasts Rat muscle-derived fibroblasts
Neuroblastoma	Mouse – Neuro-2a
Osteosarcoma	Human
Ovary	Chinese hamster – CHO

Table 7-7. (continued).

Tissue	Cell line
Pancreas	Rat
Pituitary	Rat
	Bovine
Rhabdomyosarcoma	Human – RD
Synovial fluid	Human – McCoy
Thyroid	Pig

7.10.1 Vaccines

A vast majority of vaccine producers in Europe, and many others world-wide, use surface microcarriers to produce live attenuated or inactivated vaccines for human and veterinary use (Fig. 7-36a;b) [43, 44]. Recently, they have begun to produce viral vectors used in gene therapy, adenovirus, and murine retroviruses, with both lytic and non-lytic viruses being produced. As the cells are eventually killed, natural batch or fed-batch processes are used. Normally low cell densities are cultured in stirred tank cultures for this purpose. One novel application is to use diploid MDCK epithelial cells to produce influenza vaccine [45].

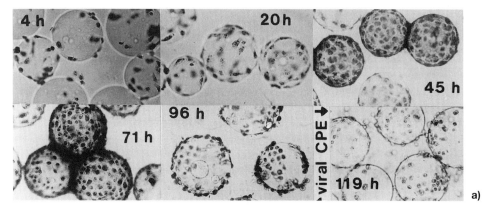

Fig. 7-36. (a) Vaccine production phases (4–45 h). Vero cell growth and virus production of Herpes simplex virus on Cytodex.

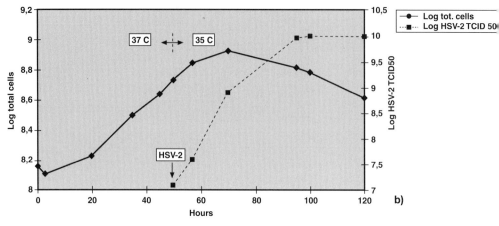

Fig. 7-36. (continued). (b) Virus production kinetics (reproduced from [43]).

7.10.2 Natural and Recombinant Proteins

A number of processes for naturally produced proteins are based on the culture of diploid cell lines on surface microcarriers in stirred tanks. Most new recombinant proteins are expressed in CHO cells. They attach and grow intially on surface microcarriers, but after some days aggregate and begin to fall off. The majority of CHO cells have been adapted for suspension culture and grown at fairly low cell densities in stirred tanks in batch or fed-batch cultures. Lately, however, some processes utilizing mar-oporous microcarriers to increase cell density have been developed. (Fig. 7-37a,b).

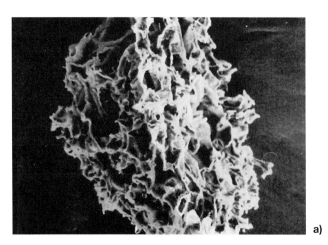

Fig. 7-37. (a) SEM micrograph from a cut of an empty Cytopore.

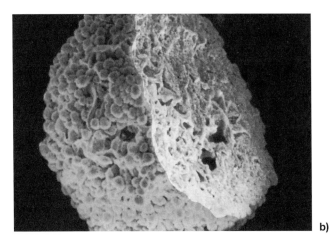

b)

Fig. 7-37. (continued). (b) of a confluent CHO culture.

This has in some cases led to increased productivity. In the report by Shirokaze et al. [46], a double productivity of r-I4 could be obtained with immobilized culture compared with suspension. Production was measured by ELISA over an 11-day period. The total productivity in suspension in 10 % calf serum was 2 mg, serum-free 1.8. In immobilized culture it was 3.8 and 3.2 mg, respectively. These run as perfusion cultures utilizing spin filters in stirred tanks. In addition, some processes are being set up using r-CHO cells in fluidized bed cultures up to 100 L reactor volumes (60 L fluidized bed volume).

7.10.2.1 Comparison of Carriers in Different Reactors (Packed Bed or Fluidized Bed Reactor)

In this experiment, a r-CHO expressing a recombinant protein was cultivated in a packed bed and a fluidized bed reactor and the productivity and functionality of the product compared (Fig. 7-38) [47]. Due to the better mixing in the fluidized bed, the cells have a better nutrient and oxygen supply (no channeling). Because of this homogeneous environment, a 10-fold increase in productivity could be observed in the fluidized bed.

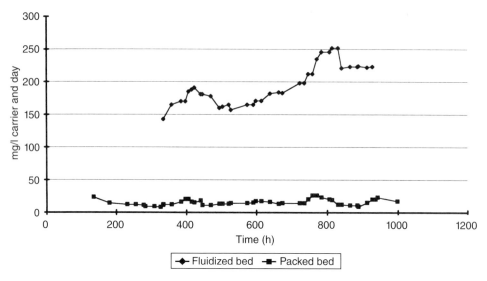

Fig. 7-38. Productivity comparison of packed bed versus fluidized bed.

7.10.3 Monoclonal Antibodies

The increased use of monoclonal antibodies in medical therapy requires cost-efficient production as the dose per patient is normally large. Packed bed technology was used for production in the past [48]. With the availability of scaleable fluidized bed technology, high productivity process can now be run over prolonged periods to produce active antibodies.

7.10.3.1 Comparison of Anti-HIV Monoclonal Antibody Productivity with Different Processes

These experiments were performed with the Xenohybridoma cell line 3D6/LC4 expressing a human anti-HIV-1 gp 41 antibody and a r-CHO cell line constructed at the Institute of Applied Mikrobiology, Vienna by co-transfecting CHO cells with two expression plasmids carrying the cloned cDNA of the heavy and light chains of the antibody. Cell number was determined by counting nuclei with a Coulter counter after treating the carrier with 0.1 M citric acid containing 1 % Triton X-100. Human IgG concentration was analyzed by anti-human IgG-ELISA. The stirred tank cultures were run as batch, fed batch, and continuous cultures using the hybridomas. The fluidized bed used hybridomas and r-CHO with Cytoline 1 and Cytoline 2 as the macroporous matrix. In addition, a stirred system using Cytoline 2 with r-CHO was run.

Table 7-8 shows two calculations of productivity: productivity per liter of carriers, and productivity per liter of reactor volume. The fluidized bed gave a 10-fold

Table 7-8. Comparison of HIV monoclonal antibody productivity with different processes.

Reactor	Cell density (cell number/ mL carrier)	Productivity/ L carrier (mg L^{-1}·day)	Productivity/ L reactor volume (mg L^{-1}·day)
Stirred tank reactor batch	6×10^5	2.5	2.5
Stirred tank reactor fed batch	$4-5 \times 10^5$	4.5	4.5
Stirred tank reactor continuous	$4-5 \times 10^5$	5.5	5.5
Fluidized bed Cytoline 1	4×10^6	40	20
Fluidized bed Cytoline 2	$5-7 \times 10^6$	110	55
Fluidized bed r-CHO, Cytoline 1	1.4×10^8	30	150
Stirred system r-CHO, Cytoline 2	6×10^7	150	37.5

increase over the stirred culture. Up to a 30-fold increase was obtained by combining CHO cells as the expression system for antibodies with fluidized bed technology and Cytoline microcarriers as the matrix for the cells.

7.10.3.2 Comparison of a Hollow Fiber Reactor with a Fluidized Bed Reactor

Comparison of a hollow fiber reactor with a fluidized bed reactor [49] showed that the former enables the production of IgA at high level (> 1 g L^{-1}) but with a negative effect on the fraction of active material. The percentage of active fraction was between 30 and 77 %. In the fluidzed bed reactor cultivated on macroporous carriers, the active fraction was higher than 80 % (personal communication).

7.11 Potential Future Applications

Applications currently being evaluated are cell expansion of blood cells (Fig. 7-39) via immobilization of hematopoietic stem cells in bioreactors, and the expansion of cytotoxic lymphocytes to generate sufficient cell numbers to be used in cell therapy. In some applications, cells are encapsulated inside a capsule (to protect them from the immune system) and then transplanted [50]. Macroporous microcarriers could easily be encapsulated when they are confluent with cells. Epithelial cells grown on Cytodex microcarriers are used for wound healing (burns) [51]. Growing cells on (degradable) microcarriers and using the entire cell/carrier complex in transplantation is also discussed in the report by Schugens et al. [52].

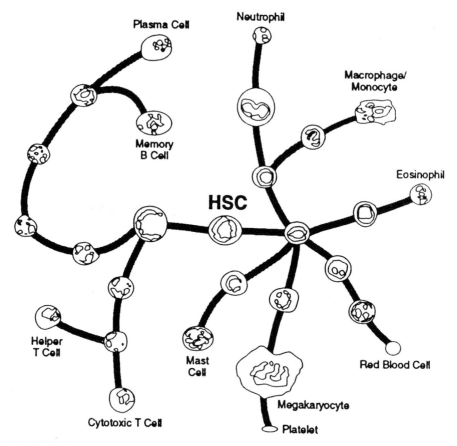

Fig. 7-39. Hematopoietic stem cell (HSC) expansion [50].

References

[1] Naveh, D., *Japanese Association for Animal Cell Technology,* 1995.
[2] Griffiths, J. B. (1992) in: *Animal Cell Culture,* Freshney, R. I. (Ed.), 2nd ed., Oxford University Press, 1992, pp. 47–93.
[3] Maroudas, N. G., *J Theoret Biol,* 1975, *49,* 417–427.
[4] Carrel, A., *J Exp Med,* 1923, *38,* 407.
[5] Butler, M. (1988), in: *Animal Cell Biotechnology,* Vol. 3, Spier, R. E., Griffiths, J. B., (Eds.), Academic Press, Inc, pp. 284–300.
[6] Keller, J., Dissertation, ETH, Zürich, 1991.
[7] Lehmann, J. H., Piehl, G. W., Schulz, R., *Dev Biol Stand,* 1987, *66,* 227–240.
[8] Spier, R. E., Whiteside, J. P., in: *Animal Cell Biotechnology,* Vol. 4, Spier, R. E., Griffiths, J. B. (Eds.), Academic Press, pp. 123–132.
[9] Reiter, M., Zach, N., Gaida, T., Blüml, G., Doblhoff-Dier, O., Unterluggauer, F., Katinger, H., in: *Animal Cell Technology: Developments, Processes and Products,* Spier, R. E., Griffiths, J. B., MacDonals, C. (Eds.), 1992, Butterworth-Heinemann Ltd, Oxford, pp. 386–390.

MAbs. However, they are a relatively inexpensive (and hence attractive) option for making MAb fragments where the only required function is antigen binding, for example in enzymatic catalysis and immunoaffinity applications. Recombinant MAbs have been expressed in various mammalian cells. Myeloma cell lines appear to offer the best expression, but Chinese hamster ovary (CHO) is also widely used for production of therapeutic MAbs because it is such a microbiologically clean and well-characterized cell line. Expression in yeast has also been attempted [6] but the yields were low and glycosylation of MAbs in yeast will almost certainly be different from that which occurs in human cells.

Glycosylation can be an important issue for any expression system used to make a MAb for therapeutic use. There are significant differences in Mab glycosylation patterns between mammalian expression systems, and even between the same cellular expression system under different cell culture conditions. These can give rise to rapid clearance or antigenicity *in vivo*.

8.3 Purification

From a regulatory point of view, biological products and chemical entities have been approached somewhat differently from each other and this has had important consequences for the purification and characterization of biological products. Increasingly, as higher-purity protein products are developed and more discriminating protein characterization methods are perfected, this distinction is disappearing. Formerly it was accepted that it might not be possible to fully define a biological product and it followed from this that the consistency and control of the manufacturing process was critical to ensure product consistency. Now the emphasis is shifting to the concept of the well-characterized molecule which allows for demonstration of product consistency despite changes in purification process. This, in turn, allows greater flexibility of purification method during product development and diminishes the need to establish a suitable purification scheme early which can be scaled-up with identical performance.

8.3.1 Initial Considerations

The following points must be considered when devising a purification scheme for a MAb.

8.3.1.1 Intended Use

The first consideration in the choice of purification strategy is the required specification: what purity is required for a given application? If the MAb is to be used as a therapeutic or *in vivo* diagnostic (imaging) agent, then not only must it be of high purity, but it must also be free of potentially harmful trace contaminants such as viruses, bacterial toxins, cytokines, and DNA (though regulatory requirements are relaxing on the last). These requirements also apply to MAbs used in the immunoaffinity purification of therapeutics. In these cases the concept of purification factor is largely inappropriate, and the performance of a purification step is measured in terms of clearance factors for defined contaminants. The required clearance factors are usually so large that they can only be achieved using chromatographic methods of purification. The performance of precipitation methods is inherently limited by interstitial supernatant in the precipitate. They are, however, often used in the early stages of purification schemes, followed by chromatographic steps.

For *in vitro* diagnostic or research tools there may only need to be minimal [7–9] or no [10] purification. Important components of the specification of the purified product in this case will be the affinity of the antibody and absence of cross-reactivity in an assay.

8.3.1.2 Culture Method

The earliest MAb preparations were made in mouse ascitic fluid and this method of production produces a relatively concentrated feedstock which may be preferable for production of material on a small scale for diagnostic or research purposes. Larger amounts are now usually produced in a cell culture system which may give a very dilute product requiring concentration before further processing. The type of production system used has another profound implication for downstream processing as it determines the nature of contaminants present.

8.3.1.3 Contaminants

Mouse ascitic fluid has a high protein concentration [11] of which ~ 40 % may be the MAb of interest [12]. It will contain other mouse IgG, high levels of lipids, and possibly also infectious agents from the mouse. A summary of the types of contaminant which might be encountered in MAb cell culture systems is shown in Table 8-1, although cell culture supernatants may vary widely in the type of contaminants they contain. It is an advantage if the cells can be grown in serum-free medium as this will mean that the problem of separating bovine IgG (introduced as a component of fetal or new-born calf serum) from the MAb does not have to be addressed. There are likely to be other proteins, such as albumin and transferrin, present, as well as many small molecular weight nutrients, waste products, and lipids. Many cell lines, even human ones, may have been fused with, or exposed to, other mammalian cell lines at some point in their development and infectious agents (particularly viruses) which may have been derived from this source must be considered. Some cell lines may have been transformed with a virus such as EBV. If a product is to be used therapeutically, extensive screening of the cell lines for adventitious agents is obligatory and this will give invaluable information for designing and vali-

Table 8-1. Potential contaminants in cell culture supernatants.

Component	Potential contaminants	
	Serum-supplemented medium	Serum-free medium
Water	Trace elements Organics Endotoxins	Trace elements Organics Endotoxins
Powdered media	Minimal nutrients Phenol red Endotoxins	Full nutrients Phenol red Endotoxins

Proteins	From animal sera:
	  Albumin (50–60 %)
	  Immunoglobulins ($\sim$ 10 % in
	    whole serum, < 0.1 % in fetal
	    calf serum)
	  Protease inhibitors ($\sim$ 10 %)
	  Transferrin (2–5 %)
	  Lipoprotein (1–2 %)
	  Peptide hormones (< 0.1 %)
	  Misc. proteins (20–35 %)
	  Proteases
	(Total 30–50 mg mL^{-1})
		Albumin (0.4–0.5 mg mL^{-1})
		Transferrin (0.03 mg mL^{-1})

Other components	Viruses from animal sera
	Lipids:
	  Cholesterol
	  Triglycerides
	  Phospholipids
	Steroids
	Vitamins
	Sugars
	Trace elements
		Inorganic salts
		Glucose
		Vitamins
		Lipids:
		  Lecithin
		  Cholesterol
		2-mercaptoethanol
		Amino acids

Hybridoma cells	Secreted proteins (50–200 µg mL^{-1}) Cellular proteins/debris Nucleic acids Viruses	Secreted proteins (50–200 µg mL^{-1}) Cellular proteins/debris Nucleic acids Viruses

dation of the purification process. As cell culture produces a relatively dilute MAb preparation, water may be regarded as a major contaminant.

8.3.1.4 Scale

The scale at which the final purification is to be run is important in determining the purification system which may be used. There are very few, if any, purification methods which cannot be scaled-up. However, methods differ enormously in their ease of scale-up. For example, one would not normally consider slab gel electrophoresis capable of practical scale-up. Precipitation methods are simple to apply in the laboratory, since bucket centrifuges are a standard item of laboratory equipment, but on scale-up the only practical solution may be a flow-through centrifuge, which at large scale is a complex and very expensive item. On the plus side, it may also be used for a cell separation/clarification process step. In contrast, the scale-up of chromatographic methods is relatively straightforward. The consideration of future process scale-up at the development stage should include an appraisal of GMP and engineering problems [13] at the final scale where appropriate.

8.3.1.5 Cost

The cost consideration in choosing a MAb purification scheme has a number of components associated with it. First there are equipment costs, for example centrifuges, chromatography columns, pumps, monitors, and ultrafiltration equipment. Second, the cost of the consumables used in the purification process, which in the case of affinity chromatography media for example can be considerable. Third, the time and cost of developing a purification process. In the case of a MAb for therapeutic use, years of investment in process development and validation go into bringing a product to market, when the investment can be recouped. Anything which can shorten this lead time is worth considering, sometimes even at the expense of process equipment costs. The researcher who may be simultaneously developing several MAbs cannot afford to spend a lot of time developing purification methods for each one. These considerations tend to favor generic methods for MAb purification, methods which can be translated from one product to the other with a reasonable degree of confidence not only that they will work, but that they will need little optimization, and one can easily build on the experience of others.

Examples would include ammonium sulfate precipitation and protein A affinity chromatography. Other chromatography techniques are likely to need significant modification for each MAb due to factors such as the isoelectric point, surface hydrophobicity, and pH stability of the individual MAb.

8.3.2 The Purification Scheme

The purification scheme can be divided into three stages. First is removal of cell debris, which may also include removal of excess water. The second stage is the major purification from gross contaminants in the feedstock. Finally, the third stage may be a polishing step if this is necessary for the intended use of the product. Although a simple purification scheme is highly desirable, most therapeutic applications of MAbs require several steps to produce the necessary degree of freedom from contaminants to meet regulatory requirements. The type of purification step which may be used at each stage is summarized in Table 8-2.

The early stages in a purification strategy will aim to achieve both concentration and purification. For some methods of MAb production, the first step will be clari-

Table 8-2. Purification techniques which may be used at different stages of a process scheme.

Stage	Clarification	Primary purification/capture	Further purification	Polishing
Purification technique	Precipitation Centrifugation Filtration Aqueous two-phase partition	Affinity chromatography Ion-exchange chromatography Hydroxylapatite chromatography Thiophilic chromatography Immobilized metal ion chromatography Precipitation Hydrophobic interaction chromatography	Ion-exchange chromatography Hydrophobic interaction chromatography Gel filtration chromatography	Gel filtration chromatography Diafiltration Ultrafiltration

fication by removal of cells and cell debris, although this may not be necessary, for example, if the culture supernatant has been produced in a hollow fiber cell culture system. The next stage will be the major purification step and will often also be a concentration step, unless a previous clarification step has achieved the required degree of concentration.

The number of steps included in the third stage of further purification and final polishing will depend on the intended use of the MAb. These steps will be designed to reduce contaminants such as DNA, host cell protein, endotoxins, process chemicals, and others such as protein A which may have been introduced early in the purification scheme. Usually, different stages will incorporate different techniques to maximize discrimination between the Mab and the contaminants. For example, if protein A affinity chromatography is used at stage two, an ion-exchange column can be used subsequently to separate the MAb from leached protein A by exploiting differences of pI between the proteins [14]. Polishing steps might include a buffer exchange by gel filtration or diafiltration.

Other steps may be introduced at any stage specifically to prevent virus contamination of the product. Any purification scheme for a therapeutic product will have to include more than one such step and again these must be different in virus reduction mode to maximize their effectiveness. Specific steps often used include solvent/detergent inactivation, which is effective for lipid-enveloped viruses, and virus filtration which is most effective for medium-sized or large viruses [15]. Viral clearance may also be demonstrable during capture chromatographic steps and some elution conditions, such as low pH, may give useful levels of inactivation of some viruses [16].

Most MAbs referred to in the literature are sourced either from human or rodent cell lines, or heterohybridomas of these, although other types of expression system are increasingly being used, particularly for immunoglobulin fragments. The majority of the cell lines secrete IgG antibodies, although a substantial minority produce IgM, and occasionally references to other classes are found. The purification methods discussed below are generally applicable to IgG molecules; where they are equally or more appropriate for other classes of antibody, this is specifically mentioned.

8.3.2.1 Clarification/Concentration

Downstream purification of MAbs produced in mammalian cell lines is simpler than the purification of many other products of cell culture processes as the required protein is secreted from the cell in a soluble form. Thus, cellular disruption is not required. Centrifugation or filtration are the two most common methods of removal of cells and naturally occurring cell debris from tissue culture supernatant [17–19]. With continuous-harvest culture systems, such as hollow fiber cell culture, filtration can be performed in-line during supernatant collection. An aqueous two-phase extraction procedure has also been used [20].

Most of the commonly used cell culture systems for MAbs result in cell suspensions in which the MAb is in solution at very low concentrations. At industrial scale this can amount to thousands of liters; hence it is desirable on economic and practical grounds to reduce the process volume at the earliest possible stage. It may in fact be essential for the performance of a precipitation step. Concentration can be done by ultrafiltration, which at large scale can also be a very expensive piece of equipment, but if one uses a tangential-flow ultrafiltration system, the same equipment (if not the same membrane) can also be used for cell separation. The reduction of process volume can be achieved very efficiently using a chromatographic capture process. Ion-exchange capture is an obvious choice, though many MAbs will not bind to either cation or anion exchangers under physiological conditions of pH and ionic strength, so the cell culture supernatant has to be conditioned by a combination of dilution and pH adjustment, or by diafiltration. Conditioning is generally not required for protein A or protein G affinity capture, though the disadvantage here is the much greater cost of the media, which may have a very limited lifetime under these process conditions. Conventional column chromatographic capture is not the preferred method for handling solutions containing particulates, which will foul the column bed and greatly reduce the gel lifetime. Methods have now been devised which allow cell-containing feedstocks to be loaded directly onto chromatography columns, thus combining clarification and purification in one step. These include expanded bed adsorption [21] and fluidized bed adsorption [22], and such systems could be applied to the production of MAbs.

8.3.2.2 Chromatography

Chromatography is a widely used technique for the purification of MAbs [11,23]. Affinity chromatography using protein A and protein G media is widely used for purification of MAbs, since it can easily achieve clearance factors of > 1000. However, in the case of MAbs for therapeutic use, one must be sure that the affinity ligand which leaches off the matrix in small amounts does not end up in the product. This is usually achieved by adding a chromatography stage after the affinity step, typically ion exchange. A conventional biochemical high-purity standard is sufficient for most of the other applications for MAbs, which again leads one towards chromatographic methods. For many research reagent applications a clarified cell culture supernatant, or at most the product of a simple precipitation purification step, is all that is required. However, the cruder a preparation is, the more likely it is to be unstable due to protease activity, either from the parent cells or from bacterial contamination. The major advantages of chromatography are that it can be readily scaled-up and is one of the easier techniques to operate to GMP.

Affinity Chromatography

This is a very powerful purification technique which can give high single-step purifications [24]. The most common types of affinity ligands in use for MAb production are immobilized bacterial cell wall proteins, e.g., staphylococcal protein A [25] or streptococcal protein G [26]. These proteins are available commercially both as the free proteins and immobilized on a variety of chromatographic supports. One of the limitations of this technique may be the affinity of the ligand for immunoglobulins of different species or of different class within one species. For example, mouse IgG1 MAbs have low affinity for protein A and, within human IgG subclasses, IgG1 binds to protein A but IgG3 does not. This problem may be overcome in some cases by enhancing binding with increased ionic strength of the loading buffer and the MAbs can then be eluted at near-neutral pH.

These bacterial proteins interact primarily through the Fc region of the IgG molecule, although both protein A and protein G have also been shown to interact with regions in the Fab part of the molecule and thus they are also used for purification of IgG fragments [27,29]. Both protein A and protein G are available in recombinant forms which possess advantages over the native protein, for example the deletion of the albumin binding site. These ligands give a good purification, especially if the feed stock contains no other IgG than the MAb of interest. However, there is always a small degree of leakage of the protein ligand and, for therapeutic products, further purification will be required. The other often-cited disadvantage of an immobilized protein ligand is its inability to withstand the kind of sanitization regimes frequently used in GMP applications. Nevertheless, these ligands can be used repeatedly with applications of a chaotropic agent such as 6 M guanidinium chloride with no loss of capacity [16].

Although proteins A and G are the most frequently used ligands for affinity chromatography of MAbs, the disadvantages mentioned above have led to the development of alternatives; for example the specific antigen for the desired antibody [30–32], a small molecular weight, non-protein affinity ligand marketed as Avid

AL [33], which also interacts with the Fab and Fc regions of the IgG, and Protein L, another bacterial cell wall protein, which has been used for preparation of kappa light chain [34]. Each of these has their own set of advantages and disadvantages. Using the specific antigen neatly overcomes the problem of purifying a MAb from a mixture containing other IgGs. On the other hand, a new chromatography gel has to be created for each MAb to be purified, which will be costly, and the conditions used to elute the MAb from the gel may have to be as harsh as the elution conditions for proteins A or G. Also, the antigen may be toxic at very low concentrations requiring a high level of clearance of any leached ligand.

Elution of the MAb from immobilized protein A or protein G is usually by a change in pH. Human MAbs require a low elution pH, although some murine MAbs can be eluted from protein A at near-neutral pH. A pH as near to neutral as possible should be used to minimize protein denaturation and, in any event, the pH of the recovered solution should be neutralized as quickly as possible. In this respect, Avid AL has an advantage as elution at neutral pH is possible, although it may have other disadvantages in binding charged substances from tissue culture medium such as phenol red [35]. A method for elution of MAb from protein G at alkaline pH has been reported [36].

Ion-exchange Chromatography
This is another powerful, well-established technique which separates molecules on the basis of charge [37]. Proteins can acquire either a positive or a negative charge by manipulation of pH either side of their pI and will then bind to immobilized groups with the opposite charge. Desorption from an ion exchanger is usually by increasing ionic strength but may also be by changes in pH.

Despite the structural commonality of all MAbs, they show a wide range of pIs. Thus, both cation- and anion-exchange chromatography can be used for the purification of individual Mabs [38]. When anion-exchange chromatography is used for IgG [12,39] or IgM [9] purification, a mixture of proteins are bound to the column and are selectively desorbed by gradient elution. This can be an inefficient use of the chromatography resin. In addition, phenol red, which is often present in cell culture supernatants, will bind to the anion exchange gel and will further decrease the capacity and could interfere with elution. Therefore it may have to be removed from the feedstock by pre-treatment. Cation-exchange chromatography for IgG [40,41] or IgM [42] can be manipulated so that only the MAb will bind, decreasing the unit volume of gel required. In this case elution will need less critical control, making scale-up easier. Ion exchange may also be used in a non-adsorptive mode in which the contaminants are bound to the gel while the MAb elutes in the flow-through [43].

A mixed mode ion-exchange resin [44] has been used for murine and rat IgG MAbs and good purity and recovery have been reported. This system has also been used for IgM [45] and for bispecific Mabs [46].

Although ion exchangers have advantages in GMP applications where they can be sanitized under extreme acid or alkaline conditions, when used as a primary capture step, this purification method may need separate optimization for each MAb and the cell culture supernatant may need manipulation (for example in ionic strength) before it can be applied to the gel. As molecules are separated on the basis of charge and charge distribution, it may be difficult to completely eliminate contaminants with similar pI to the MAb.

Other chromatography systems
These systems, which have been used for purification of MAbs, include thiophilic chromatography [47–49], hydroxylapatite [50,51], immobilized histidine ligand [52], and immobilised metal affinity chromatography [31,53]. Hydrophobic interaction chromatography is also used as a primary capture step [54] as well as a later step in purification schemes [12,55,56].

Mode
On the larger scale, a fast chromatographic throughput is required due to the large volumes to be processed. Chromatography supports such as zirconium oxide [67] and Poros™ [58] have been developed to allow operation under increasingly high flow-rate without significantly increasing back-pressure. Factors to consider in setting an upper limit for flow-rate in any chromatography system include the rate of diffusion into the bead pores for a gel (often rate limiting), and the rate of the binding reaction for any capture process (usually relatively fast). With this in mind, chemistries on membrane surfaces have also been developed and applied [8,59,60].

8.3.2.3 Other Purification Systems

Although chromatography is widely used as a purification tool, precipitation systems particularly ammonium sulfate [12,61] and polyethylene glycol (PEG) [62] precipitation or a combination of the two [63] are also used, often in the first stage of processing, but sometimes also as a major purification step. Precipitation is particularly applicable to IgM preparation and to purification of more concentrated ascites fluid as it is easier to perform on a small scale. Preparative isoelectric focusing is also used for some applications, but is not very amenable to scale-up [64].

8.3.3 Large Scale

There is no easy definition of large scale in the production of MAbs. The defining issues may be the use of the antibody and the effective dose.

8.3.3.1 Scale-up

Factors which have to be taken into consideration when planning scale up include:

- reproducible performance
- operation of process equipment to required hygiene and safety levels
- effect on processing time and buffer volumes

Any purification scheme other than those for research purposes will be developed on a small scale but finally operated on a larger production scale. There must be reproducibility of process performance over this scale-up for three major reasons. First, it is important that investigations with purified product carried out early in the development are relevant to the final product. Second, it is usually desirable to carry out the majority of the validation work on a smaller scale than the final process scale.

Table 8-3. Retention times of product peaks during purification of MAbs from two cell lines.

(i) Cell line 1; small scale

	Peak retention time (min)		
	C1[a]	C2 Product	C3
Mean	610.2	85.1	45.7
SD	2.0	2.3	2.5
n	14	14	12

(ii) Cell line 1; 5 × scale-up; linear dimensions preserved

	Peak retention time (min)		
	Column 1[b]	Column 2	Column 3
Mean	617.1	86.3	54.1
SD	1.4	0.9	0.7
n	4	4	4

(iii) Cell line 2; small scale

	Peak retention time (min)		
	Column 1[a]	Column 2	Column 3
Mean	612.9	154.5	52.0
SD	2.5	5.3	2.8
n	8	8	8

(iv) Cell line 2; 5 × scale-up; linear dimensions preserved

	Peak retention time (min)		
	Column 1[b]	Column 2	Column 3
Mean	614.3	155.2	61.8
SD	4.0	2.7	3.0
n	4	4	4.0

[a] Retention time corrected for load time.
[b] Retention time corrected for load time and changes in pump wash.

Third, the financial viability of the project will need to be predicted from yields at the earlier, small scale and if the yield changes over scale-up this may adversely affect the project.

Reproducibility of performance can be measured in several ways, including by yield or levels of contaminants in the product. With a chromatographic process, one measure of performance is the retention times of the peaks eluted from the columns. In a purification process that we have developed which consists of three chromatographic steps, these retention times have been very consistent over five-fold scale-up (Table 8-3), and it is this level of consistency which is a major argument for the use of chromatographic processes for large-scale MAb production. Discussion of the parameters which have to be taken into consideration when planning a large-scale chromatography system can be found elsewhere [65–67].

Factors which might affect the hygiene and safety of the product and which might change over scale-up include the design of liquid handling equipment, such as valves and pumps, and also the nature of the materials in contact with the process flow. Changes in processing time with scale-up may cause changes in the nature of the product if the length of exposure to denaturing conditions is changed. They may also affect the costing of the product through antibody losses and increased overhead expenses.

8.3.3.2 GMP and Validation

GMP and validation issues would fill a chapter on their own. Suffice it to say that it is essential to follow the relevant regulatory guidelines such as FDA *Points to Consider in the Manufacture and Testing of Monoclonal Antibody Products for Human Use* [68] or *CPMP Notes for Guidance: Production and Quality Control of Monoclonal Antibodies* [69] and to resolve the points raised for each particular MAb. These will certainly include viral safety, residual levels of DNA, host-cell protein, process chemicals, and the presence of any other adventitious agents in the product. Sufficient control of the process to demonstrate reproducibility, including an evaluation of the effect of expected run-to-run changes in process parameters, will also be required.

8.3.3.3 Control and Automation

Increasingly, the issue of control of the production process in all protein purification schemes – including those for MAbs – is being addressed by replacing human operators with an automated system [70–72]. Provided that the systems can be shown to be reliable, a large amount of data can be collected for each purification to provide assurance that the process proceeded as expected and was within operating limits. Unattended operation is sometimes also possible. Chromatographic processes lend themselves particularly well to this level of automation and this may be one reason for their increasing adoption for production-scale purification of protein therapeutics.

8.3.4 Future Developments

It seems likely that future generations of MAbs will be produced via a route that makes increasing use of genetic manipulations. Already MAbs have been produced which have had affinity tails grafted onto the protein sequence to facilitate purification [73,74]. MAbs can now be produced by raising antibodies in an experimental animal and then grafting the gene sequence for the immunoglobulin hypervariable region onto human IgG frameworks. Human antibodies can also be generated from a non-immunized phage display library [75], which is a useful technique for antibodies to toxic antigens. Such MAbs or MAb fragments may be produced in a variety of cell lines, both mammalian and non-mammalian [31,76], although where glycosylation is important in determining functionality a mammalian cell line may be preferable. For some applications, production of immunoglobulin fragments may be sufficient or even desirable, and in these cases use of a genetic construct is the preferred route [28]. For example, it has been suggested that the use of F(ab)$_2$ fragments in an ELISA would reduce non-specific interactions [77]. Fab and F(ab')$_2$ fragments could be used to target radioisotopes to tumors; their small size in relation to intact immunoglobulin should allow greater tumor penetration and they are also cleared rapidly [29]. Bispecific MAbs could also have applications in therapeutics, for example forming a bridge between a tumor cell and a cytotoxic cell and thus inducing the destruction of the tumor cell [31].

Chromatography or similar adsorption processes will continue to be widely applied to the purification of such MAbs. Expanded bed adsorption chromatography is being used for MAb purification [78] and other new systems, such as displacement chromatography [79] and novel small molecular weight ligands [80], may be developed in the future for the processing of MAbs.

The increasing capability for full characterization of the MAb product will have important consequences for purification, and developments in this area are discussed in the next section.

8.4 Characterization

8.4.1 The Need for Characterization

The level of characterization required for a monoclonal antibody will inevitably reflect the intended use. Whereas for a research application the basic information relating to the class, antigen specificity, and any cross-reactivity may be all that is needed, for commercial applications – and in particular where a therapeutic use is envisaged – a substantial amount of detailed molecular and functional data will be needed. We will concentrate in this discussion on the latter category.

Biotherapeutics derived from natural sources are intrinsically heterogeneous (as for example with the immunoglobulin products used in passive immunotherapy which are derived from thousands of individual human plasma donations and where an absolute characterization of the product pool would be impossible). Even where a single active agent was know, as for example with clotting factors or hormones, there was always a range of microheterogeneity due to the polymorphic expression of the product in a start pool derived from multiple individuals.

Such products required, in order to ensure efficacy and safety, a lot-by-lot testing of batches within acceptable *in vivo* models, which was costly and which has become less and less ethically acceptable. In contrast, chemical drugs have been characterized predominantly upon their physico-chemical properties with modern and rigorous molecular methodologies being applied to the analysis of purity, homogeneity, and potential contaminants.

With the advent of biotechnologically derived agents and advances in high-resolution analytical tools it has become possible to define more accurately the efficacious agent in biological parenterals. As a result, in November 1995 the FDA issued revised guidelines in terms of what it described as 'well-characterized' or 'well-specified products'. These products are those for which a range of robust validated physico-chemical and functional assays could be applied and where the structure and activity of the agent was well understood. Monoclonal antibodies were one such group of biotechnological products specified by the FDA and in the present discussion of the characterization of monoclonal antibodies this strategy will be referred to. In February 1997, a further revision of the *Points To Consider for the Manufacture and Testing of Monoclonal Antibodies for Human Use* was issued [68].

8.4.2 Aspects of MAb Characterization

8.4.2.1 Primary Sequence

Knowledge of the cDNA sequence and the corresponding post-translational amino acid sequence are crucial to the characterization of the MAb. The sequences of rodent and human myeloma-derived antibodies had been much studied and were available for comparison when assessing the sequences of new agents. Data on human MAbs were more limited and the means for their expression less amenable in general to large-scale production, though consensus human immunoglobulin sequences were also available for most of the subclasses and Gm allotypes [81]. Therefore, it was comparatively straightforward to assess whether or not the MAb under study conformed to the normal consensus sequence of its species immunoglobulin. However, with the advent firstly of chimeric and then humanized antibodies which have been specifically altered either at particularly immunogenic or functionally important sequences, the checking of the sequences at these crucial regions of the molecule became an important control in the assuring the successful exploitation of these MAbs. Indeed, the changes in three-dimensional structure and the effects on immunogenicity and efficacy were not always as predictable as had been at first

expected and much work was required in comparing these humanized antibodies to their original non-human immunoglobulin counterparts [82].

With the advent more recently of engineered antibodies and antibody fragments, the genes that code for the antigen-binding domains may have been expressed through several different host systems during their manipulation before expression of the intact MAb or engineered fragment has been arrived at in its production-scale host-cell line. For instance a pair of heavy and light chain variable region genes may have been initially selected from a bacterial host system using filamentous phage, then expressed at a larger scale after selection to ensure *in vitro* antigen recognition. Following this they will have been combined with constant region genes from the desired immunoglobulin class and subclass and eventually presented in a mammalian cell line for large-scale production and detailed study of *in vivo* efficacy. At each stage silent errors in transcription or translation may have occurred, only to show up as potential problems once the intact molecule is evaluated in its intended target or model system.

It is thus essential to have the product of the Master Cell Bank characterized in detail in terms of both the cDNA sequence and the translated protein product. Any changes in the sequence of the MAb should be investigated during the development of a clinically relevant antibody, and tests should be performed to confirm that the products of each large-scale culture conforms to that of the Master Cell Bank on which the bulk of the characterization has been performed. The techniques to perform these analyses are well documented and known in the laboratory but require automation realistically to tackle the substantial task to the necessary level of assurance and within a short timescale.

Where an immunoconjugate is the intended final product, whether it is a radionuclide complexed within a chelating ligand or a fusion protein with a larger entity such as a molecular toxin or hybrid effector molecule, the task of post-translational sequence analysis becomes even more important in order to check that the complexation has occurred, measure the proportion of unmodified antibody molecules or free ligand that remain, and identify any undesirable side reactions that may have occurred. These are particularly important checks if the complexed molecule is potentially toxic. Radio-labeled monoclonals and antibody fragments frequently use a small number of suitable radio-isotopes, indium-111, iodine-125 and -131, technetium-99, and yttrium-90. These molecules are usually attached to antibodies by cross-linkers such as 2-iminothiolane [83] or chelators such as diethyl triamino penta-acetic acid.

8.4.2.2 Sequencing Strategies

cDNA

DNA sequencing technologies (such as the Sanger method) have been well defined and indeed automated systems for DNA sequencing have been available for several years. The assignment of primary protein sequence deduced from the cDNA sequence is now more easily performed than direct protein-sequencing methodologies. However, the gene sequence alone tells one nothing of the post-translational modifications that occur in proteins after transcription of the mRNA. Although some potential glycosylation sites are detectable from the cDNA sequence, there is no guarantee that these sites will actually be occupied in the mature protein. Hence it is necessary while initially identifying the primary sequence from DNA methodologies to supplement this with selective peptide sequencing methods in order to obtain the full picture of the protein structure.

Amino Acid

For antibodies, the size of the light and heavy chains precludes direct protein sequence analysis, such has been possible for the smaller protein biotherapeutics. Instead, sequencing must be preceded by a controlled hydrolytic process and the resultant peptides separated and collected prior to analysis. Due to the distribution of suitable target amino acid residues digestion using proteolytic enzymes, primarily trypsin, has been the strategy most commonly adopted [84]. Use of specific chemical digestions (such as with cyanogen bromide) have also been reported [85]. Limited digestion might reveal altered sequences as anomalously migrating peptides in electrophoretic profiles. N-terminal sequences blocked by cyclization of the glutamine residues can be treated with pyroglutamate aminopeptidase to remove the pyroglutamyl residue, or the blocked peptide can be analyzed by mass spectrometry (MS) after cleavage from the protein and the sequence deduced from the mass [86], again taking into account the possible multiplicity of peptides that could arise from partial blockage, affecting only a subpopulation of the antibody molecules present.

Initial studies have followed the work on the sequencing of normal immunoglobulin in requiring the separation of the heavy and light chains and their independent digestion and sequencing. More recently, probably due to the excellent separation capabilities of modern reverse-phase HPLC, the digestion of whole molecules of antibody have been demonstrated [87]. This can yield 90 or more peptides which must then be separated and sequenced.

8.4.2.3 Peptide Mapping

Whereas the complete primary structure confirmation should be made once for a 'reference' lot of MAb, there is a role for limited sequence analysis as a batch-to-batch consistency test and possibly as a stability-indicating method. Proteolysis can be achieved by enzymatic means usually using TPCK-treated HPLC-purified trypsin or other sequencing grade enzymes (Lys-C, Asp N, V8 protease). Alternatively, partial chemical degradation may be useful, although only CNBr cleavage at methionine or BNPS skatole cleavage at tryptophan are sufficiently specific to be of value. Rather than obtaining exhaustive sequences, these peptides can be assigned and changes resulting from chemical modification inferred often merely by amino acid analysis or most recently by MS. In a recent review of peptide mapping [88], the advantages of computer-assisted modeling and the use of photodiode array detection to provide full UV spectral information are discussed (Fig. 8-1) as well as the application of capillary electrophoresis for peptide separation and the

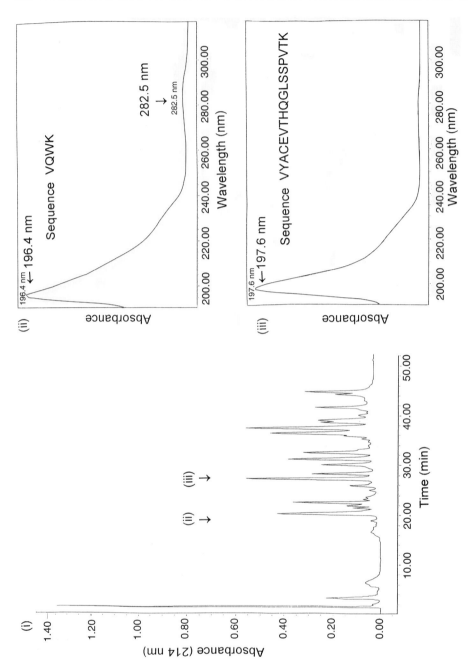

Fig. 8-1. Value of spectral analysis in peptide mapping. The reverse-phase peptide map (i) of the tryptic digest of a human monoclonal antibody light chain is shown. The spectra of two peptide components are shown (ii and iii) and compared with their known sequences. The contribution at 280 nm due to tryptophan is detected in (ii). (Sequences confirmed courtesy of Dr Will Mawby, Bristol University, UK.)

identification of the peptides by MS. A paper has appeared from a consensus of pharmaceutical firms in the field [89] on the validation of peptide mapping methods. Among the important issues they highlighted were the use of spiking with synthetic peptides to establish quantitative recovery of peptides applied in the profile generated, and the use of a digest of reference antibody to be compared with the profile of the test sample. Additionally, more conventional validation issues such as establishing reagent quality and column consistency were discussed. A fully validated peptide map has been deemed to be an acceptable tool for the assessment of the genetic stability of biotechnological medicines. Precipitation of insoluble peptides out of solution may result in irreproducible peptide maps and the presence of denaturants such as guanidinium chloride or urea may help. By running peptide maps of digests pre- and post-treatment with disulfide-reducing agents the presence of the disulfide bonds can be inferred. Automated peptide mapping has been reported recently [90] with the automation helping to reduce inter-digest variability.

8.4.2.4 Mass Spectrometry of MAbs

Mass spectrometry of the whole MAb molecule has been documented using electrospray ionization [91] and also by matrix-assisted laser desorption ionization-time of flight (MALDI-TOF) [92]. Mass spectrometry has been widely used to infer the nature of heterogeneity in peptides resulting from N-terminal blockages, amino acid oxidation, or degradation. Another area where MS has helped in recent years is in determination of glycosylation heterogeneity. Immunoglobulins contain a common Fc glycosylation in the C_H2 domain which is a complex type N-glycan, usually asialylated. However, differences in the processing following the trimannose core has been documented on a number of ocassions and this topic will be discussed further later on. Mass spectrometry is unable to distinguish between isomeric forms of sugars, but can reveal heterogeneity due to the presence or absence of the respective saccharide subunits [93].

8.4.2.5 C-Terminal Sequencing

The C-terminal sequence (of at least three residues) can be determined directly either by manual or automated C-terminal sequencing [94]. Several different chemistries are marketed for direct sequencing and each claim a specific capacity for determining a wide range of the amino acids that may potentially be present. Analysis has been either by reverse-phase HPLC of the cleaved amino acid or by MALDI-TOF mass spectroscopy. Alternatively, enzymatic release of the C-terminal amino acids using carboxypeptidase can be used [95].

Cases of the post-translational clipping of C-terminal lysine residues on antibody heavy chains have been documented for both plasma-derived immunoglobulin and monoclonals [96,97]. In these cases, components corresponding to the full-length and truncated C-terminal peptides are unlikely to co-migrate and the different forms have been separated by ion-exchange chromatography. Mass spectrometry may again be useful in assigning such components given the known C-terminal sequence, the theoretical peptide masses being calculated and compared with experimental results.

8.4.2.6 Secondary and Higher Structure

Details of the secondary and higher structure can be obtained by using techniques such as circular dichroism [98], intrinsic fluorescence [99], and differential scanning calorimetry [100], although immunoglobulins are still too large to be accessible to meaningful analysis by nuclear magnetic resonance. Such studies are useful not only to ensure a native molecule but also to detect changes induced on storage. Precise and consistent readings are required that will allow accurate comparison of runs performed under different conditions, necessitating sophisticated software. Rather than seeking absolute identities for the individual profile components the changes observed between samples under different conditions are indications of shifts in the degree of definite structural motifs, e.g., alpha-helix or beta-sheet. With molecules such as MAbs, where the generic structures are known it may be unnecessary to go too far down the road of absolute assignments of secondary and higher structure. More limited data along with accurate and specific functional and bioassays can be used to infer the integrity of the overall higher order structure.

8.4.2.7 Functional Activity

Another important way in which MAbs can be characterized is in their specificity and cross-reactivity. Indeed, specificity may be the primary criterion for a research application. The specificity of MAbs is assessed by immunoassays, of various types including for example serological, ELISA and RIA. Methodologies for these techniques are well known [101]. Increasingly, popular are methods based upon surface plasmon resonance. This technique allows quantitative data on binding specificity and kinetics to be obtained rapidly [102]. A bioassay is needed which is specific for the natural antigen and which, if possible, mimics the *in vivo* mode of action [103]. Data from such functional assays combined with that from the physical characterization methods together provide information on the nature of the MAb, which is invaluable in demonstrating comparability during process changes, development and scale-up.

Not only should MAbs be shown to be reactive with a specific antigen of interest but it is also necessary to establish the cross-reactivity, if any, of the MAb. Cross-reactions need not be restricted to molecules with similar structures to the antigen of interest, but indeed may occur with composite antigens forming a similar binding site from the association of functional groups on unrelated molecules. It is required even at an early stage in clinical trials that the potential for cross-reactivity with a number of tissues be addressed. It is important when using histochemical studies to ensure that the reactivities observed are likely to be relevant in the *in vivo* state and are not merely artefacts of the cytological preparative methods.

8.4.2.8 Stability

There is a need for limited but less time-consuming methods for the detection of any changes in the antibody throughout cell culture, harvest, processing, and formulation and indeed, across shelf life. Such techniques can be a mixture of old and new; with traditional methods such as IEF and SDS gel electrophoresis being supplemented by their capillary electrophoretic counterparts [104,105], and chromatographic methods similarly indicating structural integrity. The IEF profile or fingerprint of each MAb is characteristic and modifications to glycosylation, oxidation and deamidation of amino acid residues can all give rise to differences in the IEF profile [106,107], although some heterogeneity may be present from the beginning of culture (for instance, differences in the C-terminus due to clipping). However, not all changes in the structure of the MAb are necessarily detected by one method, and rather the approach taken is to adopt a battery of analytical techniques which are robust, quantitative and accurate in order to arrive at an overall assessment of the integrity of the antibody [108]. SDS–PAGE is a good method for showing differences in the composition of light and heavy chains as a result of differences in glycosylation, peptide integrity, and antibody class and subclass. The advantages of analyzing the antibody under both non-reducing and reducing conditions indicates changes that are both non-covalent or covalent in nature. For instance, published work on OKT3 revealed a covalent modification resulting in cross-linkage of the heavy and light chains during storage under an oxidizing atmosphere [109]. Most regulatory documents still require analysis of the MAb by traditional methods using high-sensitivity silver staining and Coomassie dyes; however, the newer methods have several advantages. The capillary form of these techniques allows more ready quantitation of components by monitoring of peptide backbone absorbancies which are less sensitive to differences in amino acid composition and alleviate differences based solely upon variation in the interaction of protein with dyes without resorting to time-consuming calibration of densitometric traces (Fig. 8-2). Most CE machines allow the automated processing under PC control of a number of samples and this again saves resources and simplifies data storage and retrieval. Free-flow electrophoresis and micellar electrophoresis utilizing the tendency for hydrophobic interaction with the detergent micelles have also been applied to the quality control of MAbs [105] being shown to be precise, quantitative, and reproducible.

Well-proven techniques such as HPLC size exclusion chromatography (SEC HPLC), used in the standardization of plasma-derived immunoglobulins, are also of value for the analysis of MAbs. This technique, when run at near-neutral pH, gives a good indication of the level of aggregation and any fragmentation which may occur in the product over time. Several monoclonals have been documented to show anomalous behavior on some SEC HPLC matrices [110] and so a typical profile for a given MAb should be compared rather than absolute retention times for individual components. This technique is a valuable tool in stability studies and there is evidence from plasma immunoglobulins that relationships between degradation at elevated temperatures and the given storage temperatures are useful predictors of antibody instability [111], although the regulations are clear that shelf life must be based upon real-time storage at the designated storage temperature.

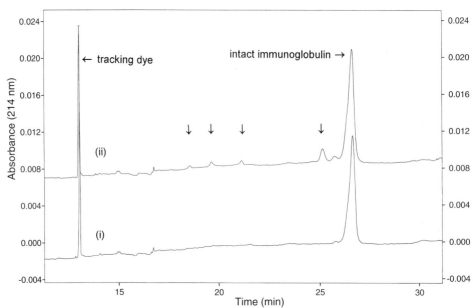

Fig. 8-2. Capillary electrophoresis to monitor antibody stability. Comparison of the SDS capillary CE absorbance profiles of frozen control antibody (i) and antibody stressed by storage at ambient temperature for 20 weeks (ii). Additional peaks in the stressed sample are indicated by arrows. Samples run on an eCAP SDS14-200 kD application kit using a Beckman PACE 5510 CE system fitted with a 47 cm neutral gel-filled capillary.

The use of elevated temperatures or other extreme conditions to provide degraded antibody material can be useful both to establish the detection limits and suitability of the techniques used as stability indicating methods, and may also assist in the prediction of the longer-term degradative changes which can occur in proteins [112]. The expression of recombinant human MAbs in SCID mice has been suggested as a means of testing the *in vivo* stability and half life of molecules intended for therapy [113].

Other powerful chromatographic techniques which have been demonstrated as useful in the characterization of MAbs include cation exchange chromatography [96], hydrophobic interaction chromatography [114], mixed bed ion-exchange chromatography [44], and chromatofocusing [115] or preparative IEF. These techniques, where applied successfully to differentiate between intact and degraded immunoglobulin molecules, have the advantage that they can yield mini-preparative scale quantities of the isoforms which greatly assists in their characterization. They also are – in principle – quantitative, rapid, and capable of automation.

Some evidence for proteolytic activity in harvests from hybridoma cell culture has been reported [116,117]. Indeed, one study has indicated a degradation of intact MAb over the course of cell culture over several weeks as revealed by non-reducing SDS–PAGE [118]. Hence, it is important to monitor harvests, intermediate bulk solutions and purified final product for the presence of proteases by the most sensitive methods available. However, electrophoretic heterogeneity may also be due to incomplete antibodies resulting from disulfide exchange between chains [119]. Such instability has been reported previously and these profiles may be characteristic for each antibody (and in our experience do not show signs of increasing degradation, as would be indicative if proteases were active), though some authors have suggested that these forms may increase in intensity with protein modification due to deamidation/oxidation [107]. One report of the presence of half antibody in IgG [120] was addressed by engineering the susceptibility to self dissociate out of the MAbs by site-directed modification. Similarly, a tendency for IgG degradation via a copper ion-mediated cleavage was reported for both monoclonal and polyclonal IgG and this again could be combatted by engineering the site of modification [121].

Formulation may play an important role in preventing the degradation of MAbs. In their work on OKT3, Rao and colleagues [109] demonstrated the value of an inert

Table 8-4. Some literature reports of degradative changes in monoclonal antibodies.

MAb/Clone name	Degradation observed	Reference
HER-2 Humanized MAb	LC deamidation at Asn30 C-terminal clipping of HC Lys450 Tyr to Gln at HC[a] cyclic imide formation at HC Asp102	[97] [96] [87] [124]
TB/C3 Mouse MAb	Unspecified proteolysis	[118]
Unnamed	Deamidation monitored by IEF	[123]
IgG4 CB72.3	80 kD[a] HL component due to Ser241	[120]
17-1A	C-terminal clipping in HC	[97] [97]
E-25 humanized anti-IgE	Isomerization at Asp32 in LC at room temperature	[97]
OKT3 mouse MAb	Deamidation Asn386, Asn423 in HC Oxidation at Met34 in HC, C-terminal clipping of HC Deamidation at Asn156 Some oxidation at Met174 in LC Covalent cross-link of LC Tyr46 and HC Cys105	[125] [97] [126] [97] [109]
OKT4a humanized	Cleavage at Asp270 HC, also some cleavage at Ser220, Thr250, Thr335, and Thr350. Trace of cleavage at Ser203 in LC	[97]
RSHZ19 humanized anti RSV	Oxidation of a Met residue C-terminal clipping of HC N terminal pyroglutamate in HC	[108]
Campath 1H humanized	Copper-induced cleavage at Lys226-Thr227 in HC at elevated temperature	[121]

[a] Rather than a degradative change this report describes the predominance of a variant sequence antibody over culture time due to differential expression of two clones.

atmosphere in preventing covalent cross-linking between chains, whereas in the same paper they report that an inert atmosphere had little effect on preventing deamidation changes. The formulation of biotechnological therapeutics has been reviewed [122] in terms of the likely modifications which may occur including isoaspartate formation, deamidation of asparagine and glutamine, and oxidation of cysteine and methionine residues. These types of modification have been monitored in MAbs [109,123]. Table 8-4 outlines the types of modification which have been reported for monoclonals under development.

8.4.2.9 Glycosylation

The topic of MAb glycosylation has occupied many reviews in its own right and the effects of variations in glycosylation in terms of functional efficacy have been discussed for a number of examples [127]. While it is clear that absence of glycosylation results in the loss of Fc-mediated effector functions [128] and reduced *in vivo* half-life [129], the influence of differences in the glycosylation pattern applied to the tri-mannose core is less clear. For some MAbs studied by certain functional assays there appears to be a role for full galactosylation of the core glycan. For instance, Kumpel et al. [130] showed that MAb possessing a higher proportion of galactosylated (G1 and G2) glycans had higher potency in lymphocyte antibody-dependent cell cytotoxicity (ADCC) than those with lower overall galactosylation. However, those antibodies with primarily monogalactosylated glycan appeared to have normal *in vivo* half life [131]. For another MAb there appears to be no effect on the functional assay results with galactose removal [132]. Certainly it is possible to influence glycosylation in terms of the species and type of host cell and the culture conditions [126,133,134]. For instance, initial studies of baculovirus-infected insect cells demonstrated non-mammalian-type terminal glycosylation processing, but later work with carefully selected strains seems to indicate that this can be overcome [135].

A number of technologies are available for analyzing glycans. Analysis without removal of the glycan from the protein can be possible by using lectin blotting [136], however this technique may not be as readily quantitative as other methods that are applied to glycans released either by enzymatic deglycosylation using PNGase F or chemical deglycosylation using hydrazine [137]. Once isolated, glycans can be analyzed by size-exclusion chromatography [138], high-performance anion-exchange chromatography [139], HPLC [140], or electrophoretic means [141]. Mass spectrometry has proven a useful tool in the characterization of glycans and glycopeptides [125].

Much has been published on the role of IgG glycosylation in disease and in addition to a conserved C_{H^2} domain N-glycosylation site both variable region glycosylation and O-glycosylation of MAbs has been reported for some mouse monoclonals [142]. Complete glycans (i.e., those possessing galactose and sometimes also sialic acid) can be produced by human lymphocytic cells lines. However, other cell lines popularly used in biotechnology may not be capable of this and so the glycosylation desired is an important consideration during the choice of host cell [135]. The glycosylation found on most human/mouse heterohybridomas has been shown to be of a rodent type [143] and to lack bisecting N-acetylglucosamine. In comparative studies using hamster (CHO), mouse (NS0), and rat (Y0) cells producing the humanized Campath-1H antibody differences in the level of galactosylation, fucosylation, and bisecting N-acetylglucosamine were found [144] and, whereas monocyte-mediated killing was unaffected, variations were found in the potency as measured by ADCC. In addition, cell lines from some species may introduce immunogenic glycosylation such as the Gal 1-3 Gal disaccharide [145].

For one IgM monoclonal a change from ascitic fluid to stirred cell culturing conditions had a dramatic effect on the *in vivo* half-life [146]. For an antibody destined for human replacement or prophylactic therapy where a normal *in vivo* half-life is required, then it is advisable to use a host cell capable of producing glycans as similar as possible to those of the native B cell. However, where antibody fragments are used (as is often the case in immunoscintigraphy or cancer therapy) then a short half-life is not a problem and the glycosylation capability of the host cell line is less vital.

8.5 Conclusions

We have reviewed the areas of the purification of monoclonal antibodies and their characterization, and given our perspectives on some of the matters discussed. The strategy applied for both downstream processing and analytical assessment of a MAb is dependent upon the use for which it is intended. Whereas for a research tool this may be the minimum to allow use of the immunoglobulin as a reagent in subsequent studies, for a MAb destined for a therapeutic application the workload will be considerable as the regulatory requirements are clearly defined.

Purification protocols rely largely upon initial capture steps (mainly affinity methods), although for the clinical applications multi-stage chromatography is necessary. In such applications the processes must be carefully and thoroughly validated to show consistency. Modern high-performance characterization methods together with functional activity studies are also an important part of the development process for successful manufacture of such biotherapeutics [147].

The recent concept of a well-characterized product has provided both a challenge to the analytical methods required and an opportunity to reduce the levels of pre-clinical and clinical evaluation which is necessary in order to assure consistency when performing modifications to the process of manufacture.

References

[1] Kohler, G., Milstein, C., *Nature*, 1975, *256*, 495–497.
[2] Winter, G., Milstein, C., *Nature*, 1991, *349*, 293–299.
[3] Lerner, R. A., Benkovic, S. J., Schultz, P. G., *Science*, 1991, *252*, 659–667.
[4] Brugemann, M., Caskey, H. M., Teale, C., Waldmann, H., Williams, G. T., Surani, M. A., Neuberger, M. S., *Proc Natl Acad Sci USA*, 1989, *86*, 6709–6713.
[5] Teng, N. N., Lam, K. S., *Proc Natl Acad Sci USA*, 1983, *80*, 7308–7312.
[6] Wood, C. R., Boss, M. A., Kenten, J. H., Calvert, J. E., Roberts, N. A., Emtage, J. C., *Nature*, 1985, *314*, 446–449.
[7] Inouye, K., Morimoto, K., *J Biochem Biophys Methods*, 1993, *26*, 27–39.
[8] Menozzi, F. D., Vanderpoorten, P., Dejaiffe C., Miller A. O. A., *J Immunol Methods*, 1987, *99*, 229–233.
[9] Rothman, S. W., Gentry, M. K., Gawne, R. D., Dobeck, A. S., Ogert, R., Stone, M. J., Strickler, M. P., *J Liq Chromatogr*, 1989, *12*, 1935–1947.
[10] Venktash, N., Murthy, G. S., *J Immunol Methods*, 1996, *199*, 167–174.
[11] Manil, L., Motté, P., Pernas, P., Troalen, F., Bohuon, C., Bellet, D., *J Immunol Methods*, 1986, *90*, 25–37.
[12] Guse, A. H., Milton, A. D., Schulze-Koops, H., Müller, B., Roth, E., Simmer, B., Wächter, H., Weiss, E., Emmrich, F., *J Chromatogr A*, 1994, *661*, 13– 23.
[13] Kenney, A. C., in: *Monoclonal Antibodies: Production and Application:* Mizrahi, A. (Ed.), New York: Alan R. Liss, Inc, 1989; pp. 143–160.
[14] Hanna, L. S., Pine, P., Reuzinsky, G., Nigam, S., Omstead, D. R., *BioPharm*, 1991, *4*, 33–37.
[15] Roberts, P. L., in: *Separations for Biotechnology:* Pyle, D. L. (Ed.), Cambridge: SCI/The Royal Society of Chemistry, 1994; Vol. 3, pp. 420–426.
[16] Baker, R. M., Brady, A.-M., Combridge, B. S., Ejim, L. J., Kingsland, S. L., Lloyd, D. A., Roberts, P. L., in: *Separations for Biotechnology:* Pyle, D. L. (Ed.), Cambridge: SCI/The Royal Society of Chemistry, 1994; Vol. 3, pp. 53–59.
[17] Ostlund, C., *Trends Biotechnol*, 1986, *6*, 288–293.
[18] Birch, J. R., Thompson, P. W., Boraston, R., *Biochem Soc Trans*, 1985, *13*, 10–12.
[19] Strobel, G.-J., in: *BIOTEC 3: Therapeutics, Diagnostics, Cell Culture, Product Isolation:* Hollenberg, C. P., Sahm, H. (Eds.), Stuttgart: Gustav Fischer, 1990; pp. 85–89.
[20] Sulk, B., Birkenmeier, G., Kopperschläger, G., *J Immunol Methods*, 1992, *149*, 165–171.
[21] Chang, Y. K., Chase, H. A., in: *Separations of Biotechnology:* Pyle, D. L. (Ed.), Cambridge: SCI/The Royal Society of Chemistry, 1994; Vol. 3, pp. 106–112.
[22] Morton, P., Lyddiatt, A., in: *Separations for Biotechnology:* Pyle, D. L. (Ed.), Cambridge: SCI/The Royal Society of Chemistry, 1994; Vol. 3, pp. 329–335.
[23] Jungbauer, A., Wenisch, E., in: *Monoclonal Antibodies: Production and Application:* Mizrahi, A. (Ed.), New York: Alan R. Liss, Inc, 1989; Vol. 11, pp. 161–192.
[24] Odde, D. J., in: *Handbook of Downstream Processing:* Goldberg, E. (Ed.), London: Blackie Academic, 1996; pp. 70–89.
[25] Lindmark, R., Thoren-Tolling, K., Sjöquist, J., *J Immunol Methods*, 1983, *62*, 1–13.
[26] Björck, L., Kronvall, G., *J Immunol*, 1984, *133*, 969–974.
[27] Carter, P., Kelley, R. F., Rodrigues, M. L., Snedecor, B., Covarrubias, M., Velligan, M. D., Wong, W. L. T., Rowland, A. M., Kotts, C. E., Carver, M. E., Yang, M., Bourell, J. H., Shepard, H. M., Henner, D., *BioTechnology*, 1992, *10*, 163–167.
[28] Zhu, Z., Zapata, G., Shalaby, R., Snedecor, B., Chen, H., Carter, P., *Bio Technology*, 1996, *14*, 192–202.
[29] Proudfoot, K. A., Torrance, C., Lawson, A. D. G., King, D. J., *Protein Expression and Purification*, 1992, *3*, 368–373.
[30] Laffer, S., Vangelista, L., Steinberger, P., Kraft, D., Pastore, A., Valenta, R., *J Immunol*, 1996, *157*, 4953–4962.

[31] De Jonge, J., Brissinck, J., Heirman, C., Demanet, C., Leo, O., Moser, M., Thielemans, K., *Mol Immunol,* 1995, *32,* 1405–1412.
[32] Kazemier, B., De Haard, H., Boender, P., Van Gemen, B., Hoogenboom., *J Immunol Methods,* 1996, *194,* 201–209.
[33] Ngo, T. T., Khatter, N., *J Chromatogr,* 1992, *597,* 101–109.
[34] Tyutyulkova, S., Paul, S., in: *Methods in Molecular Biology, Vol 51: Antibody Engineering Protocols:* Paul, S. (Ed.), Totowa, New Jersey: Humana Press Inc., 1995; pp. 395–401.
[35] Shi., J. Y., Goffe, R. A., *J Chromatogr A,* 1994, *686,* 61–71.
[36] Croze, E. M., *European Patent* 0 453 767 Al, 1991.
[37] Aguilar, M. I., Hodder, A. N., Hearn, M. T. W., in: *HPLC of Polypeptides, Proteins, and Polynucleotides:* Hearn, M. T. W. (Ed.), Weinheim: VCH, 1991; pp. 199–245.
[38] Yang, Y., Harrison, K., *J Chromatogr A,* 1996, *743,* 171–180.
[39] Burchiel, S. W., Billman, J. R., Alber, T. R., *J Immunol Methods,* 1984, *69,* 33–42.
[40] Carlsson, M., Hedin, A., Inganäs, M., Härfast, B., Blomberg, F., *J Immunol Methods,* 1985, *79,* 89–98.
[41] Hakalahti, L., Vihko, P., *J Immunol Methods,* 1989, *117,* 131–136.
[42] Boonekamp, P. M., Pomp, R., *Science Tools,* 1986, *33,* 5–8.
[43] Nadler, T. K., Regnier, F. E., in: *Chromatography in Biotechnology:* Horvath, C., Ettre, L. S. (Eds.), Washington: American Chemical Society, 1993; pp. 14–26.
[44] Nau, D. R., in: *Commercial Production of Monoclonal Antibodies: A Guide for Scale-Up:* Seaver, S. S. (Ed.), New York: Marcel Dekker Inc., 1987; pp. 247–275.
[45] Chen, F.-M., Naeve, G. S., Epstein, A. L., *J Chromatogr,* 1988, *444,* 153–164.
[46] Allard, W. J., Moran, C. A., Nagel, E., Collins, G., Largen, M. T., *Mol Immunol,* 1992, *29,* 1219–1227.
[47] Knudsen, K. L., Hansen, M. B., Henriksen, L. R., Andersen, B. K., Lihme, A., *Anal Biochem,* 1992, *201,* 170–177.
[48] Nopper, B., Kohen, F., Wilcheck, M., *Anal Biochem,* 1989, *180,* 66–71.
[49] Porath, J., Maisano, F., Belew, M., *FEBS Lett,* 1985, *185,* 306–310.
[50] Smith, G. J., McFarland, R. D., Reisner, H. M., Hudson, G. S., *Anal Biochem,* 1984, *141,* 432–436.
[51] Stanker, L. H., Vanderlaan, M., Juarez-Salinas, H., *J Immunol Methods,* 1985, *76,* 157–169.
[52] El-Kak, A., Vijayalakshmi, M. A., *J Chromatogr: Biomedical Applications,* 1991, *570,* 29–41.
[53] Hale, J. E., Beidler, D. E., *Anal Biochem,* 1994, *222,* 29–33.
[54] Johansson, H. J., Daniels, I., Weitman. A., Söderström, L., Westin, G., in: *Biologicals from Recombinant Microorganisms and Animal Cells:* White M. D. (Ed.), Weinheim: VCH, 1991; pp. 409–414.
[55] Shadle, P. J., Erickson, J. C., Scott, R. G., Smith, T. M., International Patent WO 95/22389, 1995.
[56] Grunfeld, H., Moore, P., *J Immunol Methods,* 1997, *201,* 233–241.
[57] Carr, P. W., Blackwell, J. A., Weber, T. P., Schafer, W. A., Rigney, M. P., in: *Chromatography in Biotechnology:* Horvath, C., Ettre, L. S. (Eds.), Washington: American Chemical Society, 1993; pp. 146–164.
[58] McCarthy, E., Vella, G., Mhatre, R., Lim. Y.-P., *J Chromatogr A,* 1996, *743,* 163–170.
[59] Jungbauer, A., Unterluggauer, F., Uhl, K., Buchacher, A., Steindl, F., Pettauer, D., Wenisch, E., *Biotechnology and Bioengineering,* 1988, *32,* 326–333.
[60] Jungbauer, A., Unterluggauer, F., Steindl, F., Ruker, F., Katinger, H., *J Chromatogr,* 1987, *397,* 313–320.
[61] Vallera, D. A., Burns, L. J., Frankel, A. E., Sicheneder, A. R., Gunther, R., Gajl-Peczalska, K., Pennell, C. A., Kersey, J. H., *J Immunol Methods,* 1996, *197,* 69–83.
[62] Neoh, S. H., Gordon, C., Potter, A., Zola, H., *J Immunol Methods,* 1986, *91,* 231–235.
[63] Brooks, D. A., Bradford, T. M., Hopwood, J. J., *J Immunol Methods,* 1992, *155,* 129–132.

[64] Righetti, P. G., Wenisch, E., Jungbauer, A., Katinger, H., Faupel, M., *J Chromatogr,* 1990, *500*, 681–696.
[65] Chase, H. A., in: *Discovery and Isolation of Microbial Products:* Verrall, M. S. (Ed.), Chichester: Ellis Horwood Ltd., 1985; pp. 129–147.
[66] Bonnerjea, J., Terras, P., in: *Bioprocess Engineering: Systems, Equipment and Facilities:* Lydersen, B. K., D'Elia, N. A., Nelson, K. L., (Eds.), New York: Wiley, 1994; pp. 159–186.
[67] Edwards, J., in: *Handbook of Downstream Processing:* Goldberg, E. (Ed.), London: Blackie Academic, 1996; pp. 167–184.
[68] FDA, *Points to Consider in the Manufacture and Testing of Monoclonal Antibody Producets for Human Use,* 1997.
[69] CPMP, *Note for Guidance: Production and Quality Control of Monoclonal Antibodies,* 1994.
[70] Chase H. A., *J Chem Tech Biotechnol,* 1986, *36*, 351–356.
[71] Chapman, G. E., Matejtschuk, P., More. J. E., Pilling, P., in: *Separations for Biotechnology:* Pyle, D. L. (Ed.), Cambridge: Society for Chemical Industry/Elsevier, 1990; Vol. 2, pp. 601–610.
[72] Kenney, A. C., Chase, H. A., *J Chem Tech Biotechnol,* 1987, *39*, 173–182.
[73] Knappik, A., Plückthun, A., *BioTechniques,* 1994, *17*, 754–761.
[74] Pharmacia, *Science Tools from Pharmacia Biotech,* 1996, *1*, 10.
[75] Vaughan, T. J., Williams, A. J., Pritchard, K., Osbourn, J. K., Pope, A. R., Earnshaw, J. C., McCafferty, J., Hodits, R. A., Wilton, J., Johnson, K. S., *Nature Biotechnology,* 1996, *14*, 309–314.
[76] Deramoudy, F. X., Chaabihi, H., Poul, M. A., Margaritte, C., Cerutti, M., Devauchelle, G., Bernard, A., Lefranc, M. P., Kaczorek, M., in: *Animal Cell Technology: Developments towards the 21st Century:* Beuvery, E. C., Griffiths, J. B., Zeijlemaker, W. P., (Eds.), Dordrecht: Kluwer Academic Publishers, 1995; pp.469–473.
[77] Yurov, G. K., Neugodova, G. L., Verkhovsky, O. A., Naroditsky, B. S., *J Immunol Methods,* 1994, *177*, 29–33.
[78] Hjorth, R., *Trends Biotechnol,* 1997, *15*, 230–235.
[79] Jungbauer, A., Uhl, K., Schulz, P., Tauer, C., Gruber, G., Steindl, F., Buchacher, A., Schoenhofer, W., Unterluggauer, F., *Biotechnol Bioeng,* 1992, *39*, 579–587.
[80] Birkenmeier, G., Dietze, H., in: *Biotechnology of Blood Proteins:* Rivat, C., Stoltz, J.-F. (Eds.), Paris: Colloque INSERM/John Libbey Eurotext Ltd., 1993; Vol. 227, pp. 201–206.
[81] Grubb, R. E., in: *Immunochemistry:* van Oss, C. J., Regenmortel, M. R. V. (Eds.), New York: Marcel Dekker, 1994; pp. 47–68.
[82] Clark, M. R. (Ed.), *Protein Engineering of Antibody Molecules for Prophylactic and Therapeutic Applications in Man.* Nottingham, U.K.: Academic Titles, 1993.
[83] Andersson, C. J., Schwarz, S. W., Connett, J. M. et al., *J Nucl Med,* 1995, *36*, 850–858.
[84] O'Connor, J. V. O., Keck, R. G., Harris, R. J., Field, M. J., *Dev Biol Stand,* 1994, *83*, 165–173.
[85] Schenerman, M. A., Phillips, K., *BioPharm,* 1997, *10*, 20–26.
[86] Lewis, D. A., Guzzetta, A. W., Hancock, W. S., Costello, M., *Anal Chem,* 1994, *66*, 585–595.
[87] Harris, R. J., Murnane, A. A., Utter, S. L., Wagner, K. L., Cox, E. T., Polastri, G. D., Helder, J. C., Sliwkowski, M. B., *Biotechnology,* 1993, *11*, 1293–1297.
[88] Hancock,W. S. (Ed.), *New Methods in Peptide Mapping For The Characterisation of Proteins.* New York: CRC Series in Analytical Biotechnology, 1996.
[89] Allen, D., Baffi, R., Bausch, J., Bongers, J., Costello, M., Dougherty, J., Jr., Federici, M., Garnick, P., Peterson, S., Riggino, R., Sewerin, K., Tuls, J., *Biologicals,* 1996, *24*, 255–275.
[90] Nadler, T., Blackburn, C., Mark, J., Gordon, N., Regnier, F. E., Vella, G., *J Chromatogr,* 1996, *743*, 91–98.
[91] Bennett, K. L., Smith, S. V., Lambrecht, R. M., Truscott, R. J. W., Sheil, M. M., *Bioconjug Chem,* 1996, *7*, 16–22.
[92] Alexander, A. J., Hughes, D. E., *Anal Chem,* 1996, *67*, 3626–3632.

[93] Kroon, D. J., Freedy, J., Burinsky, D. J., Sharma, B., *J Pharm Biomed Appl,* 1995, *13,* 1049–1054.

[94] Bailey, J. M., Tu, O., Issai, G., Ha, A., Shively, J. E., *Anal Biochem,* 1995, *224,* 588–598.

[95] Pattersom, D. H., Tarr, G. E., Regnier, F. E., Martin, S. A., *Anal Chem,* 1995, *67,* 3971–3978.

[96] Harris, R. J., *J Chromatogr A,* 1995, *705,* 129–134.

[97] Powell, M. F., in: *Characterization and Stability of Protein Drugs:* Pearlman, R., Wang, Y. D. (Eds.), New York: Plenum Press, 1996; pp. 1–140.

[98] Mulkerrin, M. G., in: *Spectroscopic Methods For Determining Protein Structure:* Havel H. A. (Ed.), Weinheim, VCH Press, 1996; pp. 5–27.

[99] Jiskoot, W., Hlady, V., Naleway, J. J., Herron, J. N., in: *Physical Methods to Characterize Pharmaceutical Proteins:* Herron, J. N., Jiskoot, W., Crommelin, D. J. A. (Eds.), New York: Plenum Press, 1995; pp. 1–63.

[100] Chowdhry, B. Z., Cole, S. C., *Trends Biotechnol,* 1989, *7,* 11–18.

[101] Channing-Rodgers, R. P., in: *Basic and Clinical Immunology:* Stites, D. P., Terr, A. I., Parslow, T. G. (Eds.), New York, Appleton & Lange, 1994; pp. 151–194.

[102] Cooper, L. J. N., Robertson, D., Granzow, R., Greenspan, N. S., *Mol Immunol,* 1994, *31,* 577–584.

[103] Jeffcoate, S., *Trends Biotechnol,* 1996, *14,* 121–124.

[104] Guttman, A., *Electrophoresis,* 1996, *17,* 1333–1341.

[105] Pritchett, T., in: *Handbook of Capillary Electrophoresis Applications:* Shintani, H., Polonsky, J. (Eds.), London: Blackie, 1997; pp. 240–254.

[106] Gianazza, E., *J Chromatogr A,* 1995, *705,* 67–87.

[107] Hunt, G., Moorhouse, K. G., Chen, A. B., *J Chromatogr A,* 1996, *744,* 295–301.

[108] Roberts, G. D., Johnson, W. P., Burman, S., Anumula, K. R., Carr, S. A., *Anal Chem,* 1995, *67,* 3613–3625.

[109] Rao, P. E., Kroon, D. J., in: *Stability and Characterization of Proteins and Peptide Drugs; Case Histories:* Wang, Y. J., Pearlman, R. (Eds.), New York: Plenum Press, 1993; pp. 135–158.

[110] Michaelson, T. E., Løfsgaard, M. F., Aase, A., Heyman, B., *J Immunol Methods,* 1992, *146,* 9–16.

[111] Page, M., Ling, C., Dilger, P., Bentley, M., Forsey, T., Longstaff, C., Thorpe, R., *Vox Sang,* 1995, *69,* 183–194.

[112] Usami, A., Ohtsu, A., Takahama, S., Fujii, T., *J Pharm Biomed Anal,* 1996, *14,* 1133–1140.

[113] Bazin, R., Boucher, G., Monier, G., Chevrier, M. C., Verrette, S., Broly, H., Lemieux, R., *J Immunol Methods,* 1994, *172,* 209–217.

[114] Rinderknecht, E., Zapata, G. A., World patent WO96/33208, 1996.

[115] Jungbauer, A., Tauer, C., Wenisch, E., Uhl, K., Brunner, J., Purtscher, M., Steindl, F., Buchacher, A., *J Chromatogr,* 1990, *512,* 157–163.

[116] Karl, D. W., Donovan, M., Flickinger, M. C., *Cytotechnology,* 1990, *3,* 157–169.

[117] van Erp, R., Adorf, M., van Sommeren, A. P., Grinbau, T. C., *J Biotechnol,* 1991, *20,* 249–261.

[118] Mohan, S. B., Chohan, S. R., Eade, J., Lyddiatt, A., *Biotech Bioengng,* 1993, *42,* 974–986.

[119] Li, L., Sun, M., Gao, Q.-S., Paul, S., *Mol Immunol,* 1996, *33,* 593–600.

[120] Angal, A. S., King, D. J., Bodmer, M. W., Turner, A., Lawson, A. D. G., Roberts, G., Pedley, B., Adair, J. R., *Mol Immunol,* 1991, *30,* 105–108.

[121] Smith, M. A., Easton, M., Everett, P., Lewis, G., Payne, M., Riveros-Moreno, V., Allen, G., *Int J Peptide Prot Res,* 1996, *48,* 48–55.

[122] Manning, M. C., Patel, K., Borchardt, R. T., *Pharm Res,* 1989, *6,* 903–918.

[123] Moellering, B. J., Tedesco, J. L., Townsend, R. R., Hardy, M. R., Scott, R. W., Prior, C. P., *BioPharm,* 1990, *3,* 30–38.

[124] Cacia, J., Keck, R., Presta, L. G., Frenz, J., *Biochemistry,* 1996, *35,* 1897–1903.

[125] Kroon D. J., Baldwin-Ferro, A., Lalan, P., *Pharm Res,* 1992, *9,* 1386–1393.

[126] Rao, P., Makowski, M., Meyer, E., Williams, A., Baldwin, A., Ferro, A., Hanigan, E., Kroon, D., Numsuwan, V., Tran, A., Rubin, E., *BioPharm,* 1991, *4,* 38–43.
[127] Wright, A., Morrison, S. L., *Springer Series in Immunopathology,* 1993, *15,* 259–273.
[128] Lund, J., Tanako, T., Takahashi, N., Sarmay, G., Arata, Y., Jefferis, R., *Mol Immunol,* 1990, *27,* 1145–1153.
[129] Wawrzynczak, E. J., Cumber, A. J., Parnell, G. D., Jones, P. T., Winter, G., *Mol Immunol,* 1992, *29,* 213–220.
[130] Kumpel, B. M., Rademacher, T. W., Rook G. A. W., Williams, P. J., Wilson, I. B. H., *Hum Antibod Hybrid,* 1994, *5,* 143–151.
[131] Goodrick, J., Kumpel, B., Pamphillon, D., Fraser, I., Chapman, G., Dawes, B., Anstee, D., *Clin Exp Immunol,* 1994, *98,* 17–20.
[132] Boyd, P. N., Lines, A. C., Patel, A. K., *Mol Immunol,* 1995, *32,* 1311–1318.
[133] Wright, A., Morrison, S. L., *Trends Biotechnol,* 1997, *15,* 26–32.
[134] Monica, T. J.,Goochee, C. F., Maiorella, B. L., *Biotechnology,* 1993, *11,* 512–515.
[135] Jenkins, N., Parekh, R. B., James, D. C., *Nature Biotechnol,* 1996, *14,* 975–981.
[136] Sumar, N., Bodman, K. B., Rademacher, T. W., Dwek, R. A., Williams, P., Parekh, R. B., Edge, J., Rook, G. A. W., Isenberg, D. A., Hay, F. C., Roitt, I. M., *J Immunol Methods,* 1990, *131,* 127–136.
[137] Patel, T. P., Parekh, R. B. in: *Methods in Enzymology:* Lennarz, W. J., Hart, G. W. (Eds.), New York: Academic Press, 1994; Vol. 230, pp. 57–66.
[138] Kobata, A. in: *Methods in Enzymology:* Lennarz, W. J., Hart, G. W. (Eds.), New York: Academic Press, 1994; Vol. 230, pp. 200–208.
[139] McGuire, J. M., Douglas, M., Smith, K. D., *Carbohydr Res,*1996, *292,* 1–9.
[140] Guile, G. R., Rudd, P. M., Wing, D. R., Prime, S. B., Dwek, R. A., *Anal Biochem,* 1996, *240,* 210–226.
[141] Okafo, G., Burrow, L. M., Neville, W., Truneh, A., Smith, R. A. G., Reff, M., Camilleri, P., *Anal Biochem,* 1996, *240,* 68–74.
[142] Coco-Martin, J. M., Brunink, F., van der Velden de Groot, T. A. M., Beuvery, E. C., *J Immunol Methods,* 1992, *155,* 241–248.
[143] Tandai, M., Endo, T., Sasaki, S., Masuho, Y., Kochibe, N., Kobata, A., *Archiv Biochem Biophys,* 1991, *291,* 339–348.
[144] Lifely, M. R., Hale, C., Boyce, S., Keep, M. J., Phillips, J., *Glycobiology,* 1995, *5,* 813–822.
[145] Sheeley, D. M., Merrill, B. M., Taylor, L. C. E., *Anal Biochem,* 1997, *247,* 102–110.
[146] Maiorella B. L., Winkelhake, J., Young, J., Moyer, B., Bauer, R., Hora, M., Andya, J., Thomson, J., Patel, T., Parekh, R., *Biotechnology,* 1993, *11,* 387–392.
[147] Maiorella, B. L., Ferris, R., Thomson, J. et al., *Biologicals,* 1993, *21,* 197–205.

Part Two
Quality and Characterization

9 Biological Standardization of Interferons and Other Cytokines

Anthony Meager

9.1 Introduction

An increasing number of biologically active proteins have been discovered and characterized since the early 1970s. Many of these have biological activities that have encouraged their development on a large scale for clinical evaluation as biotherapeutic products. These include interferons (IFNs), interleukins (ILs), colony-stimulating factors (CSFs), and polypeptide growth factors (PGFs), which collectively are designated as 'cytokines'. What they have in common are that they: (i) induce biological activities via specific cell surface receptors; (ii) are themselves inducible in most cases; (iii) act locally within the environs of producer cells in either an autocrine- or paracrine-manner; and (iv) are active at very low concentrations both *in vivo* and *in vitro*. This last property has permitted the development of *in vitro* biological assays (bioassays) for the quantification of the biological potency of cytokines. Bioassays are the only means by which the potency of cytokines can be determined. Physico-chemical methods of analysis are essential to ensure certain aspects of the quality of cytokines, but, on their own, do not allow full characterization of these complex, biologically active proteins; neither do such methods predict or measure cytokine potency. It is therefore essential that the potency of cytokines is quantified in robust, well-designed bioassays. These must be monitored for sensitivity on an assay-to-assay basis by the inclusion, with test samples, of an appropriately defined and characterized biological standard, normally containing the homologous cytokine, in every bioassay. To this end, individual testing laboratories have developed in-house biological standards or reference reagents for the particular cytokine(s) they are developing, or have developed.

The subject of biological standardization is, however, a complex, and often difficult one to address, particularly where cytokines are concerned, principally due to the fact that most cytokines mediate a variety of biological activities *in vitro*. Thus, there are usually no 'reference' bioassays for individual cytokines. Bioassays conducted in different laboratories often depend on variable sources of somatic mammalian cells and culture reagents, and are subject to variations in both design and methodology. This leads to the definition of unitages of biological potency for cytokines being bioassay-dependent and thus, from different laboratories, the reporting of potency values in non-comparable units. To overcome this problem, which has grown in importance as cytokines have 'entered' the clinic, the World Health Organization (WHO) has instigated and coordinated international efforts to evaluate suitable, well-characterized, cytokine materials for the purpose of establishing international standards (IS) and reference reagents (RR) for individual cytokines. WHO IS and RR, which contain lyophilized cytokines, comply with WHO guidelines for the preparation, characterization, and establishment of international and other standards and reference reagents for biological substances [1], and are of proven stability on storage at –20 °C. The National Institute for Biological Standards and Control (NIBSC), Blanche Lane, South Mimms, Herts., EN6 3QG, (Tel: +44 1707 654753, Fax: +44 1707 646730) has played the major role in the development and preparation of candidate IS and RR for individual cytokines. NIBSC has, on behalf of WHO, and in collaboration with the Center for Biologics Evaluation and Research (CBER) (The National Institutes of Health (NIH), Bethesda, Maryland 20205, USA), initiated and coordinated many international and other collaborative studies aimed primarily at the evaluation of candidate IS and RR and the identification of the most suitable one (among such candidate IS and RR) to serve as the WHO IS or RR for a particular cytokine. Priorities for cytokine standardization are set by the WHO Consultative Group on Cytokine Standardization (CGCS). Following full statistical analysis of raw data received from participants in any one collaborative study, the study's organizers make recommendations to WHO CGCS and the WHO Expert Committee on Biological Standardization (ECBS) as to which they think is the most suitable candidate IS/RR, among those evaluated, to serve as the WHO IS/RR. Final adoption and establishment of WHO IS and RR is decided by WHO ECBS on an annual basis.

Once an IS or RR for a particular cytokine has been established with an assigned ampoule content in international units (IU) or reference units (RU), ampoules are made available on a request basis to enable laboratories world-wide to calibrate their bioassays and to facilitate the assignment in IU or RU to 'in-house' working standards/reference reagents. The WHO IS or RR should not be used routinely to calibrate bioassays; this should be effected with the in-house working standard/reference reagent. Such a strategy should help preserve stocks of the WHO IS or RR so that ampoules for distribution will last for many years. Eventually, ampoule stocks of a current WHO IS will be run-down to a point at which the IS will need replacing. The IS here plays an important role in that it is used as a basis for providing continuity over long periods of time for the IU of biological activity. The first WHO IS is used to calibrate accurately the second WHO IS of the same cytokine before the first IS is exhausted. This strategy requires the best achievable standards of accuracy, reproducibility, and stability. An impressive historic example of this approach is illustrated by the insulin standards where the first IS and IU of biological activity were established in 1926. Nonetheless, in 1997, over 70 years later, manufacturers, national control laboratories, physicians, and patients will use exclusively the same IU, although the current IS is the fifth IS for insulin.

For the majority of cytokines, biological standardization has been a recent endeavor, but work was started towards this end in the 1960s with the then newly available IFNs from human, mammalian, and avian sources. The biological standardization of IFNs illustrates well the principles involved and the difficulties, both practical and theoretical, that are common to the biological standardization of cytokines in general.

Therefore, this chapter will focus largely on the biological standardization of IFNs. However, an understanding of this process requires some familiarity with IFN designations and the molecular and biological characteristics of IFNs, and these are outlined in the next section.

9.2 Interferons

9.2.1 Background Information: Definitions, Designations and Characterization

The word 'interferon' was coined in 1957 to describe a substance produced by virus-infected chick cell cultures which, on transfer to fresh uninfected chick cell cultures, could elicit a protective antiviral effect [2]. Subsequently, IFNs have been induced from many types of mammalian, including human, cells and demonstrated to be active against a broad spectrum of viruses [3]. Although the protein nature of IFNs was established at a relatively early stage following its discovery [4], it was only following the introduction of large-scale production methods in the 1970s [5] and the simultaneous development of efficient purification procedures [6,7] that sufficient amounts of IFN became available for molecular characterization and clinical evaluation. Further progress followed rapidly from the late 1970s with the advent of recombinant DNA (rDNA) technology which, together with the pharmaceutical industry's desire to produce pharmacologically active proteins cheaply and growing evidence that IFN had antitumor activity [8] besides antiviral activity, led to the cloning and subsequent mass production of IFN for clinical evaluation.

The successful cloning of the IFN-α type, the main IFN produced by virus-infected human leukocytes, revealed it to be a mixture of up to 13 distinct, but molecularly closely related proteins, now known as IFN-α subtypes. Each IFN-α subtype is expressed from a separate chromosomal gene and contains 165–166 amino acids [9]. In contrast, the IFN-β type, the IFN produced chiefly by stimulated human diploid fibroblasts, was found to be a single molecular species, 166 amino acids long, which was evolutionarily distantly related (ca. 28 % homology) to IFN-α subtypes, but antigenically distinct from them [10]. Additionally, a third IFN type, known as IFN-ω, was cloned in around 1985 and also shown to be a single molecular species of 172–174 amino acids. However, IFN-ω is produced by human leukocytes and probably arose by divergence from an ancestral IFN-α subtype [11], but is now antigenically distinct from 'modern-day' IFN-α subtypes [12]. Despite the divergence of primary amino acid sequences, these three IFN types, α, β, ω, share a common three-dimensional structure of five α-helical regions arranged in a compact bundle [13]. This characteristic, together with shared biophysical and biological properties, e.g., stability of activity at acid pH, and recognition of a common class of cell surface receptors [14,15], has led to IFN-α subtypes, IFN-β, and IFN-ω, being designated collectively as type I IFN.

A distinct IFN, originally known as 'immune IFN' because of its T-cell source, was cloned in 1982 [16]. Its amino acid sequence was found to be unrelated to any of those of the type I IFNs. Unlike the latter, it was subsequently shown to have a homodimeric structure, although each protomer contained five α-helical domains in an arrangement similar to that of IFN-β [13,17]. This IFN is now known as IFN-γ and, by virtue of the acid-lability of its activity, has been classified as type II IFN. IFN-γ has also been demonstrated to bind to a class of cell surface receptors distinct from that recognized by type I IFNs [18], although the two receptor classes show some homology in their extracellular domains [19].

Biological responses to IFN are initiated by IFN binding to cell surface receptors and contingent activation of cytoplasmic signal transduction pathways, and manifested following expression of a number of 'IFN-inducible' genes [20]. Induction of antiviral action, which is dependent on such inducible protein synthesis, can now be seen as just one of many activities mediated by IFN; these activities include inhibition of cell proliferation, regulation of cell differentiation, regulation of functional cellular activities, and immunomodulation [21]. Despite molecular differences between type I (α, β, ω) – and type II (γ) -IFNs, and differences that exist in signaling pathways [22,23] and inducible genes [20], most of these activities are shared by type I and type II IFNs [21,23]. However, type II IFN appears to have some unique activities, e.g., macrophage activation [24], induction of *de novo* class II MHC antigen (HLA-DR) expression [25], and, in some cases, to differ in the degree a shared activity is elicited.

9.2.2 Characteristics of IFNs for Clinical Use

Until about 1978, partially purified leukocyte IFN, derived from the supernatants of Sendai virus-infected human buffy-coats, produced by the Finnish Red Cross Blood Transfusion Service (FRCBTS) on a non-commercial basis, was the only IFN available in sufficient amounts for clinical evaluation. Leukocyte IFN continues to be manufactured on a modest scale by FRCBTS, but is now also made commercially elsewhere. The composition of leukocyte IFN is approximately 90 % IFN-α, a heterogeneous mixture of up to 14 IFN-α subtypes of which IFN-α_1 and IFN-α_2 subtypes are major components [9,26,27] , 2–3 % IFN-β and 7–15 % IFN-ω [28]. Purification methods often utilize IFN-α-specific antibody affinity chromatographic separations, which result in IFN preparations containing mixtures only of IFN-α subtypes, but in different proportions to those of the unpurified leukocyte IFN.

To overcome the problem of limited supplies of leukocyte IFN that could be derived from human buffy-coats, the Wellcome Foundation set up production of IFN from Sendai virus-infected Namalwa cells, a B-lymphoblastoid cell line that could be grown in large suspension cultures [29]. The purified product is known as lymphoblastoid IFN, and consists of a heterogeneous mixture of IFN-α subtypes, but in different proportions to those found in leukocyte IFN [30]. The IFN-α_2 subtype is a major component of lymphoblastoid IFN [30].

IFN-α was cloned in 1980 [9] and the mass production of particular IFN-α subtypes by genetically engineered *Escherichia coli* cultures was rapidly developed. Initially, two allelic variants of the IFN-α$_2$ subtype, designated IFN-α$_{2a}$ [26] and IFN-α$_{2b}$ [31] were manufactured and simultaneously made available for clinical trials in 1981. IFN-α$_{2a}$ differs from IFN-α$_{2b}$ in only one amino acid, Lys23 instead of Arg23. A third variant of IFN-α$_2$, IFN-α$_{2c}$, with a substitution of arginine for histidine at position 34 has also been manufactured [32]. All three variants synthesized by *E. coli* lack the O-linked oligosaccharide side chain present at Thr106 in human somatic cell-derived IFN-α$_2$ [33]. Although the three recombinant IFN-α$_2$s are very closely related molecularly, *in vitro* biological and antigenic differences have been reported [34].

The development of other individual IFN-α subtypes for clinical use has been rather limited, because of patents curbing their exploitation and the dominant market share of IFN-α$_2$ products . IFN-α subtypes, such as α$_1$, α$_4$, and α$_8$, appear to have interesting properties, and may potentially be clinically useful. However, the revolution in rDNA technology has made possible the construction of many kinds of 'non-natural' hybrid IFNs [35,36] and idealized consensus sequence IFN-α molecules [37]. Thus, hybrid IFN-α molecules were produced with an N-terminal section of one IFN-α subtype and a C-terminal section of another subtype, and these have been designated, for example, IFN-α$_{1/2}$, IFN-α$_{1/8}$ [36]. These displayed interesting *in vitro* properties [35,36], but have not been, with the exception of one particular form of IFN-α$_{1/8}$, tested clinically. A recombinant IFN-α homolog, IFN-conl, having a computer 'idealized' consensus amino acid sequence based upon known sequences of IFN-α subtypes has also been developed [37]. IFN-conl has been reported to be more active *in vitro* in certain bioassays than IFN-α subtypes, including IFN-α$_2$ [38]. Both IFN-α$_{1/8}$ and IFN-conl have been found to have more moderate side effects in phase I clinical trials than IFN-α$_2$ and leukocyte IFN-α or lymphoblastoid IFN-α [36].

Fibroblast IFN-β derived from stimulated human diploid fibroblasts is both difficult to produce in large quantities and to purify [6]. However, modest amounts have been and continue to be prepared from this source for clinical use since the late 1970s. Isolation of the IFN-β cDNA in 1980 [10] led to attempts to mass produce IFN-β in *E. coli* in a way similar to that was successful for the production of IFN-α$_2$. Production by this means proved problematic as IFN-β lacks one of the four cysteine residues which form two intramolecular disulfide bridges in IFN-α subtypes, leading to mismatched pairing of its three cysteines in *E. coli* and the formation of inactive IFN-β molecules. Subsequently, one manufacturer substituted the cysteine at position 17, which is not involved in disulfide bond formation, for serine and has developed a recombinant IFN–β Ser17 product, Betaseron or Betaferon (also known as IFN-β-1-b) [39]. This product, and any other recombinant IFN-β derived from *E. coli*, lacks the N-linked oligosaccharide side chains present in human fibroblast-derived IFN-β. Glycosylated, recombinant IFN-β can, however, be manufactured using transfected Chinese hamster ovary (CHO) cells as producer cells. The carbohydrate composition of CHO cell-derived recombinant IFN-β is non-identical to that of human fibroblast-derived IFN-β [40]. It is by comparison to *E. coli*-derived recombinant IFN-β more soluble, probably on account of its hydrophilic carbohydrate side chain.

Interferon omega (IFN-ω) is a component of leukocyte IFN [28]. It has been cloned and expressed both in *E. coli* and CHO cells [41]. However, IFN-ω is a glycoprotein containing an N-linked oligosaccharide side chain, and thus production of glycosylated IFN-ω is only attainable in CHO cells [41,42].

Interferon gamma (IFN-γ), like IFN-β and IFN-ω, is an N-linked glycosylated protein when derived from human cells [43]. Production from human T lymphocytes or CHO cells is possible, but yields are low. Therefore, the major route of manufacture is by expression in *E. coli* and subsequent purification from bacterial lysate. Nonglycosylated recombinant IFN-γ is biologically active [44].

9.3 Interferon Standardization

The biological standardization of IFNs is of critical importance to both the pre-clinical testing and clinical development of IFNs. It is clear that the uniform reporting of potency values of IFN activity in an internationally defined unitage is highly desirable in research publications and is essential for clinically used IFNs to ensure correct/precise dosing. However, there are (as outlined in Section 9.2.2) several different natural IFNs, with the possibility of markedly extending this range by rDNA technological methods. This molecular heterogeneity is associated with variable biological capability, quantitatively and, in some cases, qualitatively, among different IFN molecules, and has presented an impressive challenge to those involved in the practical aspects of biological standardization.

9.3.1 Basic Principles of Biological Standardization

The global aim of biological standardization is to permit the reliable comparison of the results of bioassays performed by different individuals with different reagents and at different times [45]. Such results must therefore be valid, reproducible, and as accurate as possible. The validity of bioassays is thus critical for achieving biological standardization. Section 9.3.2 will discuss the various types of bioassays and methodology. What follows here is a consideration of the basic principles of standardization.

In terms of clinical chemistry, an assay is seen as a procedure to quantify an 'analyte'. In the present discussion, for 'analyte' read IFN or cytokine. An IFN (or cytokine) can be quantified if an assay generates a measurable parameter that reproducibly increases with increasing concentration or dose of IFN (or cytokine), i.e. in a concentration/dose-related manner [46,47]. When two or more different IFN preparations are being compared, they must behave identically in the assay for the assay to be truly valid. Assay reproducibility is essential for biometric validity, and is best achieved by minimizing, where possible, known variables and ensuring that all variables (both known and unknown) are allocated treatments randomly (see Section 9.3.3). It helps greatly if the measured parameter is exclusively related to the IFN concentration, and not to the concentrations of active impurities, i.e. that the assay has appropriate specificity. A consequence of this is that the purer the IFN preparation, the less specific an assay system needs to be, and vice versa. It should be stressed, however, the use of purified IFN does not confer specificity on an assay system. In practice, it is often difficult completely to avoid 'interference' from substances that may contaminate IFN preparations, e.g., other cytokines, endotoxin (LPS). The contribution of such contaminants to assay responses should, where possible, be evaluated. Appropriate steps should, if necessary, be taken to minimize interference.

It has been mentioned previously that when two or more IFN preparations are being quantified in the same assay, they must behave identically for the biometric validity of results. This is an essential tenet for assay calibration, where one IFN preparation serves as the standard to which other IFN preparations included in the assay must behave as if they were a more concentrated or more dilute solution of the standard preparation. Stated differently, if like is compared with like, in the same assay system, under the same conditions, then any difference in measured response from the assay system reflects only the differences in concentrations/doses. Evidence that a test IFN preparation and an IFN standard behave similarly in an assay system is best displayed by parallelism of the graphic plots of the (log) dose–response curves (Fig. 9.1). In theory, the dose–response curve of the test IFN should, if it con-

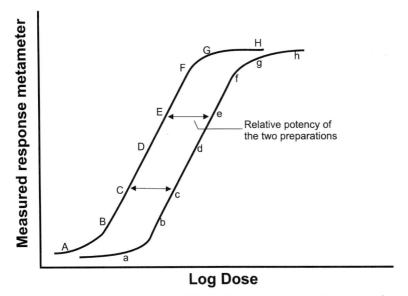

Fig. 9-1. Graphical plots of sigmoidal log dose-response curve in a comparison of two preparations that behave similarly in a bioassay system. ABCDEFGH represents the shape of a typical log dose–response curve. The potency of one preparation relative to the other is represented by the amount of one preparation which gives the same response as a measured quantity of the other, i.e., the horizontal distance between the linear part of the curves B–F and b–f. Use of three dose levels of each preparation gives the minimum information necessary to allow assessment of linearity and parallelism of the preparations.

tains IFN molecules identical to those in the IFN standard, be exactly parallel to the dose–response curve of the IFN standard [46,47]. However, it must be emphasized that although parallelism is an essential part of the evidence to prove validity of results, it is not, by itself, proof that the two IFNs compared are identical. Where parallelism of dose–response curves is evident, the *potency* of the test IFN relative to the standard IFN is represented by the amount of the test preparation which gives the same response as a known (measured) amount of the standard, i.e., the horizontal distance between the linear part of the curves A–H and a–h (Fig. 9.1). In practice, evidence of parallelism of dose–response curves may be difficult to demonstrate conclusively for reasons related to the numerous variables that are inherent in design and performance of bioassays (see below). Therefore, parallelism should be assessed statistically, both within and among assays performed on different occasions, to improve the biometric validity of results [47].

Another important aspect for standardization is the 'precision' with which results can be obtained from an assay. Precision in this context means the consistency of agreement of repeated measurements, and this may be considered as an aspect of validity. It is usual, when stating the estimate of potency of an IFN preparation, also to calculate the precision of the potency estimate, which is normally presented as fiducial limits at a given level of probability, e.g., 95 % ($P = 0.95$). The 95 % limits encompass the range of values between which the assay potency estimates could be expected to fall 95 times if the assay was done 100 times with the same precision. For clinical IFN products where precise estimation of potency is important, fiducial limits are set at not less than 64 % and not more than 156 % of the stated potency. The goal should be to achieve an estimated potency that is not less than 80 % and not more than 125 % of the stated potency. To obtain this level of precision requires careful consideration of available assay systems and designs. The range of bioassays for IFN potency estimations that are available is discussed below.

9.3.2 Bioassays for IFNs

IFNs exert a wide spectrum of different biological activities *in vitro* [21], which include antiviral-, antiproliferative-, and differentiative-actions in many cultured somatic cells. Thus, there are potentially a large number of bioassay systems that are capable of providing dose–response data for IFN potency estimations. However, historically, IFNs are thought of as antiviral agents and, as a consequence, most laboratories interested in IFN research have developed antiviral assays [48,49]. The easy availability of human and mammalian cells, especially transformed, or tumor-derived cell lines, has provided a great variety of cell 'substrates' for antiviral assay development. Additionally, there are several human and mammalian viruses that are accessible, easy to propagate, and sensitive to the antiviral action of IFNs, which enables them to be used in antiviral assays [49]. The number of cell–virus combinations is potentially vast. Nevertheless, nearly four decades of experimentation has demonstrated that only a small, select number of cell–virus combinations are really suitable and useful for antiviral assays [49]. Most human IFNs exhibit

Table 9-1. Human and mammalian cell lines used in routine antiviral assays for the quantification of human IFN.

Cell line	Origin	Morphology	Source
Hep2	Human larynx carcinoma	Epithelial	ATCC[a]/CCL231
WISH	Human amnion	Epithelial	ATCC/CCL25
A549	Human lung carcinoma	Fibroblastic	ATCC/CCL185
GM-2504	Human skin (trisomic chr 21)	Fibroblastic	HGMCR[b]/GM-2540E
GM-2767	Human skin (trisomic chr 2)	Fibroblastic	HGMCR/GM-2767B
2D9	Human glioblastoma	Fibroblastic	W. Däubener (Germany)
MDBK	Bovine kidney	Epithelial	ATCC/CCL22
EBTr	Embryonic bovine trachea	Fibroblastic	ATCC/CCL44

[a] ATCC, American Type Culture Collection.
[b] HGMCR, Human Genetic Mutant Cell Repository.

some degree of species specificity and human cells appear the best choice for the assay of human IFNs. Alternatively, some cell lines, e.g., Vero, from closely related species such as monkeys may be usable for human IFN potency estimations. IFN-α, and to a lesser extent IFN-β, are active in bovine and ovine cell lines; the bovine kidney cell line, MDBK, has been widely used. The most commonly used cell lines for antiviral assays for the titration of human IFNs are listed in Table 9-1. The viruses which are commonly used in antiviral assays include murine encephalomyocarditis virus (EMCV), Mengo virus, vesicular stomatitis virus (VSV), Semliki Forest virus (SFV), and Sindbis virus. These are all RNA viruses which replicate well in the absence of IFN, often causing a cytopathic effect in infected cells [49]. While they infect and replicate in human cell lines, these viruses are not known to be pathogenic, except in extreme circumstances, in humans, and are thus relatively safe to use.

Antiviral assays follow the basic schedule of: (i) addition of serial dilutions of IFN preparations to cell monolayers in the wells of microtitre plates; (ii) incubation at 37 °C for 20–24 h to establish the antiviral state; (iii) infection of cells with the 'challenge' virus; (iv)incubation at 37 °C for 20–24 h to produce viral manifestations, e.g., a cytopathic effect in unprotected and partially-protected cells; and (v) processing assays by addition of stain or other means to render quantifiable in a subjective/objective manner the antiviral effect of IFN [48,49]. Staining, for example, enables the dilution of IFN at which ~ 50 % of cells in the monolayer are protected, and thus alive, to be identified visually. More objective analysis may be made by stain elution and the spectrophotometric measurement of optical densities in individual wells included in the IFN titration. Optical density values (ordinate) are then plotted against the (log) dilution, or reciprocal dilution (abscissa), to generate dose–response curves. Comparison of dose–response curves of test preparations of IFN against the dose–response curve of the IFN standard of assigned potency permits calculation, normally by computer-assisted means, of the potencies of the test preparations of IFN (see Section 9.3.1 for discussion of biometric validity). Potency values should be expressed in international units (IU).

Besides antiviral assays, the potency of IFN preparations can be assessed by a variety of other quantitative bioassays. For example, IFN-α, IFN-β, and IFN-ω, but not IFN-γ, have antiproliferative activity in the human Burkitt's lymphoma cell line, Daudi [50, 51]. The Daudi cell line has appeared the most sensitive among a variety of human tumor-derived cell lines, [^{3}H]thymidine incorporation being the most common means of quantifying the effect of type I IFNs. Dose–response curves are generated by plotting c.p.m. incorporated versus reciprocal dilution and analysed in a similar way to those generated by antiviral assays. In addition, IFNs are cell-differentiating agents, inducing the expression of several cell surface- and intracellular-proteins. Cell surface proteins include class I MHC antigens (all IFNs induce/augment expression) [52], class II MHC antigens (only IFN-γ induces *de novo* expression) [53], intercellular adhesion molecule-1 (ICAM-1, -IFN-γ is the best inducer) [54] and the 9-27 (Leu13) antigen (type I IFNs are the best inducers) [55]. Usually, the up-regulated expression of cell surface antigens is detected using specific monoclonal antibodies. These bind to 'fixed' cells at the end of incubation with IFN, and are themselves quantified by the addition of an anti-murine IgG-enzyme conjugate and routine ELISA procedures. The amount of color developed is proportional to the expression of cell surface antigen and this is related to the dose of IFN. Again, dose–response curves are generated; this time absorbances or optical densities are plotted against the reciprocal dilutions. Such assays are often referred to as bio-ELISA or bioimmunoassays [54].

IFNs also induce a spectrum of intracellular proteins. Where these are enzymes, e.g., indolamine 2,3-dioxygenase (IDO), the effect of IFN may be quantified by measurement of enzymatic activity, the conversion of L-tryptophan to *N*-formyl-kynurenine in the case of IDO [56]. Where they are inert, these IFN-induced proteins are detected by specific antibodies. One such example is the MxA protein, an antiviral protein involved in resistance to influenza virus, which is highly inducible in type I IFN-stimulated cells [57]. Alternatively, use can now be made of IFN-inducible gene promoters. For example, the IFN-inducible Mx gene promoter may be linked to firefly luciferase coding gene sequences, and this construct used to transfect cells. Selected transfected cells containing Mx-Luc then respond to IFN by producing luciferase which, with an appropriate luciferase assay reagent, produces fluorescence, that may be measured spectrophometrically [58].

Lastly, IFNs appear to counteract the actions of other cytokines, and the inhibitory effect can be the basis of quantitative assessment of IFN activity. For example, type I IFNs inhibit the stimulatory effect of interleukin-5 (IL-5) or granulocyte-macrophage colony-stimulating factor (GM-CSF) on the proliferation of the human leukemia cell line TF-1 [59].

In summary, there is a variety of bioassays, with different measurable parameters, that are suitable and useful for the quantification of IFN potency. Having such a largesse of methods and activities to test may help to understand aspects of IFN biology that are likely to be involved in disease processes on the one hand, and that underly beneficial/adverse effects in the IFN therapy of human diseases on the other. However, this diversity of methods and IFN actions is also a cause of concern for the standardization of IFNs, since there are difficulties in establishing standards suitable for use in all types of bioassay. For instance, is an IFN standard whose potency is defined solely by antiviral activity really suitable for calibrating the antiproliferative activity of IFN?

9.3.3 Design of Bioassays

Most bioassays are now carried out using 96-well microtiter plates. Such bioassays utilize a systematic design related to the row and column structure of these plates. This has a variable influence on cell and virus growth and thus the positions within the 96-well plate in which test preparations are placed in relation to the standard often influence the shape of dose–response curves and potency determinations. Where there is replication of dilution series of the same preparation of IFN, statistically significant differences are likely to be observed between the replicate series, even when these replicates are placed in neighboring rows or columns of the same microtiter plate and started with aliquots from the same stock solution [60].

Besides positional effects, non-random variability of responses in bioassays in general may result from a variety of reasons. These include:

1. The variable accuracy and effectiveness of multi-channel pipets for serial dilution of test and standard preparations. These can introduce untoward biases which contribute to uneven responses.
2. The inherent difficulty of addition of a constant number of cells to the individual wells of a microtiter plate . Homogeneous single-cell suspensions should be used, but this may not be easy to achieve with naturally adherent cells, and settling effects within a cell suspension can lead to greater numbers of cells being transferred to microtiter (plates which are last in the series to be filled than those which are first in the series. Point (1) may also apply if multi-channel micropipeters are used to transfer the cell suspension to microtiter plates.
3. The diverse factors which affect the maintenance of cell monolayers or suspension in microtiter plate wells, and the replication of viruses (where used in antiviral assays). Variation in cell culture materials, especially calf serum, of which different batches can have markedly different properties, can have different effects on cells. Other inadvertant variations may also affect cell responsiveness. For example, the temperature and pH controls in the incubators in which the microtiter plates are placed may vary from time to time.
4. Microbial contamination of the cell line, especially by mycoplasma. This may have profound effects on cell responsiveness and the degree of these effects may vary from occasion to occasion.
5. Inadvertant variation introduced by staining procedures, or other procedures required to process the assays, both within an assay and among assays carried out on different occasions. These variations may reflect variable accuracy and effectiveness of multi-channel micropipets in the delivery of stain or other mate-

rial to microtiter plate wells. The condition of cells at the end of the assay, or at the time of assay processing, can introduce further variation. This may reflect uneven distributions of cells among columns or rows of microtiter plate wells, or uneven growth or survival of cells in wells. For example, it has been observed in many assays that utilize microtiter plates that the growth and survival of cells may be abnormal in the outer wells compared with that in the inner wells.

Thus, the design of assays and the control of factors affecting assay responses are of great importance for ensuring accurate and reproducible estimates of relative potency. Experience accumulated from the data obtained in international collaborative studies to evaluate candidate IFN and cytokine standards has shown that certain assay designs are to be preferred. It has been found that designs in which each preparation is tested on several plates in any assay with independent serial dilutions of the preparation on each plate have generally given more accurate and reproducible estimates of relative potency [60]. Moreover, where several preparations are tested on each of several plates, designs in which each preparation is placed at different positions on the different plates should be considered to counteract positional effects. A duplicate of one test preparation should be included in each assay to provide a check on the control of extraneous factors. Complete randomization over all factors other than dose of IFN (or cytokine) which affect assay responses may not be feasible. However, careful design can permit some measure of the influence of such factors and more valid use of classical methods of analysis.

9.4 Interferon Standards

Attempts to prepare reference preparations or standards for IFNs began soon after their discovery in 1957. The main work in this area was carried out in what was then the Medical Research Council (MRC) laboratory sited in Hampstead, London. (This facility subsequently became The National Institute for Biological Standards and Control in 1975, and transferred to South Mimms, Hertfordshire, in 1987). Initially, lyophilized reference preparations for chicken IFN (British Research Standard, catalog number 62/4) and monkey (simian) IFN (catalog number 63/1), prepared from rather crude IFN preparations, were developed. The first two figures in the catalog number indicate the year in which the reference preparation was made, i.e., 62/4 = 1962. Later in that decade a further preparation of chicken IFN (67/18) and new reference preparations of human leukocyte IFN, including the MRC Research Standard B (69/19), were produced. Reference preparations for mouse IFN, rabbit IFN, and human fibroblast IFN were subsequently developed in the U.K. and the U.S. during the 1970s . With hindsight, it is now seen that the development of these human, mammalian, and avian IFN reference preparations was carried out in complete ignorance of the structure of IFN molecules and of their molecular heterogeneity. This undoubtedly has had an impact on more recent attempts to develop IFN reference preparations, leading to issues concerning the definition of international unitages and their continuity (as discussed more fully below).

In 1969, the International Association of Biological Standardization held an International Symposium on Interferon and Interferon Inducers at which it was recommended that specific preparations of IFNs for human, chick, mouse, and rabbit, to which an agreed unitage of activity had been assigned, be adopted as Research Standards. From then on, these MRC Research Standards, as well as others prepared by the NIH, Bethesda, were made available and used in assays of activity of IFNs.

The next decade saw the first clinical trials with human IFN. Although the results of these trials were generally disappointing, interest in IFNs continued to increase with the demonstration of their antitumor activity. By 1977, the WHO recognized an urgent need to establish International Standards for IFNs [61]. On the recommendation of WHO ECBS, a study group to review the status of the then current research standards was convened in September 1978 (Woodstock, Illinois, U.S.A.) by the NIH and co-sponsored by WHO. The study group proposed that as Research Standards for human leukocyte, mouse, rabbit, and chick IFNs (adopted in 1969) [62] had been widely accepted internationally, and as the subsequent NIH standard for human fibroblast IFN had also been used extensively in many countries, that they were considered also as International Standards [63]. The findings of the study group were considered at the 30th ECBS meeting in November 1978 and the following five preparations were adopted as International Reference Preparations (IRP) of IFNs:

1. Preparation 69/19 (MRC Research Standard B) as the IRP of IFN, human leukocyte, 5000 IU of activity per ampoule.
2. The NIH preparation G023-902-527 as the IRP of IFN, human fibroblast, 10 000 IU per ampoule.
3. The NIH preparation G002-904-511 as the IRP of IFN, mouse, 12 000 IU per ampoule.
4. The NIH preparation G019-902-528 as the IRP of IFN, rabbit, 10 000 IU per ampoule.
5. Preparation 67/18 (proposed replacement British Research Standard B) as the IRP of IFN, chick, 80 IU per ampoule.

It should be pointed out that the IU for each IFN type and species were non-equivalent, and non-correlatable. Therefore each IRP was to be used only to calibrate future national standards or laboratory standards for the same type/species of IFN contained in the IRP [63].

In addition to the IRP of IFN, human leukocyte, 69/19, a further NIH preparation of similar material designated G023-901-527 with assigned potency of 20 000 IU (against 69/19) was made available [63].

As the momentum of clinical research accelerated in the late 1970s and early 1980s, breakthroughs in basic research led to the cloning of several different molecular species of IFN, and an understanding of the heterogeneous nature of human leukocyte IFN, now designated IFN-α. In 1981, cloned and highly purified IFN-α_2 subtype became available for clinical trials. Subsequently, other IFN-α subtypes, IFN-β (formerly fibroblast IFN), IFN-γ (formerly immune IFN), IFN-ω, and a variety of hybrid and consensus sequence IFN molecules were produced (see Section 9.2.2). These developments led to the call for new candidate IFN standards to be prepared from the highly purified, cloned human IFNs which, following evaluation by inter-national collaborative studies, would serve as International Standards of individual IFN types or subtypes. This call was pursued enthusiastically on behalf of WHO by the successors of the 1978 study group. Three international collaborative studies carried out during the 1980s [64–67] generated data that led to the establishment of International Standards of human IFN-α_1, human IFN-α_{2a}, human IFN-α_{2b}, human lymphoblastoid IFN, human IFN-β, human IFN-β [ser^{17}], and human IFN-γ (Table 9.2). However, the IRP for IFN human leukocyte (69/19) was used, with one excep-tion, as the primary standard to which all new IFN-α materials were calibrated. This broke a central principle in biological standardization in that like was not compared with like. The material in 69/19 was a heterogeneous mixture of up to 13 different

Table 9-2. International standards for IFNs established by the WHO Expert Commitee on Bio-logical Standardization.

Preparation no.	Interferon	Defined activity (IU/ampoule)	Year established	Source
67/18	Chick IFN	80	1978	NIBSC[a]
69/19[b]	Human leukocyte IFN	5 000	1978	NIBSC
Ga23-901-532	Human lymphoblastoid (Namalwa) IFN	25 000	1984	NIH
83/514	Human recombinant IFN-α_1	8 000	1987	NIBSC
Gxa01-901-535	Human recombinant IFN-α_{2a}	9 000	1984	NIH
82/576[b]	Human recombinant IFN-α_{2b}	17 000	1987	NIBSC
Gb23-902-531	Human IFN-β (fibroblast-derived)	15 000	1987	NIH
Gxb02-901-535	Human recombinant IFN-β [Ser^{17}]	6 000	1987	NIH
Gxg01-902-535	Human recombinant IFN-γ	80 000	1995	NIH
Ga02-902-511	Mouse IFN-α	16 000	1987	NIH
Gb02-902-511	Mouse IFN-β	15 000	1987	NIH
Gu02-901-511	Mouse IFN-α/β	10 000	1987	NIH
Gg02-901-533	Mouse IFN-γ	1 000	1987	NIH
G-019-902-528	Rabbit IFN	10 000	1978	NIH

[a] NIBSC, The National Institute for Biological Standards and Control, Blanche Lane, South Mimms, Hertfordshire, EN6 3QG, U.K.; NIH, Research Branch, National Institute of Allergy and Infectious Diseases, National Institutes of Helath, Bethesda, MD 20205 U.S.A.

[b] At its 40th Meeting Report [67], the WHO ECBS recommended that all *new* preparations of human IFN-α should be determined relative to 69/19, the IRP of IFN, human leukocyte. They further recommended that this should also apply to existing preparations of human IFN-α, and that 82/576, IS of human recombinant IFN-α_{2b}, should not be used for relative potency assays. It should be further noted that, because of inconsistent reporting of human IFN-α potency, the current IS of human IFN-α have been re-evaluated in a large International collabora-tive study organized by NIBSC and CBER [69].

IFN-α subtypes, plus small percentages of IFN-β and IFN-ω, whereas the material in a candidate IFN-α$_{2a}$ standard was solely IFN-α$_{2a}$. Additionally, although it was not recognized at the time, the material in 69/19 also contained a range of other cytokines, including IL-1β, TNF-α, and IL-6, which had the potential to affect the biological activity of IFN. It was thought then that the establishment of International Standards of different, well-characterized, IFN-α-type materials would assist manufacturers of IFN-α to harmonize unitages of their products. Regrettably, complications have resulted for a number of potential reasons:

- the continuity of IU from 69/19 to other IFN-α standards remains in doubt because of the dissimilarity of materials compared;
- the assigned potencies of the new IFN-α standards are possibly incorrect due to the fact that collaborative studies were too small (seven to eight participants in most studies) to generate sufficient data from which accurate potency values could be derived;
- the availability of multiple IFN-α International Standards has the potential for leading to inconsistencies in reporting the results of activity assays; and
- the IU has been defined on the basis of the antiviral activity of IFN-α and it remains unclear whether the current International Standards of IFN-α materials are suitable for the calibration of potency assays based on biological properties of IFN-α other than antiviral activity, e.g., antiproliferation or immunomodulation. Such alternative potency assays may have more validity for the activity required for the treatment of target diseases, such as cancer. Similar issues have appeared to affect calibration of assays with new IFN-β and IFN-γ International Standards.

Recognition of these potential problems with the standardization of IFNs has become more apparent in the last decade (1987–1997). Several manufacturers of IFN found that they were unable to make a choice from the list of WHO International Standards that would meet their needs for calibrating their own IFN products. In some cases it was found that, by changing from one International Standard to another, a significant readjustment of potency to an IFN product was required. Potentially, this meant that manufacturers would present a variation to the specifications of their product(s) which might be unacceptable to the regulatory licensing authorities. Confusion could occur in the clinic if for example a patient was receiving a 45 MIU IFN product one day and then, due to a readjustment for assays calibrated with a newly available IFN International Standard, the same product had to be assigned an activity of 8 MIU. In theory, this should not be possible, but that it has occurred in practice [68], illustrates a complex situation that probably relates back to the prior use of poorly characterized and purified, often heterogeneous materials used to develop the original early International Standards of IFN. For the most part, the standardization of other cytokines has not suffered these problems, due to the fact that almost all were developed from highly purified recombinant materials (normally homogeneous preparations of single cytokines) from the middle to late 1980s onwards.

With regard to the standardization of IFN-α, the WHO Consultative Group on Cytokine Standardization (CGCS) requested in 1995 that the U.K. National Institute for Biological Standards and Control (NIBSC) and the U.S.A. Centre for Biologics

Evaluation and Research (CBER) organize an International collaborative study to compare the activities and relative potencies of the several available IFN-α preparations, including those derived from human cells containing mixtures of IFN-α subtypes and those derived by rDNA methods containing single IFN-α subtypes, in different assays [69]. This international collaborative study has been carried out and the analysis of data completed. Seventeen (or a defined subset thereof) ampouled preparations of IFN-α were evaluated by 92 laboratories in 29 countries for their suitability to serve as International Standards for these materials. The preparations were titrated in a wide range of *in vitro* bioassays, including antiviral and antiproliferative assays. A report of the study's finding is in progress. Recommendations to the WHO CGCS and ECBS are expected to be made in 1998. It is hoped that past difficulties encountered in IFN-α standardization will be resolved by choosing the most suitable IFN-α preparations as International Standards.

The WHO CGCS has recently discussed similar problems relating to the standardization of human IFN-β, as were apparent for IFN-α, and has recommended that NIBSC organize a collaborative study to address the pertinent issues.

9.5 Cytokine Standards

A range of cytokine preparations has been developed at NIBSC. These, in many cases, have been evaluated in International collaborative studies in bioassays [70–72]. Analysis of the data of these studies has provided information on the suitability of these cytokine preparations to serve as WHO International Standards of different individual cytokines. Several have now (Table 9-3) been adopted by WHO ECBS and established as full WHO International Standards. However, the rate of discovery of new cytokines has been so rapid that it has been difficult to organize International collaborative studies for each and every one of them. The WHO ECBS therefore has instituted a new category of reference materials which have potency assigned on the basis of data provided by small collaborative studies, often involving two or three participating laboratories. Many different cytokine preparations have recently been accepted on this basis and serve as WHO Reference Reagents for particular cytokines (Table 9.3).

Table 9-3. International standards and reference reagents of cytokines established by the WHO Expert Commitee on Biological Standardization.

Preparation no.[a]	Cytokine	Defined activity (IU or U/ampoule)	Status	Year established
86/632	Interleukin-1α, rDNA, *E. coli*	117 000	IS	1990
86/680	Interleukin-1β, rDNA, *E. coli*	100 000	IS	1990
86/504	Interleukin-2, Jurkat cell line-derived	202	IS	1988
91/510	Interleukin-3, rDNA, *E. coli*	1 700	IS	1995
88/656	Interleukin-4, rDNA, CHO cell-derived	1 000	IS	1995
90/586	Interleukin-5, rDNA, SW25 cell-derived	5 000	RR	1996
89/548	Interleukin-6, rDNA, CHO cell-derived	100 000	IS	1994
90/530	Interleukin-7, rDNA, *E. coli*	100 000	RR	1996
89/520	Interleukin-8, rDNA, *E. coli*	1 000	IS	1996
91/678	Interleukin-9, rDNA, CHO cell-derived	1 000	RR	1996
92/788	Interleukin-11, rDNA, *E. coli*	5 000	RR	1996
95/544	Interleukin-12, rDNA, CHO cell-derived	10 000	RR	1996
94/622	Interleukin-13, rDNA, *E. coli*	1 000	RR	1996
95/554	Interleukin-15, rDNA, *E. coli*	10 000	RR	1996
89/512	Macrophage colony-stimulating factor, rDNA, CHO cell-derived	60 000	IS	1994
88/502	Granulocyte colony-stimulating factor, rDNA, yeast-derived	10 000	IS	1994
88/646	Granulocyte macrophage colony-stimulating factor, rDNA, *E. coli*	10 000	IS	1995
93/562	Leukemia inhibitory factor, rDNA, *E. coli*	10 000	RR	1996
93/564	Oncostatin M, rDNA, *E. coli*	25 000	RR	1996
87/650	Tumor necrosis factor alpha, rDNA, *E. coli*	40 000	IS	1993
87/640	Tumor necrosis factor beta, rDNA, *E. coli*	150 000	RR	1996

[a] All of these preparations are held at The National Institute for Biological Standards and Control (NIBSC), Blanche Lane, South Mimms, Hertfordshire, EN6 3QG, U.K. NIBSC also holds reference materials of no defined status for other cytokines, including Interleukin-10 (92/516), stem cell factor (91/682), flt-3 ligand (96/532), transforming growth factor beta-1 (89/514), transforming growth factor beta-2 (90/696), bone morphogenetic protein-2 (93/574), and for a number of chemokines.

9.6 Conclusions

Significant progress has been made in the biological standardization of cytokines, including IFNs. Problems, however, remain in some cases, particularly with some IFNs where a long history in the development of reference materials, initially from impure IFN preparations, and the heterogeneous nature of certain types of IFN, e.g., IFN-α, have combined to illustrate how difficult it can sometimes be to choose ideal biological standards and to ensure continuity of the international unitage. Nevertheless, it is essential to have International Standards of biologically active cytokines, since these are the only effective means of calibrating 'in-house' standards and biological assays used to measure the activity (potency) of cytokines. They are critical to the development of cytokines as biological medicinal products, both at the pre-clinical and clinical stages. It is imperative that the biological potency of such cytokine products is determined accurately in well-calibrated biological assays to ensure that patients receive safe, effective and consistent dosages. The very successful collaboration between NIBSC (U.K.), CBER (U. S.A.), WHO CGCS, and ECBS in bringing about the development and establishment of International Standards and reference reagents of the various and numerous human cytokines is expected to continue well into the next century.

Acknowledgements

I am indebted to Mrs Deborah Richards for expert typing of this manuscript.

References

[1] World Health Organization. *Tech Rep Series,* 1990, *800,* 181–213.
[2] Isaacs, A., Lindemann, J., *Proc Royal Soc B,* 1957, *147,* 258–267.
[3] Stewart, W. E. I., in: *The Interferon System:* Vienna, New York: Springer-Verlag, 1979, pp. 134–183.
[4] Fantes, K. H., in: *Purification, concentration and physico-chemical properties of interferons,* Finter, N. B. (Ed.), Amsterdam: North-Holland, 1966, pp.118–180.
[5] Cantell, K., Hirvonen, S., *Tex Rep Biol Med,* 1977, *35,* 138–144.
[6] Knight, E., *Proc Natl Acad Sci USA,* 1976, *73,* 520–523.
[7] Rubinstein, M., Rubinstein, S., Familetti, P. C., et al. *Science,* 1978, *202,* 1289–1290.
[8] Gresser, I., Maury, C. J., Brouty-Boyé, D., *Nature,* 1972, *239,* 167–168.
[9] Nagata, S., Mantei, N., Weissmann, C., *Nature,* 1980, *287,* 401–408.
[10] Taniguchi, T., Mantei, N., Schwarzstein, M., et al. *Nature,* 1980, *285,* 547–549.
[11] Capon, D. J., Shepard, H. M. and Goeddel, D. V., *Mol Cell Biol,* 1985, *5,* 768–779.
[12] Adolf, G. R., *J Gen Virol,* 1987, *68,* 1669–1676.
[13] Senda, T., Shimazu, T., Matsuda, S. et al. *EMBO J,* 1992, *11,* 3193–3201.
[14] Uze, G., Lutfalla, G. l., Gresser, I., *Cell,* 1990, *60,* 225–234.
[15] Novick, D., Cohen, B., Rubenstein, M., *Cell,* 1994, *77,* 391–400.

[16] Gray, P. W., Leung, D. W., Pennica, D. et al., *Nature*, 1982, *295*, 503–508.

[17] Ealick, S. E., Cook, S. E., Vijay-Kumar, S. et al. *Science*, 1991, *252*, 698–702.

[18] Aguet, M., Dembic, Z., Merlin, G., *Cell*, 1988, *55*, 273–280.

[19] Bazan, J. F., *Cell*, 1990, *61*, 753–754.

[20] Sen, G. C., Lengyel, P., *J Biol Chem*, 1992, *267*, 5017–5020.

[21] De Maeyer, E., De Maeyer-Guignard, J., in: *The Interferon Gene Family*, New York: Wiley Interscience, 1988, pp. 5–38.

[22] Müller, M., Briscoe, J., Laxton, C. et al., *Nature*, 1993, *366*, 129–135.

[23] Vilcek, J., Oliviera, I. C., *Int Arch Allergy Immunol*, 1994, *104*, 311–316.

[24] Le, J., Prensky, W., Yip, Y. K. et al., *J Immunol*, 1983, *131*, 2821–2826.

[25] Collins, T., Korman, A. J., Wake, C. T. et al., *Proc Natl Acad Sci USA*, 1984, *81*, 4917–4921.

[26] Goeddel, D. V., Leung, D. W., Dull, T. J. et al., *Nature*, 1981, *290*, 20–26.

[27] Kauppinen, H.-L., Hirvonen, S., Cantell, K., *Methods Enzymol*, 1986, *119*, 27–35.

[28] Adolf, G. R. *Virology*, 1990, *175*, 410–417.

[29] Phillips, A. W., Finter, N. B., Burman, C. J. et al., *Methods Enzymol*, 1986, *119*, 35–38.

[30] Zoon, K. C., Miller, D., Bekisz, J. et al., *J Biol Chem*, 1992, *267*, 15210–15216.

[31] Streuli, M., Nagata, S., Weissmann, C., *Science*, 1980, *209*, 1343–1347.

[32] Dworkin-Rastl, E., Dworkin, M. B., Swetly, P. J., *Interferon Res*, 1982, *2*, 575–585.

[33] Adolf, G. R., Kalsner, I., Ahorn, H. et al., *Biochem J*, 1991, *276*, 511–518.

[34] von Gabain, A., Lundgren, E., Ohlsson, M. et al., *Eur J Biochem*, 1990, *190*, 257–261.

[35] Alton, K., Stabinsky, Y., Richards, R. et al., in: Production, characterization and biological effects of recombinant DNA derived human IFN-α and IFN-γ analogs: De Maeyer, E., Schellekens, H. (Eds.), Amsterdam: Elsevier, 1983, pp. 119–128.

[36] Horisberger, M., Di Marco, S., *Pharmacol Ther*, 1995, *66*, 507–534.

[37] Ozes, O. N., Reiter, Z., Klein, S. et al., *J Interferon Res*, 1992, *12*, 55–59.

[38] Blatt, L., Davis, J., Klein, S. et al., *J Interferon Cytokine Res*, 1996, *16*, 489–499.

[39] Mark, D. F., Lu, S. D., Creasey, A. A. et al., *Proc Natl Acad Sci USA*, 1981, *81*, 5662–5666.

[40] Utsumi, J., Mizuno, Y., Hosoi, K. et al., *Eur J Biochem*, 1989, *181*, 545–553.

[41] Adolf, G. R., Frühbeis, B., Hauptmann, R. et al., *Biochim Biophys Acta*, 1991, *1089*, 167–174.

[42] Adolf, G. R., Maurer-Fogy, I., Kalsner, I. et al., *J Biol Chem*, 1990, *265*, 9290–9295.

[43] Rinderknecht, E., O'Connor, B. H., Rodriguez, H. et al., *J Biol Chem*, 1984, *259*, 6790–6797.

[44] Rinderknecht, E., Burton, L. E., in: *Biochemical characterization of natural, and recombinant, IFN-gamma;* Kirchner, H., Schellekens, H. (Eds.), Amsterdam: Elsevier, 1984, pp. 397–402.

[45] Bangham, D. R., in: *Assays and Standards:* Gray, C. C., James, V. H. T. (Eds.), London: Academic Press, 1983, pp. 256–297.

[46] Bliss, C., in: *The Statistics of Bioassays:* New York: Academic Press, 1952, pp. 256–297.

[47] Finney, D. J., in: *Statistical methods in Biological Assay,* 3rd edition, London: Charles Griffin Co Ltd, 1978.

[48] Grossberg, S., Jameson, P., Sedmark, J., in: Came, P., Carter, W. (Eds.), Berlin: Springer, 1983, pp. 23–43.

[49] Meager, A., in: *Quantification of interferons, by anti-viral assays and their standardization;* Clemens, M. J., Morris, A. G., Gearing, A. J. H. (Eds.), Oxford: IRL Press, 1987, pp. 129–147.

[50] Adams, A., Strander, H., Cantell, K., *J Gen Virol*, 1975, *28*, 207–217.

[51] Nederman, T., Karlström, E., Sjödin, L., *Biologicals*, 1990, *18*, 29–34.

[52] Hermodsson, S., Strannegard, O. J., Jeansson, S., *Proc Soc Exp Biol Med*, 1984, *175*, 44.

[53] Gibson, U. E. M., Kramer, S. M., *J Immunol Methods*, 1989, *125*, 105–113.

[54] Meager, A., *J Immunol Methods*, 1996, *190*, 235–244.

[55] Deblandre, G., Marinx, O., Evans, S., et al., *J Biol Chem*, 1995, *270*, 23860.

[56] Daubener, W., Wanagat, N., Pilz, K., et al., *J Immunol Methods*, 1994, *168*, 39–37.

[57] Ronni, T., Melen, K., Malygin, A. et al., *J Immunol,* 1993, *150,* 1715–1726.

[58] Canosi, U., Mascia, M., Gazza, L. et al., *J Immunol Methods,* 1996, *195,* 55–61.

[59] Mire-Sluis, A. R., Page, L. A., Meager, A. et al., *J Immunol Methods,* 1996, *195,* 55–61.

[60] Gaines Das, R. E., Meager, A., *Biologicals,* 1995, *23,* 285–297.

[61] Memorandum, *Bulletin WHO,* 1978, *56,* 229–240.

[62] *International Symposium on Standardization of Interferon and Interferon Inducers:* Basel, New York: Karger, 1969, p. 328.

[63] Interferon Standards: a memorandum, *J Biol Standardization,* 1979, *7,* 383–395.

[64] WHO Report on the Standardization of Interferons. *WHO Tech Rep Ser,* 1983, *687,* (Annex 1), 35–60.

[65] WHO Report on the Standardization of Interferons. *WHO Tech Rep Ser,* 1985, *725,* 28–64.

[66] WHO Report on the Standardization of Interferons. *WHO Tech Rep Ser,* 1988, *771,* (Annex 1), 37–87.

[67] WHO Expert Committee on Biological Standardization 45th Report. *WHO Tech Rep Series,* 1995, *858,* 6, 18.

[68] Paty, D. W., Li, D. K. B., UBC MS MRI Study Group, et al., *Neurology,* 1993, *43,* 662–667.

[69] Mire-Sluis, A., Gaines Das, R., Zoon, K. et al., *J Interferon Cytokine Res,* 1996, *16,* 637–643.

[70] Meager, A., Gaines Das, R. E. J., *J Immunol Methods,* 1994, *170,* 1–13.

[71] Mire-Sluis, A., Gaines-Das, R., Thorpe, R., *J Immunol Methods,* 1996, *194,* 1–12.

[72] Mire-Sluis, A. R., Gaines-Das, R., Thorpe, R. et al., *J Immunol Methods,* 1995, *179,* 117–126.

10 The Strategic Role of Assays in Process Development: A Case Study of Matrix-Assisted Laser Desorption Ionization Mass Spectroscopy as a Tool for Biopharmaceutical Development

T. J. Meyers, P. G. Varley, A. Binieda, J. A., Purvis and N. R. Burns

10.1 Introduction

Without a robust, efficient production process the development of any protein as a pharmaceutical product is not feasible. To achieve such a process it is obviously necessary to be able to monitor the quality, quantity, and consistency of the product. Similarly it is widely appreciated that the finished product attributes of purity, potency, identity, and stability comprise the backbone of any specification. The measurement of these attributes requires appropriate assays. Perhaps what is less well appreciated is that the development of the assays that will be applied to the specification of a licensed, finished product is intimately linked to the development of the production process itself. Both process and assay development go through series of inter-dependent iterative cycles of development and that the quality of any manufacturing process and its resultant product is fundamentally linked to the quality of the analytical support during its development. The only certain feature of the purification process as it moves from research through clinical evaluation to full-scale, market production is that it will change. Without the analytical tools to provide steering through the effects of these process changes, there is little hope of successfully developing a product. It has often been said of biologicals that 'the process is the product', by which it is meant that the manufacturing process defines the physical attributes of the product; if this is so, then it is equally true that the process can be equally well defined as the analytical scrutiny to which it has been subjected. In this paper we illustrate these points through data on the development, validation, and application of a single assay, Matrix-assisted laser desorption ionization-time of flight spectroscopy (MALDI-TOF), in the context of the process development for a specific protein.

The complexity of protein-based pharmaceuticals is such that no single analytical method is capable of describing all the quality attributes of a product of this class. This has resulted in the quality of these products being controlled by a combination of very tight regulations,on the manufacturing process and the establishment of an assay 'portfolio' in which a battery of assays are used to describe the quality of the final product. However, recent advances in analytical methodologies are moving the emphasis away from the regulation of the manufacturing process towards final product analysis [1]. Despite this shift in emphasis there are two main reasons why a well-characterized production process is required for protein pharmaceutical development. First, no process will deliver a truly homogeneous product as a variety of post-translational modifications introduce microheterogeneity. Second, as a product moves through clinical evaluation changes to the process will occur which will inevitably lead to changes in the composition of the product. This molecular variation may be reflected in the pharmacokinetics, toxicity, or clinical efficacy of the product. Therefore, not only is there a need to employ the increasingly sophisticated analytical methods to the final bulk and finished product, but also in process development and process characterization.

At the forefront of the improvement in analytical methods has been mass spectroscopy, particularly electrospray (ES-MS) [2] and matrix-assisted laser desorption/ionization time-of-flight mass spectroscopy (MALDI) [3]. Since the molecular mass of a protein is dictated by its amino acid composition, the accurate determination of molecular mass can act as a valuable confirmatory identity test. ES-MS is a higher-resolution method than MALDI (precision figures range from 0.01–0.005 % and 0.1–0.05 %, respectively) [4,5]. However, MALDI has a number of advantages over ES-MS: notably its high sensitivity, relative insensitivity to buffer components in the sample; the tendency not to be subject to compound specific suppression effects; the ease of sample preparation and analysis; and the relatively low cost of the instrument [6,7]. These factors all contribute to making MALDI particularly useful for process development and identity testing.

This paper describes a formal validation study on MALDI and the application of this technique in the context of the process development and quality control of a recombinant protein intended for clinical use. The protein in question is BB-10010, which is an engineered variant of Macrophage Inflammatory Protein-1α (MIP-1 α). BB-10010 differs from wild-type MIP-1 α by a single amino acid which dramatically alters its association equilibrium, thereby rendering it soluble in physiological buffers [8]. BB-10010 is currently being clinically evaluated as an adjunct to chemotherapy in cancer. More recently the discovery that MIP-1 α prevents HIV proliferation [9] has heightened expectations for the molecule.

10.2 Materials and Methods

Each of the variant proteins were purified from culture supernatant derived from recombinant yeast strains containing vectors with genes that encode a 69-amino acid protein based upon MIP-1 α [8]. The yeast were cultured and expression induced and the variants prepared as described previously [10,11]. The fidelity and homogeneity of the BB-10010 was confirmed independently by ES-MS, N-terminal sequencing, SDS–PAGE and reversed phase HPLC. All samples were stored in phosphate-buffered saline at a concentration of 2 mg mL^{-1} prior to use. The molecular weights of all samples were determined by direct calibration with bovine insulin as the internal calibrant, molecular weight 5733.6 Da (Sigma Chemical Co., Poole, Dorset, U.K.). All solvents or other chemicals used in this work were of HPLC

Table 10-1. Summary of molecular weight data for analysis of protein BB-10010 and its variants.

Parameter	BB-10010 (7668.6 Da)[a]		Variant A (7712.6 Da)[a]		Variant B (7685.5 Da)[a]	
	All data	Selected data[b]	All data	Selected data[b]	All data	Selected data[b]
Experimental mean (Da)	7669.3	7668.5	7715.4	7713.8	7691.7	7678.0
Standard deviation (Da)	7.04	5.55	6.99	4.4	14.7	4.78
No. of observations	48	39	48	43	48	39
CV (%)	0.09	0.07	0.09	0.06	0.19	0.06

[a] Theoretical molecular weight.
[b] Contains only data selected using pre-defined criteria that the peak width at half-height is < 1.5%.

grade (BDH, Poole, Dorset, U.K.). Table 10-1 details the MIP-1 α variants used for the validation.

MALDI was performed on a LaserMAT 2000 mass spectrometer (Thermo BioAnalysis, Hemel Hempstead, Hertfordshire, U.K.) [12]. All samples were analysed in 0.5 µL of α-cyano, 4-hydroxycinnamic acid matrix solution concentration 10 mg mL in 70% acetonitrile, 30% water, 0.1% trifluoroacetic acid, and 2 pmol of bovine insulin internal calibrant. Purified proteins were analyzed at concentrations of 0.2 and 0.02 mg mL^{-1} prepared by dilution of a 2 mg mL^{-1} protein stock solution with water. A 0.5-µL aliqout of each protein sample was applied to each sample slide, giving loadings of 13 pmol and 1.3 pmol, respectively.

10.2.1 Validation Study

Parameters studied for validation were specificity and precision (repeatability). Specificity was validated by demonstrating the ability of MALDI to measure and distinguish accurately between the molecular weights of BB-10010 and single amino acid variants of BB-10010 (shown in Table 10-1). The repeatability was examined by assaying BB-10010 on 12 slides with bovine insulin as the internal calibrant. All four target sites on each slide, six containing 13 pmol of BB-10010 and six containing 1.3 pmol of BB-10010, were analyzed. A spectrum was deemed acceptable if the width of the peak at half-peak-height was < 1.5% of the protein's molecular weight.

10.2.2 Statistical Analysis

SAS Version 6.08 was used to perform all the statistical analysis. A Wilcoxon rank sum test was performed between the mean molecular weight for the BB-10010 and each of the BB-10010 variants. From the Wilcoxon Rank Sum test, a significant difference at the 95 % confidence level between the BB-10010 and each BB-10010 variant was applied to show assay specificity.

10.2.3 In-process Analysis

MALDI has been used to characterize and develop the BB-10010 production process. Samples from fermentation harvests and expanded bed adsorption chromatography (EBA) were 0.2 μm filtered to remove any cell debris that was present. The fermentation harvest and in-process samples were diluted with water, where required, to allow a minimum of ~1.3 pmol to be loaded in 0.5 μL. Waste fractions, where necessary were concentrated using C18 Sep Pak cartridges (Waters, Milford, MA) by eluting the sample in 90 % acetonitrile, 10 % water, and 0.1 % trifluoroacetic acid and drying down the eluate by vacuum centrifugation.

10.3 Results

10.3.1 Identity Test – Validation

Figure 10-1 shows typical spectra obtained for BB-10010 and the two BB-10010 variants. The results are summarized in Table 10-1. Data for each protein are presented in two columns, the first column summarizes all the data. The second column summarizes the results from the data that meet the pre-defined criteria that the width of the sample peak at half-weight is < 1.5 % of the protein's molecular weight (115 Da for BB-10010). This is a specific, pre-defined rule which we have found to be useful in ensuring that an occasional poor-quality spectrum does not impact upon identity testing of purified BB-10010, and was included in the validation protocol. The advantage of applying this criterion to our data is illustrated in the study results. Some spectra are not included in the final analysis, resulting in an improvement in both the precision and the accuracy of the molecular weight determinations.

The basis of the validation of the identity test is to demonstrate the ability of the technique to distinguish between BB-10010 and very closely related variants. None of the mean experimental molecular weights differed from the theoretical molecular weight by more than 3.4 Da. In the case of BB-10010, the mean experimental molecular weight was 7668.5 Da, a mere 0.1 Da difference from the theoretical molecular weight. The samples all had a coefficient of variation of less than 0.10 %. A Wil-

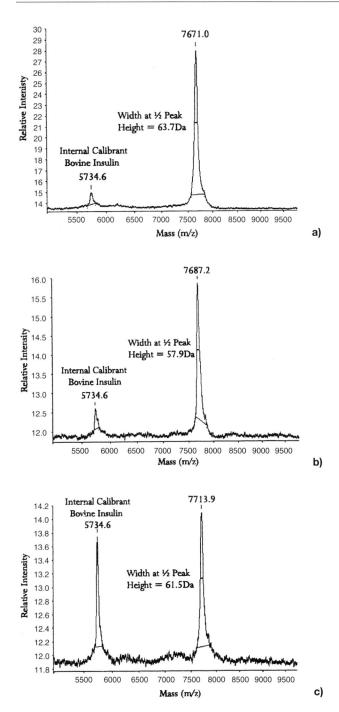

Fig. 10-1. Typical MALDI mass spectra for each of the proteins used in the validation study: (A) BB-10010, (B) Variant A, and (C) Variant B.

Table 10-2. Effect of sample loading upon mass determination of BB-10010 protein and its variants.

Loading	BB-10010 (7668.6 Da)[a]		Variant A (7712.6 Da)[a]		Variant B (7685.5 Da)[a]	
	13 pmol	1.3 pmol	13 pmol	1.3 pmol	13 pmol	1.3 pmol
Experimental mean (Da)	7668.5	7668.6	7714.5	7713.1	7687.7	7686.3
Standard deviation (Da)	5.40	5.79	4.84	3.97	5.58	3.79
No. of observations	17	22	20	23	20	19
CV (%)	0.07	0.08	0.06	0.05	0.07	0.05

[a] Theoretical molecular weight.

coxon rank sum test was used to compare mean experimental calibrated results of BB-10010 against each of the BB-10010 variants. The P value was 0.05 in each case. This indicates that the molecular weights of BB-10010 and its variants can easily be distinguished one from another. This is illustrated by the calculation of the 95 % confidence interval for the BB-10010 data. Using the number of determinations as $n = 10$, this was calculated as the mean molecular weight $\pm$ 3.97 Da. This confidence interval implies that we can be 95 % confident that the population mean for BB-10010 is between the values 7664.5 to 7672.5 Da, when 10 measurements are made. On this basis we routinely perform 10 measurements at five different concentrations across the working range of the spectrometer (for this protein) on a single sample in a formal BB-10010 identity test.

Table 10-2 compares the results from 1.3 pmol and 13 pmol loadings of BB-10010 and the two BB-10010 variants. These amounts approximate the extremes of the working range for the spectrometer for this particular protein. There are no significant differences in the consistency or absolute values of the molecular weights determined at each of the sample loadings. The molecular weight determination is therefore independent of sample loading across the working range of the instrument.

10.3.2 Applications In-Process

A key advantage of the MALDI technique is its relative robustness in the presence of buffer salts and other contaminants. The technique is therefore able to be used to analyze proteins during their production without excessive sample preparation which may in itself affect the analysis. Figure 10-2 shows a spectrum of a typical fermentation harvest from a BB-10010 pilot-scale production run. Table 10-3 shows data concerning the repeatability of such measurements. The fermentation harvests had an overall mean experimental molecular weight of 7668.8 Da, differing from the theoretical molecular weight by 0.2 Da. The Wilcoxon rank sum test compared the mean experimental calibrated results for the BB-10010 used in the study against the overall mean for the fermentation harvests. The P value was 0.762.

Table 10-3. Summary of the analysis of BB-10010 fermentation harvests using the original expression vector.

Parameter	Data from eight production runs
Experimental mean (Da)	7668.8
Standard deviation (Da)	2.96
No. of observations	41
CV (%)	0.04

The samples had an overall coefficient of variation of 0.1 %. The mean and coefficient of variation were similar to the data obtained for BB-10010 during the validation study.

Subsequent to the validation of MALDI as an identity test, the technique was applied to the development and characterization of the BB-10010 production process. This involved the characterization and optimization of the process to ensure the consistent production of optimal amounts of high-quality material. An understanding of the process on the molecular level also allows the impact of changes to the process upon product quality to be assessed. This information can then be used to ensure that any equivalence issues are minimized. One such change required to facilitate scale-up during BB-10010 development has been the use of an alternative expression vector. Figures 10-2 and 10-3 show typical fermentation harvests from the production of BB-10010 using the different expression vectors. Two additional peaks of larger mass than BB-10010 can be seen in the spectrum in Fig. 10-3. These peaks were not observed with the original expression vector (Fig. 10-2). MALDI was used to confirm the clearance of these impurities from the BB-10010

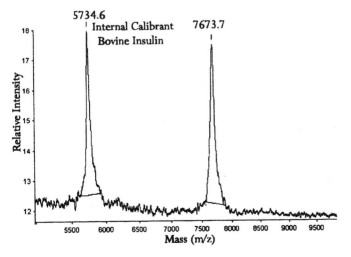

Fig. 10-2. Typical MALDI spectrum of a fermentation harvest from the BB-10010 production process. This material was produced from the original expression vector.

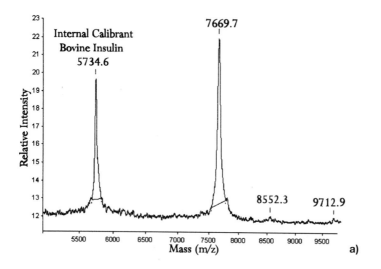

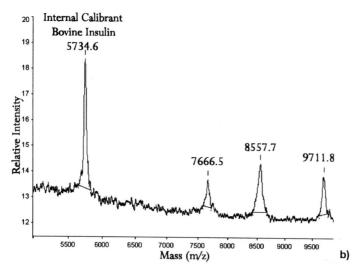

Fig. 10-3. MALDI mass spectra from the EBA stage of the BB-10010 production proces using the new expression vector. (A) Column load (fermentation broth). (B) column flow-through.

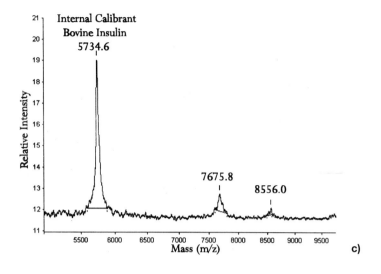

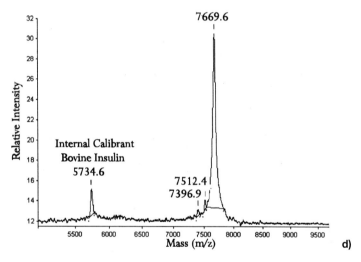

Fig. 10-3. (continued). (C) column wash fraction. (D) BB-10010 product eluate.

product stream (Fig. 10-3). Although the low quantities involved precluded an accurate mass-balance, the presence – and indeed removal – of these impurities was confirmed by examining the discarded fractions. This results in a product eluate from both processes which are indistinguishable.

The MALDI spectrum of the EBA eluate in Fig. 10-3 has a main peak that corresponds to intact BB-10010 and small peaks that correspond in mass to N- or C-terminally degraded protein. This indicates that contaminating exo-proteases are present in the product stream. Figure 10-4 shows how MALDI was used to establish that these proteases are separated from the product during the course of the EBA chromatography. After incubation of the column load and wash fractions at 25 °C for 10

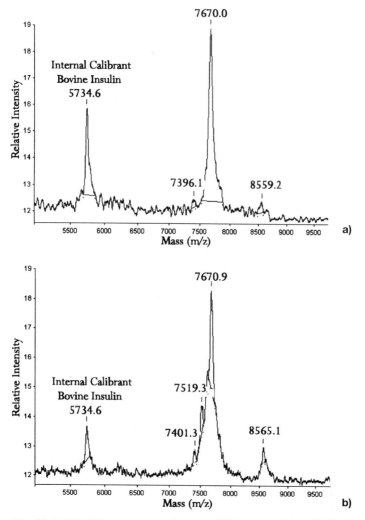

Fig. 10-4. MALDI mass spectra from the EBA stage of the BB-10010 production process. (A) column load material; (B) column wash fraction.

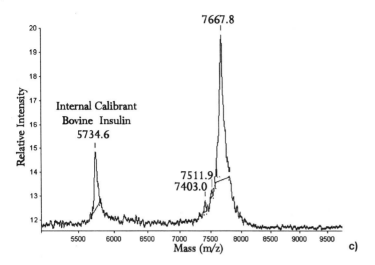

Fig. 10-4. (continued). (C) BB-10010 product eluate.

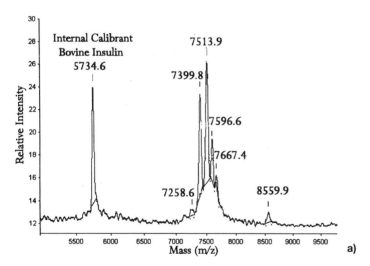

Fig. 10-5. MALDI mass spectra of the samples shown in Fig. 10-4 following a 10-day incubation at 25 °C. (A) column load material.

days the formation of degraded BB-10010 demonstrates that proteases are present in these fractions. The degradation is most apparent in the column wash which no longer contains intact BB-10010. The largest peak in this fraction corresponds to the loss of three residues from the termini (N or C) of the molecule. This indicates that the bulk of the contaminating proteases are contained in this fraction. Interest-

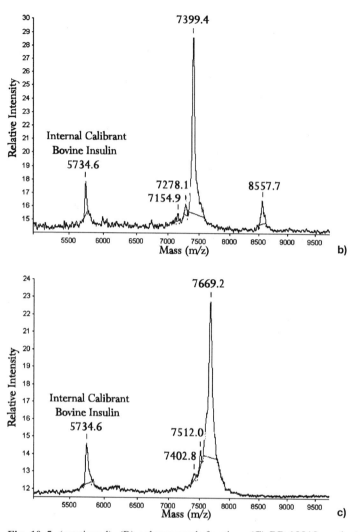

Fig. 10-5. (continued). (B) column wash fraction; (C) BB-10010 product eluate.

ingly, in both the flow-through and wash fractions, the larger contaminating peak (~ 8560 Da) remains unchanged, indicating that the proteases do not affect this impurity. In the BB-10010 product eluate there is no significant difference between the levels of degradation present before (Fig. 10-4) and after the 10-day incubation period (Fig. 10-5). The results therefore demonstrate the removal of proteases from the product stream by this stage of the process.

The application of MALDI in BB-10010 process development is further illustrated by the spectra in Fig. 10-6. This shows analysis of samples from the hydrophobic interaction chromatography (HIC) stage of the production process. The spectrum

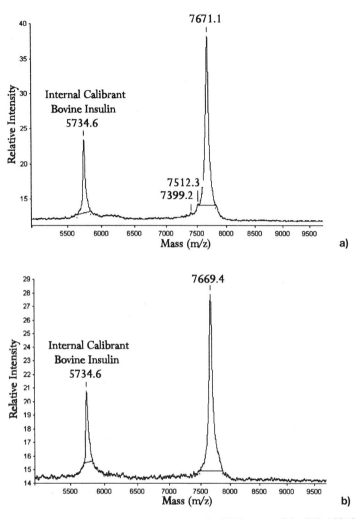

Fig. 10-6. MALDI mass spectra across the HIC stage of the BB-10010 production process. (A) column loading; (B) BB-10010 product fraction.

of the product eluate contains a single peak with a molecular weight that corresponds to intact BB-10010. Degraded or truncated BB-10010 which is present in the column load is eluted in the waste fraction. This illustrates how MALDI has been used to design and optimize the HIC stage of the process in order to separate full-length BB-10010 from any BB-10010 degradation products.

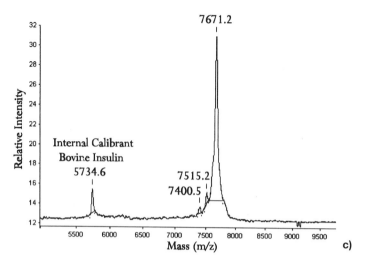

Fig. 10-6. (continued). (C) column waste fraction.

10.4 Discussion

Mass spectroscopy has been at the forefront of recent progress in the analysis and characterization of proteins as pharmaceutical products. As these techniques have become more routine and increasingly accessible, their application in the field of analysis of protein–based pharmaceuticals has focused mainly upon the characterization and understanding of final bulk or finished product. This is illustrated here by the validation study which demonstrates the ability of MALDI to be used as a simple, but effective, identity test for BB-10010. The primary application of the technique is, however analysis of the BB-10010 production process. Here, the ability of the technique directly to analyze the product in crude mixtures, such as fermentation harvests, is utilized to make MALDI an effective tool for development and analysis of the production process.

The data obtained from the validation study show the inherent assay variability which defines the accuracy of the technique. The results demonstrate that when 10 spectra are acquired with an average molecular weight of within 4 Da of 7668.6 Daltons, we can be greater than 95 % confident that the analyte has the molecular weight expected for BB-10010. This is the basis of a powerful identity test.

During this work we have considered the limitations of the technique. For example, it is possible theoretically to have a completely unrelated compound with a similar molecular weight. Alternatively, the protein could be altered by either amino acid substitutions (in itself unlikely), or chemical modifications that result in a mass change of < 4 Da. Any such change would result in the incorrect identification of an analyte as BB-10010. However, most amino acid mutations or chemical modifications would result in a mass change of > 4 Da. For example, only 7.9 % of all the possible single amino acid mutations in BB-10010 would result in a mass change of less than 4 Da (analysis not shown). The combined probability of a change occurring and resulting in a mass change of < 4 Da is therefore very low. For this reason we believe that the MALDI identity test is a valuable and worthwhile addition to the portfolio of assays used to characterize BB-10010. The probability of incorrect identification could be reduced by using electrospray ionization mass spectroscopy or even tandem mass spectroscopy techniques. These techniques do not however display all the unique advantages of MALDI, namely a combination of sensitivity, high tolerance to buffers and salts, and speed and ease of analysis. For example, the use of electrospray would require the protein to be subjected to a desalting step to remove buffer contaminants prior to analysis. Moreover, the exclusive use of tandem mass spectroscopy would prohibit rapid analysis and data interpretation. This would cause problems with effective real-time analysis of BB-10010 production process intermediates, which was the primary goal of this work.

The data obtained from the fermentation harvest are comparable in quality with those from purified material from the validation study. The BB-10010 fermentation harvest contains more contaminants than at any other point during the production process. The results therefore demonstrate how MALDI can provide accurate molecular weight determinations, despite the presence of contaminating buffers, salt, and other small molecules. This is the key advantage of MALDI over other types of mass spectroscopy. This work was subsequently extended to characterize and optimize each stage of the BB-10010 production process using MALDI as the principal analytical tool. The technique has been applied to monitor routinely all stages of the production process to ensure that it is functioning correctly and is under control. The turn-around time of typically 10 minutes, and speed and ease of data interpretation also contribute to make MALDI an extremely convenient and effective in-process monitoring tool.

This work demonstrates how MALDI has been an important factor in determining the quality and consistency of the BB-10010 product and illustrates how product quality relates to, and is best addressed by, process development. The characterization of the final product, although becoming more important as analytical methodologies improve, does not make the detailed analysis of the production process redundant. In fact, it could be argued that the reverse is true. As ever more sensitive methods of analysis are used, any shortcomings in product quality will be more readily identified which in turn will require that more detailed in-process analysis is carried out. MALDI, with its ability to perform fast accurate analysis in relatively crude mixtures, will be at the forefront of this analytical strategy.

References

[1] Henry, C., *Anal Chem,* 1996, *68,* 674A–677A.
[2] Fenn, J. B., Mann, M., Meng, C. K., Wong, S. F., Whitehouse, C. M., *Science,* 1989, *246,* 64–71.
[3] Karas, M., Bhar, U., Geisseman, U., *Mass Spectrom Rev,* 1991, *10,* 335–358.
[4] Nguyen, D. N., Becker, G. W., Riggin, R. M., *J Chromatogr A,* 1995, *705,* 21–45.
[5] Geisow, M. J., *Trends Biotechnol,* 1992, *10,* 432–441.
[6] Stults, J. T., *Curr Opin Struct Biol,* 1995, *5,* 691–698.
[7] Kaufmann, R., *J Biotechnol,* 1995, *41,* 155–175.
[8] Hunter, M. G., Bawden, L., Brotherton, D. et al., *Boold,* 1995, *86,* 4400–4408.
[9] Cocchi, F., DeVico, A. L., Garzino-Demo, A., Arya, S. K., Gallo, R. C., Lusso, P., *Science,* 1995, *270,* 1811–1815.
[10] Clements, J. M., Craig, S., Gearing, A. J. H., Hunter, M. G., Heyworth, C. M., Dexter, T. M., Lord, B. I., *Cytokine,* 1995, *4,* 76–82.
[11] Patel, S. R., Evans, S., Dunne, K., Knight, G. C., Morgan, P. J., Varley, P. G., Craig, S., *Biochemistry,* 1993, *32,* 5466–5471.
[12] Mock, K. K., Sutton, C. W., Cottrell, J. S., *Rapid Commun Mass Spectrometry,* 1992, *6,* 233–238.

11 Quality Control of Protein Primary Structure by Automated Sequencing and Mass Spectrometry

Philip J. Jackson and Stephen J. Bayne

11.1 Introduction

Protein primary structure can be defined in two ways. The simpler definition refers to the linear sequence of amino acid residues in a protein, as specified by the gene sequence which codes for it. This definition assumes that each position in the sequence is occupied by one of the 20 amino acid residues specified by the genetic code and is used to provide a unique description of a particular protein in the databases.

The more complex definition of primary structure includes all of the covalent bonds within a protein, taking into consideration co- and post-translational modifications such as glycosylation, methylation, and disulfide bond formation. Excluded in both definitions are noncovalent interactions which, in addition to covalent bonding, govern the three-dimensional structure of a protein.

This chapter deals with the analysis of protein primary structure at both levels, focusing on applications in the quality control of peptides and proteins, whether synthetic, native, or recombinant, for use in therapeutics, antibody production, and crystallography, for example. Linear sequence determination may be sufficient to verify a protein's identity, characterize contaminants, or detect proteolytic degradation. However, since it may be structurally or functionally crucial for the desired product to have undergone accurate bioprocessing or synthetic modification, more complex characterization strategies, involving both chemical and mass analysis, are also covered.

11.2 Automated Edman Degradation

The idea of analyzing a protein by the sequential chemical cleavage of amino acid residues from the N-terminus was first published in 1930 (for a historical review, see [1]). The major breakthrough in the chemistry occurred in 1950 when Pehr Edman (1916–1977), then of the University of Lund in Sweden, published a manual chemical process, popularly referred to as Edman degradation [2]. This chemistry has remained largely unaltered over the past 48 years, except for automation, adaptations to the original reagent formulations, and increased sensitivity [3].

At present, there are a number of different types of N-terminal sequencing instruments in laboratories around the world, supplied by several manufacturers. As would be expected, there are differences between these instruments, relating to hardware, reagent formulations employed, and means of sample presentation. However, as far as chemistry, operating principles, and limitations are concerned, all of the instruments currently in use (some are 14 years old) are similar enough to allow coverage in a single generic description.

11.2.1 Chemistry

Before sequencing is started, the protein sample must be immobilized on to an inert support material, which is then placed inside a heated reaction cartridge on the instrument. Depending on the sample type or instrument, the target protein may be adsorbed on to a glass-fiber filter [4], a proprietary membrane disk [5], a biphasic sorbant column [6], or a poly(vinylidene difluoride) (PVDF) membrane [7,8]. In all of these cases the sample is retained on the support by noncovalent interactions. For such samples the instrument must incorporate chemicals and delivery regimes which are specifically designed not to cause solubilization and subsequent loss from the reaction cartridge by elution. Sample retention, particularly on glass-fiber supports, can be augmented by the addition of a polymeric quaternary amine carrier known as Polybrene [4].

It is also possible to immobilize proteins covalently (via their amino groups) to a PVDF-type membrane functionalized with isothiocyanate or via carboxyl groups after carbodiimide activation, to an amino-functionalized membrane (see Section 11.2.2).

In essence, all automated protein sequencers (also referred to as sequenators) consist of three subsystems as summarized in Fig. 11-1. The first is the reaction cartridge

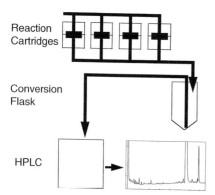

Fig. 11-1. The three subsystems of a modern, automated protein sequencer. Samples are placed in the reaction cartridges where they are subjected, one at a time, to the coupling and cleavage steps of Edman degradation. After cleavage, residues (as ATZ derivatives) are transferred to the conversion flask and subsequently injected (as PTH derivatives) on to a reverse-phase HPLC system for identification and quantitation.

which typically operates in the temperature range 45–55 °C and holds the protein sample. Its construction is instrument-specific but mainly designed according to the particular means of sample presentation. A single instrument may in fact possess more than one reaction cartridge (the maximum at present is four) to increase sample throughput by allowing additional samples to be loaded and sequenced automatically overnight or over a weekend.

The first chemical step in protein sequencing (Fig. 11-2) is termed coupling and occurs with the entry of both the Edman reagent, phenylisothiocyanate (PITC), and the coupling base into the reaction cartridge. The PITC is stored and delivered as a 0.5–10 % (v/v) solution in *n*-heptane or acetonitrile. Typically, only a few microliters are delivered at a time, just sufficient to wet the sample support. The solvent is then driven away by a stream of argon or nitrogen, leaving the PITC, which is volatile only under reduced pressure, in contact with the sample. The coupling base, a tertiary amine such as triethylamine, diisopropylethylamine, or *N*-methylpiperidine

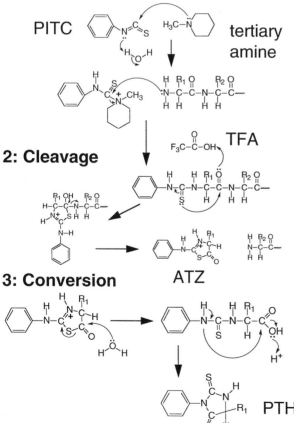

Fig. 11-2. Edman degradation chemistry.

is stored in aqueous solution, sometimes with methanol, ethanol, or propanol as co-solvent (several formulations exist) but delivered in some instruments as a vapor over a period of 3–8 minutes. This delivery method is, as mentioned above, to mini-mize sample wash-out by what would be an effective protein solvent. Liquid base delivery is used in instruments where either the applied volume is precisely metered so that no flowthrough occurs or the sample is covalently immobilized.

One function of the coupling base is to buffer the environment within the reaction cartridge at high pH (9–10) to promote the deprotonation of the protein's terminal α-amino group so that it can react, by nucleophilic addition, with the PITC. The other function is catalytic in that a PITC adduct can also form and subsequently react more rapidly with the terminal amino group by nucleophilic substitution, regenerating the coupling base in the process. Evidence for this mechanism is provided by the obser-vation that coupling efficiency in Edman degradation increases with the leaving-group potential of the tertiary amine [9].

In a typical sequencing cycle, the PITC and coupling base deliveries are repeated two or three more times to drive the reaction as near to completion as possible, then all remaining base is blown from the reaction cartridge by argon or nitrogen. The vast excess of PITC, together with any reaction by-products, are extracted by the delivery of nonpolar solvents such as ethyl acetate, *n*-heptane, 1-chlorobutane, either singly or in combination, or acetonitrile/toluene mixtures. Again, the rationale is to maximize the extraction of everything except the sample, which is now derivatized at its N-terminus with a phenylthiocarbamyl (PTC) group.

The next step to occur in the reaction cartridge is known as 'cleavage' whereby the N-terminal residue side chain forms part of a cyclic derivative after reaction with anhydrous trifluoroacetic acid (TFA) which is delivered either as vapor in a stream of argon or nitrogen, or as a precisely measured pulse of liquid. When vapor delivery is employed, the sequencing chemistry is referred to as 'gas phase' and the optimum duration of this step is 10–20 minutes. The ability to deliver liquid TFA allows this step to be speeded up to 5–10 minutes; however, the delivery volume must be accu-rately controlled to prevent sample wash-out. The cleavage step is complete when the derivative, an anilinothiazolinone (ATZ) undergoes a rearrangement and detaches from the rest of the polypeptide chain, leaving a new N-terminus. In the case of pro-line, the cleavage reaction is relatively slow and leads to the carry-over of a large proportion (about 30%) of the signal from this particular residue into the next cycle. This signal lag accumulates through subsequent cycles, essentially increasing the background and curtailing the potential length of sequence read. Therefore, if the protein sequence is already known, it is usual to incorporate doubled cleavage times into the appropriate cycles to enhance sequencing efficiency for proline-containing samples. Extended cleavage steps are not employed at every cycle because this would increase the total exposure time of the sample to TFA, leading to a rise in unwanted side reactions (see Section 11.2.3).

After cleavage, the ATZ derivative, which possesses the original amino acid resi-due side chain, is extracted with nonpolar solvents, for example, ethyl acetate, 1-chlorobutane, or acetonitrile/toluene and transferred to the conversion flask, the sec-ond subsystem of a protein sequencer. The nature of the side chain, however, modi-fies the extractability of the ATZ such that no single solvent is optimal for all residues. In addition, the sample support medium exerts an influence with the consequence that certain residue signals may be slightly depressed and carried-over. This type of lag, seen more commonly with histidine and arginine, does not accumulate through subsequent cycles, in contrast to proline-associated lag.

At this point, the new N-terminus is ready for the next round of coupling and cleavage. Meanwhile, in the conversion flask at 60–70 °C, the ATZ solution is blown down to a few microliters, then reconstituted in 25 % (v/v) aqueous TFA, which drives a rearrangement to the more stable phenylthiohydantoin (PTH) derivative. Despite the overall stabilization, side reactions which essentially lead to signal reduction can occur. For this reason a scavenger, dithiothreitol at 0.01 % (w/v) is included in the conversion reagent. This additive has a positive effect on the final recoveries of all PTH derivatives, and is most apparent in the case of lysine.

After conversion, which takes 10–20 minutes, the aqueous TFA is evaporated from the conversion flask to leave the PTH ready for reconstitution in a suitable solvent, for example 10–20 % (v/v) aqueous acetonitrile, for injection on to a high-performance liquid chromatograph (HPLC), the third protein sequencer subsystem. Like the ATZ, the PTH retains the original amino acid residue side chain and this determines its elution time from a monofunctional C_{18} reverse-phase column. The mobile phase consists of an aqueous/organic solvent gradient system, with the PTH derivatives detected and quantitiated spectrophotometrically at 269 nm (Fig. 11-3). Each degradation cycle generates a separate chromatogram, therefore a polypeptide sequence is read by interpreting successive chromatograms as shown in Fig. 11-4.

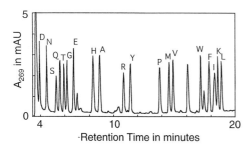

Fig. 11-3. Reverse-phase HPLC of PTH amino acid derivatives. Peaks are identified by the standard one-letter code for amino acids. PTH-Cys is missing because cysteine is only identifiable during sequencing after sample alkylation. In the chromatography system shown here, PTH-pyridylethyl-*S*-cysteine for example, elutes between PTH-Tyr and -Pro. PTH-Lys elutes at the hydrophobic end of the chromatogram because it is ε-PTC-derivatized by the Edman reagent and therefore carries an additional phenyl group.

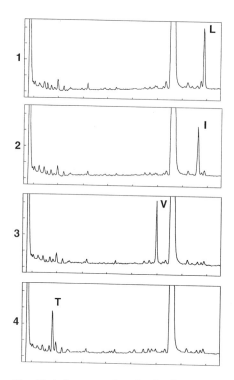

Fig. 11-4. Sequence data for the first four residues of bovine β-lactoglobulin. These are Leu, Ile, Val, and Thr.

11.2.2 Sample Preparation

11.2.2.1 Compatibility with Edman Chemistry

Once an appreciation of the chemistry of Edman degradation has been gained, the necessary criteria for protein or peptide sample preparation are easily rationalized. While it is crucial that the sample should not have undergone any deleterious chemical changes during production, media which are formulated to preserve conformational stability or biological activity are not necessarily compatible with sequence analysis by Edman degradation.

One particular type of chemical modification renders a sample resistant to Edman degradation. As described in the previous section, after reactions involving the Edman reagent PITC and the coupling base, the N-terminal α-amino group becomes derivatized. Such derivatization is only possible by virtue of the amino group's nucleophilicity. If this property is eliminated by prior derivatization, then Edman degradation cannot begin. In other words, the sample is N-terminally blocked.

A large proportion, probably more than 50 % according to current experience [10], of eukaryotic proteins are naturally blocked by co- or post-translational processing.

If the blocked N-terminus is known to be the cyclic, glutamine-derived residue pyrrolidone carboxylic acid (pyroglutamic acid), de-blocking prior to sequencing can be achieved routinely by incubation with pyroglutamate aminopeptidase (5-oxoprolyl peptidase, EC 3.4.19.3) which allows Edman degradation from residue 2 onwards [10]. Unfortunately, for other commonly encountered N-modifications, for example formylation, acetylation, myristoylation, and methylation, there are no reliable, specific de-blocking protocols. An acylpeptide hydrolase (*N*-acylamino-acid releasing enzyme, EC 3.4.19.1) [11] can be employed for the removal of α-*N*-acylated N-terminal residues, as the name suggests. Unfortunately, there are known to be problems with the stability of the enzyme and its low activity on polypeptides larger than 10–15 residues [11]. Chemical methods using, for example, trifluoroacetic acid vapor [12] or concentrated hydrochloric acid in methanol [10] carry the risk of sample fragmentation by nonspecific peptide bond cleavage. Sequencing of N-terminally blocked samples requires controlled proteolysis and subsequent isolation of the peptide fragments (Section 11-5). Even then, the N-terminal fragment is still refractory to Edman chemistry but may be verified by mass spectrometry (Section 11-4).

Samples which are not naturally blocked at the N-terminus should not be exposed to chemicals or conditions which might lead to blockage. Amino-reactive compounds such as aldehydes, carboxylic acid anhydrides, active esters, and cyanates should not be present at any stage of sample preparation. Furthermore, components of buffers, solvents, etc. which are liable to contain amino-reactive contaminants should be used at sufficient levels of purity. For instance, it is common practice to pre-treat urea solutions with a mixed-bed ion exchange resin, e.g., Amberlite MB-1, to remove ammonium cyanate. However, as cyanate is continually formed from urea in aqueous solution, sequencing efficiency of samples stored in the presence of urea may be compromised.

Artefactual N-terminal blockage problems can often be managed by keeping protein concentrations as high as possible, or processing and storage at low temperature. Maintaining the pH well below the pK_a of the α-amino group (i.e., below 9), to reduce its nucleophilic potential by protonation, may also be helpful. However, one should be aware that in this environment the protein may be subject to other problems like denaturation, aggregation, and precipitation.

For the coupling reaction of Edman degradation, N-terminal protonation has to be inhibited by maintaining a high pH. In this situation, the terminal amino group is particularly susceptible to blockage by suitably reactive sample matrix components, especially since the temperature is now also elevated. Sequencing efficiency may similarly be compromised if the ability of the coupling base to deprotonate the N-terminus is inhibited by the presence of buffer compounds of $pK_a < 9$ in the sample matrix. For this reason, matrix components such as acetate, phosphate, citrate, *N*-[2-hydroxyethyl]piperazine-*N'*-[2-ethanesulfonic acid] (HEPES), 2-[*N*-morpholino]ethanesulfonic acid (MES), and other commonly used buffers of this type must be removed before sequencing. Fortunately, protein sequencer manufacturers have addressed this problem by providing support materials which allow routine sample desalting during immobilization [6,8].

Desalting is also required to remove PITC-reactive matrix components, e.g., thiols, ammonia, primary and secondary amines (such as Tris), amino acids, and ampholines. Apart from effectively reducing the PITC concentration and therefore the coupling efficiency, these compounds form stable PITC adducts which are extracted with the ATZ derivatives subsequently to appear on the PTH analysis chromatograms. In some cases, the adducts co-elute with particular PTHs, for example the ammonia adduct with PTH-Asp, causing potential difficulty in reading the sequence. The most common experience with these adducts is that they are present in such high concentrations, by protein sequencing standards, that their signals obscure those of the PTHs. It should also be noted that volatile ammonium salts cannot be completely removed by lyophilization, even if this is repeated several times. Tertiary amines for use as the coupling base in Edman degradation are extensively purified: nonsequencing grade products typically contain sufficient contaminating ammonia and primary or secondary amines to cause problems.

Other nonvolatile matrix components which require removal prior to sequencing are multivalent metal salts, detergents, guanidinium salts, carbohydrates, and glycerol. These compounds are deleterious to Edman degradation by various mechanisms including coordination with reaction intermediates, promotion of sample wash-out, reaction with amino groups, and inhibition of sample penetration by the reagents.

11.2.2.2 Manipulation of Samples in Solution

In terms of final sample application to a sequencing support matrix, with or without desalting as necessary, specific protocols are provided by instrument manufacturers for use by sequencer operators. The aim of this section is to provide guidelines for the manipulation of samples before presentation for sequencing so that unbiased data may be obtained.

Sequencing for research purposes frequently involves sample amounts at low- or even sub-picomole levels. In such situations, sample manipulation strategies are designed to maximize recovery by minimizing losses by adsorption on to surfaces. One such strategy is to maintain protein solutions at high concentration which, as stated in the previous section, also reduces the extent of N-terminal blockage by matrix contaminants. However, when samples are concentrated by centrifugation either through an ultrafiltration membrane or *in vacuo*, it is important to not allow them to dry completely. If this happens, sample recovery tends to be reduced.

For quality control of bulk and formulated protein or peptide products, sample amounts tend not to be limiting. However, issues relating to the reconstitution of freeze-dried material and further dilution still need to be addressed. Clearly, to assess product purity by any analytical method it is of paramount importance for the sample to be representative. For this reason the solvent selected to reconstitute a freeze-dried sample should dissolve all polypeptide components completely. The potential problem of sample component bias caused by the use of an inappropriate solvent is illustrated in Fig. 11-5. Furthermore, in a situation which may be more specific to sequencing, the solvent may also have a profound effect on the relative amounts of different polypeptide components within a sample which become immobilized on a sequencing support matrix.

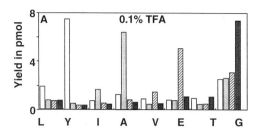

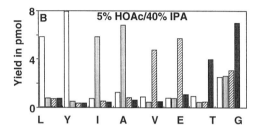

Fig. 11-5. Sequence data for an equimolar mixture of bovine β-lactoglobulin (sequence: Leu, Ile, Val, Thr) and a synthetic decapeptide (sequence: Tyr, Ala, Glu, Gly). Duplicate lyophilized samples were reconstituted in (A) 0.1 % (v/v) trifluoroacetic acid in water and (B) 5 % (v/v) acetic acid, 40 % (v/v) propan-2-ol in water before sequencing. (A) shows a significantly lower recovery of β-lactoglobulin compared with (B), illustrating the importance of using appropriate solvents to obtain representative sequence data.

Typical quality control sample loadings on to a protein sequencer are in the range 200 pmol to 1 nmol, depending on the anticipated impurity levels. It is generally not worth exceeding 1 nmol because of the effect of background (see Section 11.2.3). Consequently, if a sample requires dilution before sequencing it is worth making a careful choice of both container and solvent, again to eliminate the possibility of component bias. Polystyrene and glass tubes are notorious for polypeptide adsorption which, at low sample concentrations, may become selective. It is therefore preferable to use high-quality polyethylene or polypropylene tubes which, if required, can be purchased with passivated internal surfaces. Preferential adsorption of hydrophobic polypeptides on to tube walls can also occur when the organic solvent concentration in the sample diluent is too low. For this reason, water alone is rarely appropriate as a diluent and detergents are not included for noncovalent sequencing because of potential problems with immobilization or sequencing efficiency, depending on the type of support. A frequently used diluent is 0.1 % (v/v) TFA/10–20 % (v/v) acetonitrile in water.

11.2.2.3 Polyacrylamide Gel Electrophoresis (PAGE)

In addition to its analytical applications, PAGE is widely used as a means of preparing protein samples for sequencing [7]. In a biotechnological context, the same gel which is used to monitor the isolation or verify the purity of a protein can be used as the first step towards characterizing any impurities or confirming the identity of the target protein by Edman sequencing.

When PAGE is used for preparative purposes, several precautions need to be taken. First, it is advisable to either use the best available acrylamide, methylenebisacrylamide and sodium dodecylsulfate (SDS) for gel production or purchase high-quality, commercially made gels. This precaution not only minimizes potential N-terminal blockage by contaminants but also gives minimal staining backgrounds, tight protein bands, and optimal resolution. Freshly made gels should be aged at 4 °C for 16–24 h to allow time for all free radicals generated during polymerization to be consumed. In some laboratories it is customary to subject gels to pre-electrophoresis with reduced glutathione in the cathode buffer to scavenge free radicals and oxidants before the sample is applied [13]. It is preferable to disperse the sample in loading buffer at 40–50 °C rather than 90–100 °C as is usual for analytical PAGE. This precaution reduces the thermal degradation of tryptophan to improve its recoveiy during sequencing (see Section 11.2.3).

After electrophoresis, the proteins within the polyacrylamide gel are electro-transferred to a PVDF membrane [7] before visualization by staining. Protein bands are then excised and can be placed directly into the reaction cartridge of a sequencer. In terms of sensitivity, as a general rule if a band is visible after staining with Coomassie Brilliant Blue R250 or Naphthol Blue-Black (Amido Black) there is sufficient protein for sequencing. To control background levels, staining solutions should only be used once.

11.2.2.4 Covalent Immobilization

Progressive sample extraction, leading to a reduction in the potential length of sequence read, can be prevented by covalent attachment to the sequencing support. This approach, also called 'solid phase' sequencing, is used routinely in instrument systems which are designed specifically for covalent sequencing but can also be valuable for particular applications on instruments which normally operate with non-covalently immobilized samples. Peptides shorter than about 30 residues which are largely hydrophobic or with an increasing hydrophobic bias towards the C-terminus tend to be more rapidly desorbed during sequencing from a noncovalent support than peptides with predominantly polar or ionic residues. Consequently, where there is a specific need to confirm the identity of residues further towards the C-terminus than is possible with noncovalent immobilization, covalent sequencing would be appropriate. A more frequent application of covalent sequencing, however, is the confirmation of residues that have undergone post-translational modification by, for example, phosphorylation or glycosylation. These modifications, being hydrophilic, prevent the extraction of their ATZ derivatives during sequencing with the normal nonpolar solvents and result in gaps in the sequence. Covalent immobilization enables the use of more polar solvents such as methanol to extract these modified residues without the risk of also desorbing the sample from its support.

As mentioned in Section 11.2.1, covalent immobilization requires pre-derivatized membrane supports. These are currently available commercially in the form of kits which also contain the appropriate reagents and buffers [14]. A widely used protocol is to immobilize a sample via the C-terminal and side-chain carboxyl groups to an amino-functionalized support by inducing amide cross-links with 1-ethyl-3-(3-dimethylaminopropyl)carbodiimide (EDC). The sample obviously needs to be free of any matrix components which might compete in the reaction such as ammonia, amines, and carboxylic acids. Appropriately for hydrophobic samples, however, is tolerance of high concentrations of strong detergents such as SDS which can be removed by extensive washing after immobilization. During sequencing aspartic acid, glutamic acid, and the C-terminal residues tend to be detectable at lower levels than the other residues because only those ATZ molecules from the total population which are not immobilized are extractable for conversion to PTHs. Provided that the amount of sample is not limiting, this problem would be outweighed by those factors which dictate the use of covalent immobilization.

The PTH derivatives of phosphorylated residues, by virtue of their double negative charge, co-elute with the injection artefacts in the normal PTH chromatography systems. Confirmation or identification of phosphorylation sites by covalent sequencing is usually undertaken by using ^{32}P-radiolabeled samples and collecting the methanol-extracted ATZ derivatives for scintillation counting [10]. For glycosylated residues, the PTH derivatives can be accommodated on a modified chromatography system [15]. The identification of modified amino-acid residues in proteins and peptides typically requires the application of at least several complementary technologies, two of which are chemical degradation and mass spectrometry.

11.2.3 Data Analysis

As will be apparent later in this section, protein sequencing is not generally regarded as a strictly quantitative method. Nevertheless, the first important point about sequence data analysis and interpretation is that quantitation should never be ignored. The amount, expressed in nanomoles or picomoles, of sample loaded into the instrument should ideally be known, having been determined by a validated assay or amino acid analysis.

Once sequence data have been obtained and quantitated by comparison with PTH calibration standards, it is always observed that the amount of the N-terminal residue is less than the original amount of sample loaded. This amount, termed the 'initial yield' is typically in the 30–70 % range and dependent on several factors relating to the sample, its means of isolation and presentation for sequencing; also the instrument hardware, optimization, and chemicals.

A major influence on initial yield is the degree of N-terminal blockage. Proteins with completely blocked N-termini will obviously give no sequencing signal whatsoever, only an amino acid background related to the amount loaded and the molecular weight, as will be discussed below. Partial blockage is probably unavoidable and occurs during both polypeptide isolation and sequencing. An N-terminal glutamine residue is susceptible to spontaneous cyclization to form pyrrolidone carboxylic acid, under both acidic and alkaline conditions [10]. While such conditions can be avoided during sample preparation, especially if the N-terminus is known to be glutamine, they occur within a protein sequencer as part of Edman degradation. Assuming that all possible precautions to minimize blockage have been taken during sample preparation, side reactions in addition to glutamine cyclization occur during sequencing. Some are specific to serine, threonine, and cysteine which have reactive side chains, while others are caused by reagent contaminants and therefore minimized by the use of specially purified sequencing chemicals.

In addition to chemical blockage, initial yield is determined by the amount of N-terminal PTH derivative eventually recovered for detection after all the reaction steps, solvent extractions, and transfers through the instrument subsystems. Not only is none of these steps 100 % efficient but also the ATZ and PTH derivatives are subject to varying degrees of degradation. Serine and threonine undergo β-elimination to form dehydro-derivatives which then partially react with dithiothreitol, the oxidant scavenger present in the conversion reagent. Consequently, the final recoveries of PTH-Ser and PTH-Thr are 30–40 % and 60–70 % respectively relative to PTH-Ala or PTH-Leu, for example, which are considerably less labile. PTH-Trp is recovered at about 10 % for sample loadings > 50 pmol, decreasing to zero for loadings < 5 pmol because of the lability of the indole ring system. Owing to the reactivity of the thiol group, PTH-Cys is only detected during sequencing in exceptional circumstances, giving rise to a blank chromatogram at a cysteine residue position. The only way to identify cysteine residues in a sequence is by prior S-alkylation of the sample by one of the many protocols available (e.g., see [10,16]). The remaining two residues to undergo significant degradation are asparagine and glutamine. These deamidate to form the corresponding acids so that PTH-Asp and PTH-Glu always accompany PTH-Asn and PTH-Gln respectively. It is worth noting, however, that partial deamidation can also occur during bioprocessing, isolation, and storage. Furthermore, the extent of deamidation in all cases is highly dependent on the identity of neighboring residues, in terms of both linear sequence and three-dimensional structure [17].

As Edman degradation progresses there is a general downward trend in the recovery of each successive residue. The amount of each residue recovered can be plotted against its position in the sequence. On a semi-logarithmic scale a straight line may be fitted to the data by linear regression, the slope of which gives the average repetitive yield, expressed as a percentage as shown in Fig. 11-6A. This graph may be modified to exclude the labile residues, as in Fig. 11-6B, to obtain a more realistic impression of sequencing efficiency, especially if these residues bias the slope of the line. An alternative way of eliminating bias associated with the recovery characteristics of different residues is to calculate repetitive yields for individual residues which occur more than once in the sequence, using the equation

$$RY = [(Y_B/Y_A)^{1/B-A}] \times 100$$

where RY is the repetitive yield, Y_A is the amount of a residue the first time it occurs in the sequence (position A), and Y_B is the amount of the same residue as it occurs further along the sequence (position B).

Repetitive yield is not only dependent on protein sequencer performance but also polypeptide chain length and amino acid composition. A protein such as bovine

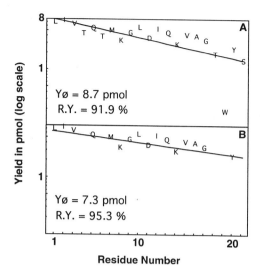

Fig. 11-6. Repetitive yield determinations from sequence data for the first 21 residues of bovine β-lactoglobulin. Linear regression calculations were based on (A) all residues and (B) all residues except serine, threonine, and tryptophan which occur at lower recoveries and, in this case, bias the slope of the line.

β-lactoglobulin (mol.wt. 18.4 kDa) typically sequences over the first 20 residues with a repetitive yield of 93–96 % when immobilized noncovalently (see Fig. 11-6). Under identical conditions a 20-residue peptide typically sequences with a repetitive yield in the range 80–85 %. Therefore, as polypeptide chain length decreases the tendency for sample wash-out increases during noncovalent sequencing. This tendency is enhanced in relatively hydrophobic samples and, as described in Section 11.2.2.4, may be circumvented by covalent sequencing. Alternatively, for quality control of specific samples, custom-designed solvent delivery regimes which minimize wash-out, usually with increased chemical background as the trade-off, can be used.

Other composition-related factors to impact negatively on repetitive yield concern the number of serine, threonine, nonalkylated cysteine, glutamine, and proline residues occurring in the sample over the stretch of sequence which is subjected to Edman degradation. The amino acid side chains of serine, threonine, and cysteine are susceptible to trifluoroacetylation during the cleavage reaction. As sequencing progresses, any residues which are derivatized in this way undergo O→N (serine, threonine) or S→N (cysteine) acyl shift on reaching the N-terminus, with a consequent increase in the population of blocked polypeptide molecules [3]. The other blockage event occurring in the presence of TFA is glutamine cyclization, as discussed above. These side reactions can all be minimized by optimizing the cleavage time, however, proline requires an extended exposure to TFA for efficient cleavage (see Section 11.2.1). Consequently, whenever a proline is encountered in the sequence, carry-over into the next cycle occurs with the result that the proline and all subsequent signals are effectively reduced, lowering the overall repetitive yield. These observations mean that the sequencing efficiency of individual samples is idiosyncratic and that repetitive yield values cannot be interpreted without the relevent sequence data.

A sample giving an initial yield of 100 pmol for an N-terminal alanine, for example, gives an alanine signal at residue 20 of 37.8 pmol at a 95 % repetitive yield. This residue 20 signal would be reduced to 4.6 pmol at a repetitive yield of 85 %; however, even this signal is 100-fold higher than the theoretical minimum detectable level for modern sequencing systems. The length of sequence read depends not only on the actual residue yield over a number of degradation cycles but also the amino acid background level. The sequence only becomes unreadable when the yield becomes indistinguishable from the background. As is the case for repetitive yield, background levels are sample-dependent and influenced by molecular weight and amino acid composition. In general, longer polypeptides give rise to higher backgrounds because there are more peptide bonds per mole to undergo spontaneous acidolysis during exposure to TFA in the reaction cartridge. Samples rich in serine, threonine, aspartic acid, and cysteine (nonalkylated) tend to fragment more readily as a result of cyclization reactions involving the side chains, followed by rearrangements which lead to peptide bond fission [9].

Unlike repetitive yield, the background level increases as a function of the amount of sample loaded and is independent of the actual amount giving rise to a sequencing signal, i.e., the initial yield. These factors have two consequences. First, the length of sequence read does not increase in proportion to the amount of sample loaded into the sequencer. The same protein sequenced with different amounts loaded should give initial yields in the same proportions and identical repetitive yields. However, higher loadings produce higher backgrounds which converge with the progressively falling sequence signal after fewer degradation cycles. Second, samples of the same protein sequenced with identical loadings but exhibiting different degrees of N-terminal blockage give rise to the same repetitive yield and background level. Lower initial yields mean that this background level is again reached after fewer cycles. These relationships between initial yield, repetitive yield, sample characteristics, and background are illustrated in Fig. 11-7. Apart from ensuring that instrumentation

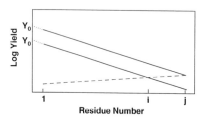

Fig. 11-7. The relationship between theoretical initial yield (Y_0), repetitive yield, background, and length of sequence read. When two samples containing the same amount of a particular polypeptide are sequenced, the repetitive yields (i.e., the slope of the solid lines) and backgrounds (broken line) should be identical. If one sample gives a lower sequencing yield as a result of a higher level of N-terminal blockage, the length of sequence read is curtailed, stopping at residue i intead of j.

is fully optimized and operated using the appropriate sequencing protocol for a particular sample, two further means of maximizing the length of sequence read are concerned with controlling background levels. The first is to minimize contamination which might otherwise produce a chemical or amino acid background (see Section 11.2.2.1). The second strategy is to sequence proteolytic fragments derived from proteins whose molecular weight or composition might otherwise generate excessive background levels (see Section 11.5).

From the above discussion, it is clear that quantitation in Edman degradation is subject to the idiosyncracies of individual polypeptides and amino acid residues. Consequently, for the majority of sequencing applications, quantitation is secondary to qualitative residue identification. As stressed in the first paragraph of this section, quantitation is, however, still valuable.

Comparison of the actual amount of sample loaded into the reaction cartridge with the initial yield can provide an indication of whether the sequence belongs to a major component of the sample or, if this is N-terminally blocked, a minor component. This information may be useful in the identification of contaminants during isolation process monitoring.

For quality control, the initial yield expected for a particular protein product is established as part of method validation. As an alternative to quoting the N-terminal residue yield, extrapolation to the y-axis of a repetitive yield graph gives an estimate of the amount of sample available for sequencing before the start of Edman degradation, the theoretical initial yield, or Y_0. This parameter may provide a better indication of the original sample amount than the yield of the N-terminal residue when this forms one of the low recovery PTH derivatives. On the other hand, the theoretical initial yield can be biased by the means of repetitive yield calculation, as exemplified by comparing Fig. 11-6A and 11-6B. The method used to represent the data should be determined empirically according to which is more sensitive to deviations from the standard. Such deviations can result from either incorrect bioprocessing or degradation during isolation or storage. N-terminal blockage by, for example, reaction with a reducing sugar is detectable by a decreased initial yield. Proteolytic cleavage might produce a frayed N-terminus, with signals from residues 2 and possibly 3 appearing with residue 1 (sequence 'preview'). Initial yield data may then be used to set acceptance criteria on levels of contaminants such as truncated species which are closely related to the desired product. Quantitating the relative amounts of different polypeptide components of a sample may not be as easy because these may not be reflected precisely by the relative initial yields (see Fig. 11-5). To apply sequence analysis to characterizing mixtures requires a validation strategy in which the relative proportions of the expected components are varied over an appropriate range. In this way, factors which relate initial yields to the amounts present can be determined. Where theoretical initial yields are employed, a sufficient number of cycles to provide a representative repetitive yield line should be performed. This number would be influenced by the positions of serine, threonine, and tryptophan in the sequence and is usually in the 10–15 range.

Repetitive yield is a measure of sequencing system performance for a particular polypeptide sample. Its utility in quality control is to demonstrate that the complete analytical system including sample dissolution, immobilization on the sequencing support, instrument programming and performance, and PTH quantitation is valid for its intended purpose. Protein sequencer manufacturers use a standard protein such as bovine β-lactoglobulin to test the installation and performance of an instrument which should achieve minimum initial and repetitive yields under specified analysis conditions. Instrument verification should be carried out as part of quality control and in some laboratories this is done both before and after the product analyses. Whether or not a performance test is done using the manufacturer's test protein, verification always involves sequencing a reference sample of the product.

11.3 Carboxy Terminal Analysis

For many quality control applications, full verification of the identity and chemical structure of a polypeptide is attained without direct C-terminal analysis. In other words, determination of the C-terminal amino acid residue and adjacent residues in the C-terminal region does not necessarily provide information which cannot be deduced from a combination of Edman degradation, mass analysis, and peptide mapping. If the polypeptide is small, both complete Edman degradation from its N- to C-terminus, assuming that the N-terminus is not blocked, and unambiguous characterization by mass analysis within the accuracy limits of the instrumentation (see Section 11.4) is possible. For both types of analysis the upper mass limit for characterization of a polypeptide as a single molecule is, in practice, dependent on individual circumstances but typically about 3000–5000 Da, i.e., 25–40 residues. To overcome the limitations associated with the analysis of larger polypeptides, digestion and subsequent analysis of the isolated peptide fragments, i.e., peptide mapping or fingerprinting, is the preferred strategy. This approach to protein characterization will be covered in Section 11.5. Of relevance here is that, by cleaving a protein into more conveniently sized pieces, complete verification of its primary structure is possible without the need for specific C-terminal residue identification or sequencing from the C-terminus. Confirmation of correct C-terminal processing or determination of the extent of spurious proteolytic degradation at the C-terminus, for example, therefore requires several steps: the generation, identification, isolation, and analysis of the C-terminal peptide fragment (or fragments). Consequently, the principal reasons for direct C-terminal analysis are speed and economy.

Like N-terminal sequencing, the idea of C-terminal analysis is not new. The first methods for N-terminal and C-terminal sequencing were published in 1930 [1] and 1926 [18] respectively. While N-terminal sequencing has been in common use since Edman's publication in 1950 [2], this has not been the case for C-terminal sequencing. The reason for this is the relative inefficiency of carboxyl group activation compared with that of the amino group [1]. Because of the potential utility of C-terminal sequencing, there has been considerable commitment to developing a viable chemical degradation method [18,19]. The challenges associated with the development of a chemical C-terminal degradation method have provided an incentive for the use of enzymatic degradation with carboxypeptidases [3]. As will be discussed in the following sections, these two approaches currently have neither the sensitivity nor length of sequence read achievable with N-terminal sequencing. However, for quality control applications in biotechnology, these limitations may not be disadvantageous.

11.3.1 Automated Chemical Degradation

Two automated C-terminal sequencing systems have recently entered the protein analysis arena. The first was launched as a commercial product in 1994 [20]. The second system is, at the time of writing, undergoing its pre-launch testing phase in a number of protein sequencing laboratories and has been used in several research projects, e.g., [21]. Both systems are currently subject to progressive modifications in both chemistry and instrument programming with a view to enhancing their sensitivity and performance. Because specific details are liable to become rapidly outdated, only a general appreciation of C-terminal sequencing chemistry will be given here.

As in Edman degradation, the first step in C-terminal sequencing is referred to as coupling and similarly involves derivatization to form an active moiety. Carboxy-reactive compounds in current use are diphenylphosphoroisothiocyanate [20] and acetic anhydride [18]. After coupling, an intramolecular cyclization to form a thiohydantoin is induced at the C-terminus. One problem at this point is that, because aspartic and glutamic acid side chain carboxyl groups are also activated, deleterious side reactions involving amino, hydroxyl, and thiol groups can occur in the steps leading to thiohydantoin formation. Premature termination of sequencing by this mechanism, a major contributor to the inefficiencies associated with this chemistry, can be circumvented by prior derivatization of nucleophilic groups by for example N-phenylcarbamylation and O-acetylation, and quenching activated side chain carboxyls by amidation [21]. The next step is cleavage of the C-terminal residue as its thiohydantoin derivative which, like the phenylthiohydantoin resulting from N-terminal degradation, is identified and quantitated by reverse-phase HPLC.

Refinements to the cleavage chemistry have been introduced by the instrument manufacturers to improve efficiency. These include the use of aqueous TFA vapor to enable cleavage at proline, which is otherwise highly inefficient [20], and alkylation prior to cleavage to increase the reaction rate [19]. Another refinement involves the inclusion of thiocyanate in the cleavage reagent formulation to drive the direct formation of the pre-cleavage derivative of the following residue. In this way, C-terminal activation of the polypeptide is only done at the first cycle with the consequence that spurious fragmentation of the sample during sequencing cannot generate background signals (cf. Edman degradation) [19]. Key elements of the currently available C-terminal sequencing chemistries are shown in Fig. 11-8. As in the case of Edman degradation, overall sequencing efficiency is highly sample dependent but C-terminal sequencing is, at present, significantly less sensitive. A sequence read of 10 residues typically requires a sample loading of 1000 pmol and a realistic minimum loading to obtain three residues is about 100 pmol. Despite these current

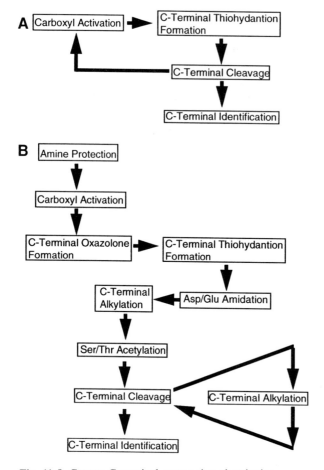

Fig. 11-8. Current C-terminal sequencing chemistries.

limitations, a significant number of both commercial and academic laboratories have invested in the available technologies.

11.3.2 Enzymatic Degradation

In view of the difficulties encountered during the development of a robust chemical C-terminal degradation method, there has been an interest in exploiting carboxypeptidases for this purpose [22]. The original strategy, summarized in Fig. 11-9, involves amino acid analysis to identify the residues released from the C-terminus of the protein sample by carboxypeptidase action. To derive sequence information, released amino acids must be identified and quantitated over a time-course on the basis that the release of each amino acid can only occur once the previous one has

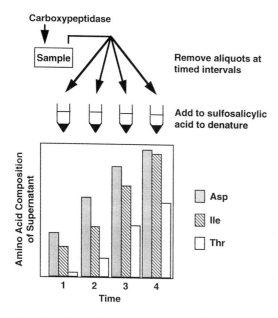

Fig. 11-9. C-terminal sequencing by using carboxypeptidase and amino acid analysis.

been removed. In practice, this method is complicated by the individual substrate specificities of carboxypeptidases. For example, carboxypeptidase A (EC 3.4.17.1) releases histidine, glutamine, threonine, and nonpolar residues rapidly from the C-terminus. Asparagine, serine, lysine, glycine, and acidic residues are released more slowly while proline and arginine are not released at all. Furthermore, the penultimate residue has an effect on the rate of C-terminal release [23]. Therefore method optimization for each polypeptide sample must be done on an individual basis by testing a range of carboxypeptidases and enzyme:substrate ratios to achieve reproducible and interpretable data. The sensitivity of this method is determined by that of the amino acid analysis method used (reviewed in [3]) and the background levels; however, a realistic minimum sample amount might be 500 pmol. A potential problem is the inability to detect and quantify ragged C-termini as a result of bioprocessing or spurious proteolytic degradation, both potentially important applications of C-terminal analysis in quality control.

11.3.3 Mass Analysis

Since gaining wider accessibility to the protein characterization laboratory, mass spectrometry has largely replaced amino acid analysis for sequencing in conjunction with carboxypeptidase digestion [24]. Instead of identifying the released amino acids over a time-course, the masses of the truncated forms of the sample polypeptide are

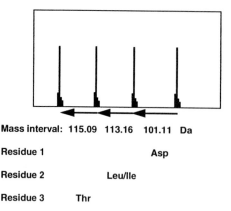

Mass interval: 115.09   113.16   101.11   Da

Residue 1   Asp

Residue 2   Leu/Ile

Residue 3   Thr

Fig. 11-10. C-terminal ladder sequencing by using carboxypeptidase and mass spectrometry.

used to deduce the C-terminal sequence. This approach, outlined in Fig. 11-10, is referred to as 'ladder sequencing' because the mixture of truncated polypeptide chains are interpreted in decreasing order of mass, starting with the mass of the complete molecule and calculating each successive mass difference to derive the sequence of residues. The advantages of ladder sequencing are that it is more rapid and sensitive than both the chemical and carboxypeptidase-amino acid analysis methods, with potential minimum sample loadings in the 10–100 fmol range [24]. An obvious limitation is that some residues and residue combinations are isobaric (see Section 11.4). Another possible limitation is that, because of the idiosyncracies of carboxypeptidase action (see above) or ionization properties of polypeptides, some truncated forms may be missing from the mass profile. Therefore, while it may still be possible to deduce the combination of residues within a particular segment of the polypeptide, their sequence will be indeterminate. Ragged C-termini may be detected by mass analysis of the sample before the addition of carboxypeptidase, provided that all molecular species readily ionize.

C-terminal sequencing can also be done by mass spectrometry without the use of carboxypeptidases. This approach takes advantage of the ability of some types of instrumentation to induce polypeptide fragmentation, producing ladder sequencing data. This and other applications of mass spectrometry will be covered in the next section.

11.4 Mass Spectrometry

Figure 11-11 shows the basic components of a mass spectrometer. The different types of instrument differ in their means of sample introduction, ionization, and mass analysis; therefore specific details will be covered in relevent sections below (see also [25–28]). All MS methods aim to generate gas phase ions with the subsequent determination of their mass-to-charge (m/z) ratios and hence their masses. The more

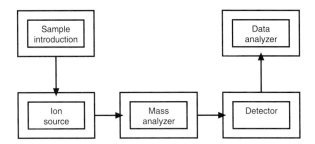

Fig. 11-11. The basic components of a mass spectrometer.

easily ions are formed (e.g., by protonation), the better the probability of obtaining a mass measurement. Every ion produces its own characteristic m/z value(s) and fragmentation enables structural diagnosis, although interpretation of the resulting spectra may require considerable skill on the part of the analyst.

Twenty years ago, the idea of using mass spectrometry (MS) for protein analysis would have been met with considerable resistance, with thoughts of huge instruments, incompatible solvents, complicated data programs, and specialized training. Many of the technologies available at that time were applicable only to relatively small molecules, as opposed to peptides and proteins, because of difficulties with volatilization and ionization.

The utility of MS in protein primary structure determination has, however, long been recognized. It is clearly possible, in principle, to deduce the amino acid composition of a polypeptide from its molecular mass and, by various fragmentation strategies, to determine its sequence. Ambiguities, however, may arise because of the isobaric residue pairs, leucine/isoleucine and glutamine/lysine, and combinations which are close in mass, e.g., arginine/glycine + proline. Additionally, MS may allow the identification and localization of co- and post-translational modifications. In fact, a well-known scientist stated at a conference that he could not see any future for Edman degradation because of the 'simplicity' of sequencing proteins and peptides on a tandem mass spectrometer. One of the drawbacks not mentioned was the large amount of physical space that was required for such an instrument and its mass limit, at that time, of around 2500 Da. The impressive accuracy of this type of mass spectrometer, with its ability to resolve isotopes, illustrates the traditional analytical chemistry focus of MS before specific strategies and hardware were developed to enable protein and peptide analysis.

An early and well-studied approach made use of peptide derivatization by N-acetylation and N,O-permethylation to enhance volatility for GC-MS [29]. This strategy enabled the analysis of peptides with up to ten residues and characteristic fragmentation patterns could be interpreted to derive sequence data, as shown in Fig. 11-12.

Applicability to polypeptide MS continued to evolve in the early 1980s with the development of fast atom bombardment (FAB) ionization, which will be described in Section 11.4.1. In the mid 1980s, BioIon AB, a Swedish company based in Uppsala, introduced a plasma desorption mass spectrometer (PDMS). This was the first compact and simple (for the non-MS specialist) protein/peptide mass analyzer

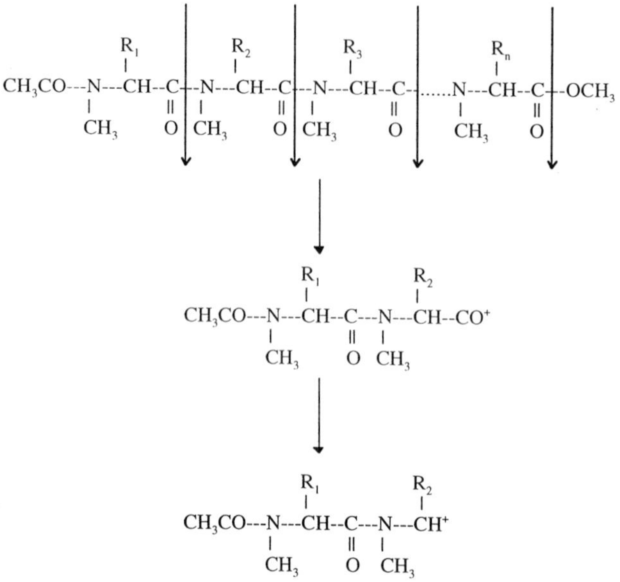

$$CH_3CO\text{---}N\text{---}CH\text{--}C\overset{R_1}{\underset{CH_3\quad O}{|}}N\text{---}CH\text{--}C\overset{R_2}{\underset{CH_3\quad O}{|}}N\text{---}CH\text{--}C\overset{R_3}{\underset{CH_3\quad O}{|}}\ldots\ldots N\text{---}CH\text{--}C\overset{R_n}{\underset{CH_3\quad O}{|}}\text{-OCH}_3$$

$$CH_3CO\text{---}N\text{---}CH\text{--}C\overset{R_1}{\underset{CH_3\quad O\quad CH_3}{|}}N\text{---}CH\text{--}CO^+ \;\;(R_2)$$

$$CH_3CO\text{---}N\text{---}CH\text{--}C\overset{R_1}{\underset{CH_3\quad O\quad CH_3}{|}}N\text{---}CH^+ \;\;(R_2)$$

Fig. 11-12. Ions formed during mass analysis by GC-MS of a permethylated peptide which has been dissociated with a collision gas (see Fig. 11-13 and 11-14).

capable of measuring molecular masses up to about 10–30 kDa (see Section 11.4.2). Continued development of 'friendly' mass analyzers resulted in the emergence of so-called hyphenated techniques, in addition to GC-MS, such as LC-MS and CE-MS. The advent of atmospheric pressure interfaces to mass analyzers permitted the use of liquid junctions (see Section 11.4.3). Simultaneously, the attainment of increasingly higher mass measurements, particularily by the use of matrix-assisted laser desorption ionization (MALDI) instruments was the result of the pioneering work of Hillenkamp and Karas in 1988, who broke the 100 kDa barrier (see Section 11.4.4).

As can be gathered from this introduction, there are several approaches through various analyzers to the measurement of mass; each has its own particular strengths. These will be addressed in the following sections describing four main types of instrument. The order of presentation should not imply any preference on our part. These four types of mass analyzer can be divided into two general categories. With respect to sample application, FAB- and electrospray (ES)-MS are congenial to hyphenated techniques such as online LC. They are, however, critical with respect to solvent composition, in that salts can disturb and often obliterate any analyte ion measurements, owing to the fact that the solvent ions 'absorb' ionization potential, inhibiting ionization of the analyte. PD- and MALDI-MS require, at present, that samples are dry when introduced into the high-vacuum region of the instrument. They are more accomodating with respect to salt content; the PDMS target can be washed after sample application to remove interfering salts, the presence of sodium ions in particular being detrimental to the quality of the spectrum obtained.

11.4.1 FAB-MS

Barber and colleagues' introduction of FAB ionization technology provided a significant step forward in the analysis of labile and non-volatile compounds, including
polypeptides [30,31]. A viscous, nonvolatile solvent such as glycerol is used as
the sample matrix but other useful matrices include thioglycerol and 2, 2'-dithio-
diethanol, which are thought to enhance sample protonation. The sample/matrix mixture is placed on to the tip of a probe which is introduced into the ion source and held
at high potential. A beam of, for example, xenon or argon atoms bombards the sample, releasing ions which are then extracted, focused and accelerated into the mass
analyzer. Ions generated by FAB tend to be the singly charged $(M+H)^+$ species,
but $(M+Na)^+$, $(M+K)^+$, and glycerol adduct ions can also occur. In most situations,
the practical upper mass limit is 3–5 kDa.

As mentioned above, FAB-MS is amenable to online coupling with HPLC by
means of a continuous-flow probe [32] and tryptic digests (see Section 11.5) have
been successfully analyzed in this way. In addition, when it is coupled to a tandem
MS, as shown in Fig. 11-13, selection of a specific peptide ion prior to further collision enables sequence analysis [33]. This type of analysis also permits distinction of
leucine and isoleucine in a peptide, owing to intra-sidechain fragmentation in addition to the series of main-chain fragment ions formed by the dissociation (Fig. 11-14)
[25,34].

FAB-MS has been successfully used for the determination of disulfide bonds in
proteins by comparing mass data before and after reduction. However, it is worth
noting that the thiol-containing matrices referred to above may lead to unintentional
disulfide bond reduction [33]. There may be advantages in using FAB-MS for the
analysis of hydrophobic peptides since there appears to be a positive correlation

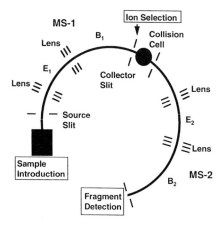

Fig. 11-13. A tandem (MS-MS) mass spectrometer. Ions formed from the sample are resolved by
the first electric (E_1) and magnetic (B_1) field regions. A selected ion is then exposed to gas at low
pressure within a collision cell prior to separation of the dissociated fragments in the second electric (E_2) and magnetic (B_2) fields.

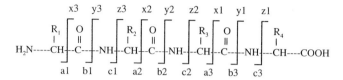

Fig. 11-14. Possible fragment ions formed by collision induced dissociation (CID, also referred to as collisionally activated dissociation, CAD) of a polypeptide during an MS-MS experiment. This type of data can be interpreted to obtain a polypeptide sequence.

between the hydrophobicity of a peptide and its ionization efficiency by FAB. This may be explained by the glycerol used to disperse the sample inducing the accumulation of hydrophobic peptides at its surface. An increased occurrence of fragment ions may be observed with more concentrated samples because of facilitated diffusion of the analyte to the surface, where collision-induced dissociation (CID) by the bombarding atoms occurs.

11.4.2 PDMS

The basic principles of PDMS are described in Fig. 11-15 [34,35]. The sample, in solution, is loaded on to a nitrocellulose-coated Mylar foil target. For best results, the target should be attached to a horizontal spinning rotor so that the solvent is rapidly removed by the centifugal effect while the analyte is retained on the nitrocellulose. Any salts present can be washed off the target with water while it is spinning on the rotor.

Figure 11-15 presents a typical mass spectrum obtained by PDMS, which in the absence of salts gives predominantly the $(M+H)^+$ ion, with fragment ions (e.g., the des-carboxyl species) seen only at high sample loads. No sequence information can be derived by PDMS unless used in conjunction with, for example, carboxypeptidase digestion which can be done on the target (see Section 11.3.3). Although mostly used for samples of molecular mass under 10 kDa, the technique has been used for some proteins of higher mass, for example growth hormone and trypsin. For characterization of polypeptide mixtures, PDMS only provides qualitative information; signal levels reflect peptide-specific ionization efficiencies, not absolute amounts. In other words, peptides giving a low relative abundance in a spectrum may be the predominant components of a mixture and vice versa.

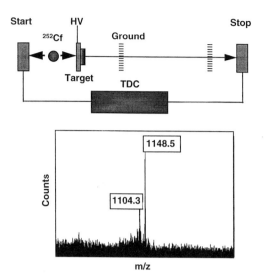

Fig. 11-15. Plasma desorption mass spectrometry (PDMS). The sample is placed on a target which is held at 10–20 kV. Ionization is induced by collisions with particles produced by nuclear fission of the radioactive californium source. Each fission event produces two particles which are ejected simultaneously in opposite directions. While one particle desorbs ions from the target, the other triggers a start signal in the time-to-digital converter (TDC). The desorbed ions are accelerated towards a grid at ground potential then travel in a field-free region before hitting a detector which simultaneously triggers a stop signal. In this type of instrument, m/z ratios are resolved by time-of-flight (TOF). The lower panel shows a PDMS spectrum of a decapeptide (1148 Da). The signal at -44 Da corresponds with the loss of a carboxyl group.

11.4.3 ESMS

The use of ESMS for the mass analysis of proteins and peptides has expanded exponentially since the pioneering work of Fenn in the late 1980s [36]. Its adaptability to online analysis, particularily with HPLC, has been of tremendous benefit to the protein scientist and peptide mapping (see Section 11.5) of recombinant proteins for quality control can be performed routinely with LC-MS.

The probable mechanism of ionization by electrospray, which is not fully understood [37], is shown in Fig. 11-16. Ions are resolved with a quadrupole mass analyzer, which scans across a specified m/z range, e.g., 100–3000 Da, filtering and allowing ions to reach the detector in 0.5 Da steps. Instruments can either be fitted with a single quadrupole or, for tandem MS, a triple quadrupole, as illustrated in Fig. 11-16 [38–40]. The top m/z limit for quadrupole mass analyzers is 2500–3000 Da. Therefore the ability to analyze proteins of higher molecular mass depends on the sample's acquisition of many charges to produce a series of ions with m/z values within the operating range of the quadrupole. A typical ESMS spectrum of a protein is shown in Fig. 11-17. The molecular mass of a protein analyzed by ESMS is cal-

A: Ion Source

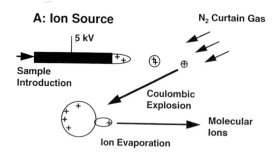

B: Mass Analyser

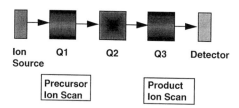

Fig. 11-16. Triple quadrupole electrospray mass spectrometry (ESMS). The ion source (A) comprises a needle at high electrical potential through which the sample flows in solution. Charged droplets emerge from the needle and solvent evaporation occurs as they are accelerated towards the entrance to the mass analyzer. The flow of nitrogen (curtain gas) aids solvent evaporation and blows uncharged liquid away. As the droplets reduce in size, a point is reached when the charge density becomes too great and coulombic explosion occurs. Molecular ions are then released in the vapor phase. The mass analyzer (B) contains a first quadrupole (Q1) which scans the m/z ratios of the ions entering from the electrospray. The second quadrupole (Q2) holds selected precursor (parent) ions in contact with gas at low pressure while CAD gives rise to product (daughter) ions which are then scanned by the third quadrupole (Q3).

culated by using a computer algorithm supplied by the instrument manufacturers, with a typical accuracy of 0.01 %.

Because mass resolution is by filtration, the number of ions reaching the detector compared with those actually generated by the ion source is typically reduced by five to six orders of magnitude. Despite this, sensitivity for protein analysis by ESMS can be around 1 picomole. The characterization of glycoproteins presents particular challenges owing to microheterogeneity and the presence of sialic acids, which both have a negative impact on sensitivity. This problem was largely overcome in a study of tissue plasminogen activator by neuraminidase treatment to remove sialic acids, and trypsin digestion followed by online HPLC to allow analysis of individual glycopeptide fragments by LC-MS [41]. HPLC also overcomes a potential problem with ESMS: suppression of sample ionization by the presence of salts which reverse-phase LC removes.

In ESMS, a useful application is single ion monitoring to scan peptides as they enter the mass analyzer for specific structural features such as glycosylation or phos-

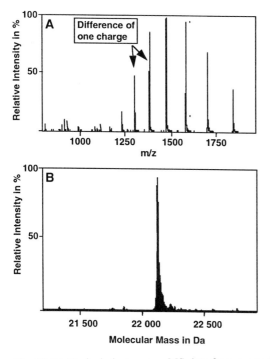

Fig. 11-17. Typical electrospray MS data from a protein. The raw spectrum (A) shows a distribution envelope of multiply charged ions, each successive ion differing by one unit of charge. A data analysis algorithm is used to deconvolute the ion series data to derive the mass of the protein (B).

phorylation sites. It is possible rapidly to alternate the potential at the entrance to the mass analyzer between normal and increased settings so that, in the latter case, CID occurs. At the same time as measuring the total mass of a peptide it is possible to record the presence of specific ions released during dissociation. For example, positive ions at 162 or 203 Da are derived from carbohydrate moieties, while a negative ion at 79 Da is phosphate.

11.4.4 MALDI-MS

This technology had its origins in the 1960s, but only with the development of suitable matrix compounds, particularly the pioneering work of Karas and Hillenkamp in the 1980s, did its forte as an analysis method for determining the molecular weights of proteins greater than 100 kDa emerge [42,43]. The principle of the method is similar to that of PDMS, as can be seen in Fig. 11-18 with a laser substituted for the radioactive source. Whereas in PDMS only one choice of surface is available, in MALDI the choice of matrix is determined to some extent by the analyte of interest. The most commonly used matrix compounds include nicotinic, dihydroxybenzoic,

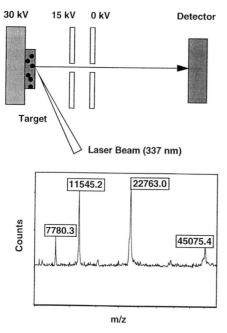

Fig. 11-18. Matrix-assisted laser desorption ionization (MALDI) mass spectrometry. The polypeptide sample is mixed with a matrix compound, e.g., sinapinic acid, before application to the target. A laser beam activates the matrix which transfers energy and ions to the sample. Desorbed ions are then accelerated across a potential gradient towards a detector for resolution by TOF, as in PDMS (see Fig. 11-15). The lower panel shows a MALDI mass spectrum of a 45 kDa protein. Accompanying the singly charged $(M+H)^+$ ion are the $(M+6H)^{6+}$, $(M+4H)^{4+}$, and $(M+2H)^{2+}$ ions with m/z values of 7780, 11454, and 22763 respectively.

sinapinic, and α-cyano-4-hydroxycinnamic acids [44]. With MALDI analysis, the most predominant signal can be attributed to the singly charged species, in contrast to electrospray analysis, and the upper limit of detection is theoretically unlimited, but practically ca. 200 kDa. Multiply charged species can also be found, as can be seen in Fig. 11-18, but their abundance is dependent to some degree on the matrix used. Tryptic digests (without prior separation by HPLC) may be analyzed by MALDI-MS and a higher resolution may be achieved by use of a reflectron, which in effect doubles the flight length. The enhanced resolution provided by the reflectron is apparent in Fig. 11-19 and also enables sequence information to be obtained from fragmentation due to post source decay [45].

On the MALDI targets, as is the case on PDMS targets, the possibility exists for performing reactions such as enzymatic digestion of the protein or peptide of interest, reduction of disulfide bonds, and other manipulations. In this way, samples for MALDI and PDMS can, in principle, be used many times.

A relatively new configuration of MALDI-MS is 'delayed extraction', in which the ionizing event is separated from the acceleration event [46]. This enables higher laser power to be used, with a consequent improvement in signal intensity and mass

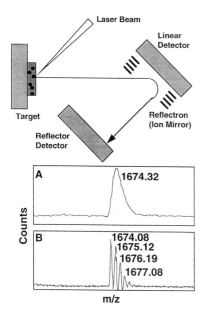

Fig. 11-19. MALDI mass spectometry incorporating a reflectron. Ions accelerated towards the linear detector can be deflected towards the reflector detector to effectively double the flight length. Panel (A) shows the signal given by neurotensin, using α-cyano-4-hydroxycinnamic acid as the matrix, in linear mode. The same sample analyzed in reflector mode (B) shows the presence of isotopes as a result of enhanced resolution.

accuracy. This configuration can function in both linear and reflector modes and is of particular benefit in the latter.

11.5 Protein Fragmentation

In the previous sections, there have been several references to protein fragmentation, which can be achieved by chemical or enzymatic means or both. Commonly used methods are summarized in Table 11-1, with more detail provided in [3,10].

A widely used application in protein quality control is in peptide mapping or 'fingerprinting'. Exploitation of the cleavage site specificities of enzymes, trypsin being the most popular, results in the generation of set of peptides which can be analyzed by HPLC (with UV detection), LC-MS, or MS without prior separation. By comparison of the pattern obtained from the test sample with that from a reference, deviations from the expected primary structure may be detectable.

Owing to the progressive decrease in signal and increase in background experienced with Edman degradation, it may not be possible obtain sequence data as far away from the N-terminus as needed. Alternatively, the N-terminus of the sample may be blocked and therefore refractory to Edman degradation. In these situations,

Table 11-1. A summary of enzymes and reagents for protein fragmentation.

Protease or reagent	Cleavage site	Conditions
endoproteinase Arg-C	–Arg–X–	ammonium bicarbonate pH 8 8 h at 37 °C
endoproteinase Asp-N	–X–Asp–	ammonium bicarbonate pH 7 4–16 h at 40 °C
chymotrypsin	–Trp,Tyr,Phe,Leu–X–	ammonium bicarbonate, Tris–HCl pH 8, 4 h at RT
endoproteinase Glu-C	–Glu–X– but slow at –Glu–Glu–	ammonium bicarbonate, Tris–HCl pH 8 4–16 h at RT
endoproteinase Lys-C	–Lys–X–	Tris–HCl pH 8 > 2 h at RT
pepsin	–Glu/Phe/Tyr/Leu–X–	0.1 % TFA, 0.1 % HCl 2 h at RT
thermolysin	–X–Leu/Ile/Phe/Val/Ala/Met–	Tris–HCl pH 7.6 (+ 2 mM $CaCl_2$) 1–16 h at RT
trypsin	–Arg/Lys–X– but not when –Arg/Lys–Pro–	ammonium bicarbonate, Tris–HCl pH 8 4–16 h at RT
cyanogen bromide	–Met–X– but very slow –Met–Ser/Thr–	70 % formic acid overnight at RT
hydroxylamine	–Asn–Gly– –Asn–Asp–	1.8 M-hydroxylamine for 3 h at 45 °C
o-iodosobenzoic acid	–Trp–X– –Tyr–X–	*o*-iodosobenzoic acid (4 g L^{-1}) in 4 M guanidinium-HCl 16 h at RT
BNPS-skatole	–Trp–X–	BNPS-skatole (10 g L^{-1}) in 50 % acetic acid 48 h at RT
acid	–Asp–X– –Asp–Pro–	70 % formic acid for 24–86 h

the only way to obtain internal sequence data is by fragmentation followed by separation of the peptides. This procedure will enable identification of the majority of peptide fragments from a protein, but will not give any information as to the order in which they occur in the protein. Thus, the protein can be digested by a second, or third method to derive overlaps which enable the complete primary structure to be deduced. This strategy clearly involves a large amount of sequencing; therefore, in the quality control environment, peptide mapping and mass analysis can provide complementary information to minimize the time and cost of Edman degradation.

Fragmentation is the obvious means of preparing specific segments of a protein for further characterization. The task of verifying modifications such as glycosylation or disulfide bonds at specific sites on a protein is simplified or even made practicable by this strategy. The choice of cleavage method depends on the sequence of the protein, which, for quality control purposes will already be known. Proteolytic enzymes are generally used in preference to chemical methods because they tend to generate smaller fragments which are more amenable to isolation by reverse-phase HPLC. Because chemical cleavage sites (e.g., Met -X and Trp-X, see Table 11-1) are relatively few in number within typical proteins, fragments can be large and may not adequately simplify the subsequent analysis. In addition, the strongly denaturing conditions used in chemical cleavage protocols frequently cause aggregation to the extent that the fragments are poorly recovered and resolved by HPLC. The problem may be overcome in some cases by the use of C_4-type reverse-phase media instead of C_8 or C_{18} and by using propan-2-ol in acetonitrile instead of acetonitrile alone as the mobile phase organic modifier.

Many proteins are resistant to enzymatic proteolysis unless placed under denaturing conditions. This resistance may be due to either tertiary structure or steric hindrance by modifications like glycosylation. The inclusion of denaturants such as urea, SDS, or acetonitrile must be tempered with the stability (activity) of the protease under those conditions. The best option is to consult data sheets provided by suppliers of proteases. It should be noted that many of the proteases in common use have their optimum activity at pH values above 7 (see Table 11-1). At this pH and above, disulfide bonds are liable to scramble. Therefore, for the determination of disulfide bond connectivities, fragments must be generated at pH < 6 using mild acid cleavage or pepsin, for example.

11.6 Summary and Future Prospects

We have described two types of instrumentation used for the structural characterization of proteins. Mass spectrometry is rapidly gaining importance as a technology which complements the more traditional Edman sequencing. It is clear that no single technology can provide all the information which is demanded: an orthogonal approach is required.

Acknowledgements

We would like to express our thanks to Jonathan MacBeath, Ian Blench, Kevin Howland, and Dorte Christensen for their valuable input.

References

[1] Doolittle, R. F., in: *Methods in Protein Sequence Analysis:* Elzinga, M. (Ed.), Clifton: Humana Press, 1982; pp. 1–24.
[2] Edman, P., *Acta Chem Scand,* 1950, *4*, 283–293.
[3] Allen, G., *Sequencing of Proteins and Peptides.* Amsterdam: Elsevier, 1989.
[4] Hewick, R. M., Hunkapillar, M. W., Hood, L. E., Dreyer, W. J., *J Biol Chem,* 1981, *256*, 7990–7997.
[5] Linse, K. D., Carson, W., Farnsworth, V., Technical Information Bulletin T-0105. Beckman Instruments, Inc.
[6] Horn, M. J., Miller, C. G., Harrsch, P. B., Woo, W., Wagner, G. W., *Presentation Abstracts,* 6th Symposium of The Protein Society, San Diego, CA, 1992. Cambridge: Cambridge University Press, 1992.
[7] Matsudaira, P., *J Biol Chem,* 1987, *262*, 10035–10038.
[8] Werner, W. E., Hsi, K.-L., Grimley, C., Yuan, P.-M., *Presentation Abstracts,* 9th Symposium of The Protein Society, Boston, MA, July 1995. Cambridge: Cambridge University Press, 1995.
[9] Tarr, G. E., in: *Methods of Protein Microcharacterization:* Shively, J. E. (Ed.), Clifton: Humana Press, 1986; pp. 155–194.
[10] Aitken, A., *Identification of Protein Consensus Sequences.* Chichester: Ellis Horwood, 1990.
[11] Farries, T. C., Harris, A., Auffret, A. D., Aitken, A., *Eur J Biochem,* 1991, *196*, 679–685.
[12] Wellner, D., Panneerselvam, C., Horecker, B. L., *Proc Nat Acad Sci USA,* 1990, *87*, 1947–1949.
[13] Dunbar, B., Wilson, S. B., *Anal Biochem,* 1994, *216*, 227–228.
[14] Coull, J. M., Pappin, D. J. C., Mark, J., Aebersold, R., Koster, H., *Anal Biochem,* 1991, *194*, 110–120.
[15] Gooley, A. A., Packer, N. H., Pisano, A., Redmond, J. W., Williams, K. L., Jones, A., Loughnan, M., Alewood, P. F., in: *Techniques in Protein Chemistry VI:* New York: Academic Press, 1995; pp. 83–90.
[16] Yarwood, A., in: *Protein Sequencing: a Practical Approach.* Findlay, J. B. C., Geisow, M. J. (Eds., Oxford: IRL Press, 1989, pp. 119–145.
[17] Wright, T. W., *Crit Rev Biochem Mol Biol,* 1991, *26*, 1–52.
[18] Tarr, G. E., in: *Methods in Protein Sequence Analysis:* Wittman-Liebold, B. (Ed.). Berlin: Springer-Verlag, 1989; pp. 129–136.
[19] Boyd, V. L., Bozzini, M. L., Zon, G, Noble, R. L., Mattaliano, R. J., *Anal Biochem,* 1992, *206*, 344–352.
[20] Miller, C. G., Bailey, J. M., *Genetic Engineering News,* 1994, *14*, 16.
[21] Nguyen, D. N., Becker, G. W., Riggin, R. M., Boyd, V. L., Bozzini, M. L., Yuan, P.-M., Loudon, G. M. *Presentation Abstracts,* 9th Symposium of The Protein Society, Boston, MA, July 1995. Cambridge: Cambridge University Press, 1995.
[22] Ambler, R. P., *Methods Enzymol,* 1972, *25*, 262–272.
[23] Ambler, R. P., *Methods Enzymol,* 1972, *25*, 143–154.
[24] Patterson, D. H., Tarr, G. E., Regnier, F. E., Martin, S. A., *Anal Chem,* 1995, *67*, 3971–3978
[25] Biemann, K., *Biomed Environ Mass Spectrom,* 1988, *16*, 99–111.
[26] Burlett, O., Yang, C.-Y., Guyton, J. R., Gaskill, S. J., *J Am Soc Mass Spectrom,* 1995, *6*, 242–247.
[27] Naylor, S., Findeis, A. F., Gibson, B. W., Williams, D. H., *J Am Chem Soc,* 1986, *108*, 6359–6363.
[28] Siudzdak, G., *Mass Spectrometry for Biotechnology.* New York: Academic Press, 1996.
[29] Biemann, K., in: *Biochemical Applications of Mass Spectrometry:* Waller, G. R. (Ed.), New York: Wiley, 1971.

[30] Barber, M., Green, B. N., *Rapid Commun Mass Spectrom,* 1987, *1*, 80–83.

[31] Barber, M., Bordoli, R. S., Elliott, G. J., Sedgewick, R. D., Taylor, A. N., *Anal Chem,* 1982, *54*, 645A–657A.

[32] Ashcroft, A. E., Chapman, J. R., Cottrell, J. S., *J Chromatogr,* 1987, *394*, 15–20.

[33] Biemann, K., in: *Protein Sequencing: a Practical Approach.* Findlay, J. B. C., Geisow, M. J. (Eds.), Oxford: IRL Press, 1989, pp. 99–118.

[34] Cotter, R. J., *Anal Chem,* 1992, *64*, 1027A–1039A.

[35] NcNeal, C. J., *The Analysis of Peptides and Proteins by Mass Spectrometry.* Chichester: Wiley, 1988.

[36] Fenn, J. B., Mann, M., Meng, C. K., Wong, S. F., Whitehouse, C. M., *Science,* 1989, *246*, 64–71.

[37] Kebarle, P., Tang, L., *Anal Chem,* 1993, *65*, 972A–986A.

[38] March, R. E., Hughes, R. J., *Quadrupole Storage Mass Spectrometry.* New York: Wiley, 1989.

[39] Bruins, A. P., Covey, T. R., Henion, J. D., *Anal Chem,* 1987, *59*, 2642–2646.

[40] Whitehouse, C. M., Dreyer, R. N., Yamashita, M., Fenn, J. B., *Anal Chem,* 1985, *57*, 675–679.

[41] Guzzetta, A. W., Basa, L. J., Hancock, W. S., Keyt, B. A., Bennett, W. F., *Anal Chem,* 1993, *65*, 2953–2962.

[42] Karas, M., Hillenkamp, F., *Anal Chem,* 1988, *60*, 2299–2301.

[43] Hillenkamp, F., *Adv Mass Spectrom,* 1989, *11*, 354.

[44] Cohen, S. L., Chait, B. T., *Anal Chem,* 1996, *68*, 31–37.

[45] Kaufmann, R., Spengler, B., Lutzenkirchen, F., *Rapid Commun Mass Spectrom,* 1993, *7*, 902–910.

[46] Vestal, M. L., Juhasz, P., Martin, S. A., *Rapid Commun Mass Spectrom,* 1995, *9*, 1044–1050.

12 General Strategies for the Characterization of Carbohydrates from Recombinant Glycoprotein Therapeutics

Gerrit J. Gerwig and Jan B. L. Damm

12.1 Introduction

The past few years have seen an impressive increase of the number of biotechnologically produced glycoprotein-drugs approved for diagnostic and/or therapeutic use in humans (see Chapter 5). The carbohydrate moieties of these pharma-glycoproteins often play an essential role in their overall biological properties, e.g., pharmacokinetic and immunogenic behaviour *in vivo*. In first instance, it is important that the oligosaccharide moiety of recombinant glycoprotein does not differ significantly from its 'natural' human counterpart. However, the glycan structures are dependent on the expression system and cell culture conditions, commiting pharmaceutical companies to monitor batch-consistency and batch-uniformity of recombinant glycoproteins. On the other hand, by investigating the biological activities and the carbohydrate chain structures of recombinant glycoproteins with altered glycans, a new approach to elucidate the function of carbohydrate chains of glycoproteins has been opened. Understanding the influence of glycosylation on the structure and function of (recombinant) glycoprotein therapeutics, in most cases, still requires the full characterization of all oligosaccharides attached covalently to that glycoprotein. This characterization of glycosylation is generally seen as a relatively difficult problem, in particular because it demands for the use of highly sophisticated instrumentation. In this chapter, we describe the types and some functions of carbohydrate moieties occurring in glycoproteins, together with some biosynthetical aspects. Furthermore, we will discuss some general strategies for the isolation, fractionation and characterization of glycoprotein glycans.

12.2 Functions of Glycoprotein Glycans

Many biological functions have been ascribed to oligosaccharides of the various classes of glycoconjugates (i.e., conjugates of carbohydrate and other biomolecules, such as proteins and lipids). From the literature data [1–5], it is evident that the functions of carbohydrate chains span the whole spectrum from effects on physico-chemical properties of the molecule to which they are linked to highly sophisticated functions in cell–cell recognition, fine tuning of biological activity, and masking of functions. However, in spite of the accumulating data, it is not possible to denominate generalized functions for (specific types and/or structures of) carbohydrate chains. Rather, for each (recombinant) glycoprotein the effect(s) of glycosylation on all the properties of the molecule have to be determined. The various reasons why it is so difficult to unravel clear functions for oligosaccharides have been presented in Chapter 5. Bearing this warning in mind, several more or less, specific functions of carbohydrates can be mentioned. In general, the carbohydrate functions described below refer to, but are not necessarily limited to, protein-linked carbohydrate chains in higher animal species.

12.2.1 Transport, Stabilizing, Protecting and Structural Functions

Already during the biosynthesis of glycoproteins the attached oligosaccharides can play a crucial role. A well-accepted function of oligosaccharide units of glycoproteins is the involvement in the initiation of correct polypeptide folding in the rough endoplasmic reticulum (rER), and in the subsequent maintenance of protein solubility and conformation. Many proteins that are incorrectly glycosylated fail to fold properly and, by consequence, fail to exit the ER and are degraded [6–8]. The carbohydrate sequences also influence the further intracellular trafficking of the glycoprotein. The best known example is the Man-6 phosphate residue present in high-mannose-type N-glycans of newly synthesized lysosomal enzymes: the Man-6 phosphate residue is the label which targets the enzymes to their final destination in the lysosomes. Deficiency in phosphorylation impedes lysosomal targeting leading to I-cell disease and pseudo-Hurler polydystrophy [9,10]. Glycosylation of proteins, especially when multiple carbohydrate chains are attached, creates a 'carbohydrate shell' around the protein. For an impression of the dimensions of oligosaccharide chains versus the protein to which they are attached, the reader is referred to Chapter 5 (3D model of hCG). The carbohydrate cover may serve to protect the protein against recognition by proteases [11] and/or antibodies [12]. The stabilizing function (with respect to denaturation) of submaxillary gland mucin, and the protective function (against auto-digestion) of gastric mucus have already been described in Chapter 5.

It is widely recognized that carbohydrates are essential for the physical maintenance and functional integrity of various structural elements in the body, such as collagens and proteoglycans [13]. Chondroitin sulfate and keratan sulfate chains, for instance, have a pronounced effect on the organization and tensile strength of cartilage [14]. Also well-known is the 'glycocalyx' covering the surface of whole cells, thus functioning as a structural and protective element. The porosity and functionality of 'borders' in the body formed by continuous membrane systems (e.g., the basement membrane and extracellular matrix of the vascular wall) is profoundly influenced by the presence and composition of glycoconjugates [15].

12.2.2 Storage Function

Glycosaminoglycans, such as chondroitin sulfate, heparin, and heparan sulfate, loca-lized in the extracellular matrix, tightly bind various growth factors, such as fibro-blast growth factors and macrophage colony stimulating growth factor [16]. The gly-cosaminoglycan-mediated sequestration of growth factors warrants the concentration of growth factors at or near their target area and prevents diffusion of these biologi-cally highly active molecules to 'unwanted' sites. Furthermore, after binding to the glycosaminoglycan chains the growth factors are less prone to proteolytic degrada-tion. Several other storage functions have been reported, e.g., the binding of comple-ment regulatory protein H to sialic acid residues on cell surfaces [17] (preventing the alternate pathway of complement activation), and the sequestration of calcium [18] and sodium [19] ions and water [20].

12.2.3 Masking Functions

Carbohydrates may serve to prevent infection by microorganisms or (unwanted) immune reactions by masking protein or other carbohydrate epitopes. Examples include: (i) the 4- and/or 9-*O*-acetylation of terminal sialic acid residues which pre-vent the recognition, binding and invasion of certain bacteria and viruses (influenza A and B) [21] that normally infect cells via sialidase-mediated recognition of term-inal sialic acids of membrane bound glycoconjugates; (ii) the presence of heteroge-neous carbohydrates on secreted mucins and in milk which inhibit the infection of the gut by microbial pathogens [22,23]; (iii) terminal sialic acids on O-linked oligo-saccharides which mask the recognition of core O-linked glycans by natural antibo-dies against T and Tn antigens [24]; and (iv) the presence of terminal sialic acid resi-dues on glycoproteins and cell surfaces which prevent recognition by the asialoglycoprotein receptor and/or uptake by macrophages via the Gal/GalNAc lectin [25].

By contrast, carbohydrate sequences may also render the cells on which they occur vulnerable to infection, noxious agents, autoimmune responses or malignant disease: (i) terminal sialic acids and specific carbohydrate sequences of glycoconjugates on mucosal surfaces serve as recognition points for various viruses, protozoa, patho-genic bacteria, chlamydiae, mycoplasma and parasites (thus mediating infection) [26,27]; (ii) particular gangliosides on the cell surface can act as receptor for several bacterial toxins [28]; (iii) certain N-linked glycans on gastric parietal cell glycopro-teins act as antigen for antibodies involved in autoimmune gastritis and pernicious anemia, causing mucosal atrophy and parietal cell loss [29]; and (iv) sialylated, fuco-sylated (poly)*N*-acetyllactosamine-glycans on tumor cells act as ligands for selectin molecules on endothelial cells, thus probably enhancing the metastatic capability of the tumor cells [30], to mention just a few examples.

12.2.4 Receptor Functions

Several examples have been reported where oligosaccharides function as ligands for a receptor, thus mediating the interaction between biomolecules (as such or imbedded in larger cellular systems, like membranes) in the body. For instance, the interaction between oligosaccharide ligands and the selectin family of receptor proteins, mediating the adhesion of (stimulated) leukocytes or platelets to activated endothelial cells [31–33], and the interaction between B cells and activated T or B cells, mediated by the CD22β lectin of B lymphocytes [34]. Furthermore, cell–cell and cell–matrix interactions during development and tissue organization are proposed to be conferred by the mutual recognition of soluble or surface-bound β-Gal-binding lectins and *N*-acetyllactosamine units of the complementary cells [10]. The role of oligosaccharides in the species-specific recognition of sperm and egg cell has already been discussed in Chapter 5. The interaction between heparan sulfate chains on cell surface proteoglycans and extracellular matrix proteins, such as fibronectin, laminin, and thrombospondin, has been implicated in cell adhesion, differentiation, spreading, or invasion [35,36].

12.2.5 Regulation of Clearance

A long-known function of glycosylation is its effect on the circulatory half-life of soluble glycoconjugates and even whole cells. In general, the carbohydrate-mediated clearance of glycoproteins (drugs) from the blood is driven by lectins (carbohydrate receptors) that occur in cells/organs that are specialized in the removal of (partially degraded and/or non-functional or harmful) molecules or cells from the circulation. An example is the removal of pathogens by the macrophage Gal/GalNAc receptor that recognizes Gal/GalNAc-terminated glycans of endogenous and exogenous origin [37]. Several types of lectins occur in the liver: the asialoglycoprotein receptor which removes glycoproteins that bear exposed Gal residues [38], and the non parenchymal terminal 4-*O*-sulfate-β-GalNAc receptor which removes certain glycoprotein hormones [39].

12.2.6 Tuning of Biological Activity

One of the best documented funtions of protein-linked carbohydrate chains is the effect on the biological activity of the protein to which they are attached. These properties are defined as 'tuning' effects. Briefly, it can be stated that the oligosaccharides, depending on their structure and number, modulate the biological activity of the glycoprotein (enzymes, hormones, etc.) in subtle ways through effects on binding affinity (of the glycoprotein to its receptor), efficiency of signal transduction, circulatory half-life, and biodistribution. Many examples can be found in the literature

[3,40]. As a specific example, the influence of the carbohydrate chains on the biological activity of hCG has been described in Chapter 5. The fact that certain terminal carbohydrate sequences on glycoconjugates act as blood-group determinants is commonly known. Easy access to the literature on the functions of glycans can be gained via the compendious review paper of A. Varki [41].

12.3 Types of Carbohydrate Chains in Glycoproteins

Glycoproteins are biomacromolecules consisting of a polypeptide backbone with covalently attached carbohydrate side chains. These molecules may contain from 0.4 % (by weight) to more than 80 % carbohydrate (see Chapter 5). Three major classes of glycan chains can be distinguished:

- *N-glycosidically linked carbohydrate chains,* in which the reducing-end sugar residue (generally *N*-acetylglucosamine) is linked to the amide nitrogen of asparagine,
- *O-glycosidically linked carbohydrate chains,* in which the linkage is formed between the reducing sugar residue and an amino acid hydroxyl group,

Table 12-1. Covalent linkages between monosaccharides and proteins.

Class	Linkage	Class	Linkage
N-linked:		*other O-linked:*	
	GlcNAc→Asn		Glc→Tyr
	GalNAc→Asn		Gal→OH-Lys
	Glc→Asn		L-Ara$f$→OH-Pro
	L-Rha→Asn		Gal→OH-Pro
O-linked:			Gal→OH-His
	GalNAc→Ser/Thr		GlcA-OH-Trp
	GlcNAc→Ser/Thr		GlcA→OH-Phe
	Man→Ser/Thr		GlcA→OH-Ser
	L-Fuc→Ser/Thr	*S-linked:*	
	Gal→Ser/Thr		Gal→Cys
	Glc→Ser		Glc→Cys
	Xyl→Ser	*glycation:*	
ADP-ribosylation:			Glc→Lys
	ADP-Rib→N^ζ-Arg		Rib→Lys
	ADP-Rib→N^δ-Asn	*amide bond:*	
	ADP-Rib→N^1-His		GlcA/GalA(6→N^α)Lys
	ADP-Rib→*O-C(O)*Glu		GlcA/GalA(6→N^α)Thr/Ser
	ADP-Rib→S-Cys		GlcA/GalA(6→N^α)Ala
glypiation:			MurNAc(3→N^α)Ala
	Man6→PO$_4$→(CH$_2$)$_2$→NH→C(O)→protein		

Linkage	Structure	Occurrence
N-Glycosidic GlcNAc-Asn		Widely distributed in animals, plants, and micro-organisms
O-Glycosidic GalNAc-Ser/Thr		Glycoproteins from animal sources
Xyl-Ser		Proteoglycans, human thyroglobulin
Gal-Hyl		Collagens
Ara-Hyp		Plant and algal glycoproteins

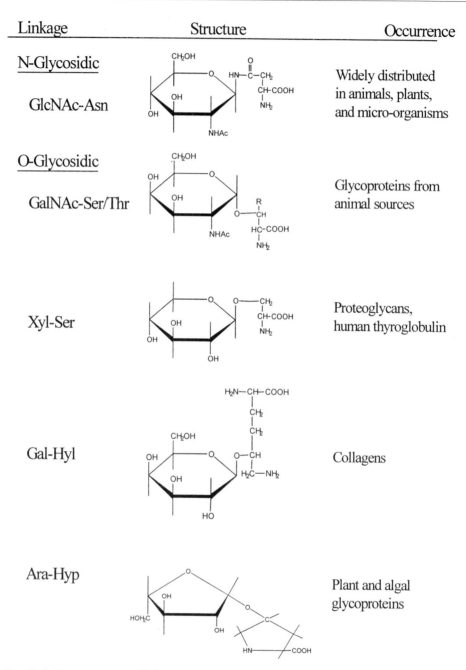

Fig. 12-1. Some common carbohydrate-protein linkage types.

- *Glycosylphosphatidylinositol(GPI)-anchors* (glypiation [42,43]). A (glyco)pro-
tein is commonly linked to the GPI anchor by its C-terminus through an amide
bond involving the amine of an ethanolamine moiety.

N- as well as O-linked carbohydrate chains can occur at one protein backbone. In
some glycoproteins, other, less common, covalent bonds involving N- or S-atoms
have been found between the ε-amino group of Lys and Glc or Rib (glycation
[44]), between Rib and the side chains of Arg, Asn, or His residues (ADP-ribosyla-
tion [45]), and between Rib and Cys (S-glycosylation [46]). The established carbohydrate–protein linkage types are compiled in Table 12-1, and some common carbohydrate–protein linkage types are illustrated in Fig. 12-1.

12.3.1 Structure of N-linked Carbohydrate Chains

In general, the N-glycans have a common pentasaccharide core, consisting of two
GlcNAc residues and three Man residues, which can be extended with additional
monosaccharide units giving rise to four types of N-glycosidic carbohydrate chains
[40,47] (Fig. 12-2):

- *Oligomannose (or high-mannose) type,* consisting of Man and GlcNAc. The
number of Man residues linked to the pentasaccharide core generally varies
between 0 and 6 in most mammalian glycoproteins, but in yeast over 100 Man
residues can be attached. The oligomannose type includes also structures with
up to three non-reducing-end terminal Glc extensions (see Section 12.4.1) and
sometimes *O*-phosphorylated Man residues.
- *N-acetyllactosamine (or complex) type,* being composed of Man, GlcNAc, and
NeuAc. The core is, in general, extended with two or more Galβl-4GlcNAc
units (sometimes repeated) and terminated with NeuAc. In addition, Fuc, Gal-
NAc, Xyl, and noncarbohydrate substituents like acetate, lactate, sulfate, phosphate, or methyl groups may occur [41]. In mammalian glycoproteins, di- to
penta-antennary branching of the core structure is regularly found.
- *Hybrid type,* combining the characteristics of the oligomannose and the *N*-acetyl-
lactosamine type. In most hybrid type (and some complex type) structures, an
extra GlcNAc residue (inter- or bisecting GlcNAc) is attached to the core β-Man.
- *Xylose-containing type,* in which the β-Man residue is substituted with β1-2
linked Xyl. Frequently, α1-3 linked Fuc and occasionally α1-6 linked Fuc is present at the Asn-bound GlcNAc residue. Small extensions on the α-Man residues
may occur. These Xyl-containing carbohydrate chains are often found in higher
plants.

D3   B
Manα1-2Manα1-6
                          4'
  D2   A   Manα1-6⟍
Manα1-2Manα1-3 ╱   ⟍Manβ1-4GlcNAcβ1-4GlcNAc-Asn
                            ╱   3   2   1
    Manα1-2Manα1-2Manα1-3╱
    D1   C   4

(oligomannose type)

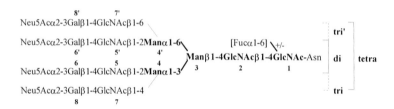

         8'   7'
Neu5Acα2-3Galβ1-4GlcNAcβ1-6⟍
                                              ┊   tri'
Neu5Acα2-3Galβ1-4GlcNAcβ1-2Manα1-6⟍   [Fucα1-6] ⟍+/-   ┊
     6'   5'   4'   ╱Manβ1-4GlcNAcβ1-4GlcNAc-Asn di │ tetra
     6   5   4   ╱   3   2   1
Neu5Acα2-3Galβ1-4GlcNAcβ1-2Manα1-3╱
Neu5Acα2-3Galβ1-4GlcNAcβ1-4
     8   7   ┊   tri

(*N*-acetyllactosamine type)

D3   B
Manα1-2Manα1-6
                          4'
  D2   A   Manα1-6⟍
Manα1-2Manα1-3 ╱   ⟍Manβ1-4GlcNAcβ1-4GlcNAc-Asn
                            ╱   3   2   1
Neu5Acα2-3Galβ1-4GlcNAcβ1-4Manα1-3╱
     6   5   4

(hybrid type)

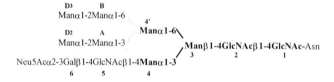

    4'
Manα1-6⟍
            ⟍Manβ1-4GlcNAcβ1-4GlcNAc-Asn
         ╱   3   2   1
Manα1-3╱   │
    4
        Xylβ1-2

(Xylose-containing type)

Fig. 12-2. Examples of the four types of N-linked carbohydrate chains of glycoproteins. The common pentasaccharide core is shown in bold face. For the *N*-acetyllactosamine type the di-, tri-, tri'- and tetra-antennary sub-types are indicated. The standard notation of the monosaccharide residues has been included.

12.3.2 Structure of O-linked Carbohydrate Chains

The O-glycosidic carbohydrate chains are classified according to the specific combination of the bond-forming amino acid and the sugar residues. In animal systems, the mucin (sub)type, in which the carbohydrate–protein linkage is formed between Gal-NAc and Ser or Thr, occurs frequently. The GalNAc residue linked to Ser/Thr may be extended with Gal, GlcNAc, or GalNAc, giving rise to at least eight established core structures [48,49] (Fig. 12-3). The peripheral sequences which are attached to the core structures are formed from GlcNAc, Gal, Fuc and NeuAc/NeuGc residues and often contain elements also occurring in N-linked oligosaccharides. Furthermore, the Gal and GalNAc residues may be sulfated. Apart from the mucin type, other types of O-linked carbohydrates are known [50]. As indicated in Table 12-1, monosaccharides other than GalNAc may be linked to hydroxy amino acids. These sugar residues may also serve as core elements for further extension [51]. O-glycosylation of glycoproteins with GlcNAc seems to be a dynamic modification of specific intra-

Type	Structure	Type	Structure
Core 1	Galβ1-3 ⟋GalNAc-Ser/Thr	Core 5	GalNAcα1-3 ⟋GalNAc-Ser/Thr
Core 2	GlcNAcβ1-6 ⟍ / GalNAc-Ser/Thr Galβ1-3 ⟋	Core 6	GlcNAcβ1-6 ⟍ GalNAc-Ser/Thr
Core 3	GlcNAcβ1-3 ⟋GalNAc-Ser/Thr	Core 7	GalNAcα1-6 ⟍ GalNAc-Ser/Thr
Core 4	GlcNAcβ1-6 ⟍ GalNAc-Ser/Thr GlcNAcβ1-3 ⟋	Core 8	Galα1-3 ⟋GalNAc-Ser/Thr

Fig. 12-3. Eight core structures of mucin-type O-linked carbohydrate chains of glycoproteins.

cellular proteins, in which addition and removal of O-GlcNAc is a highly regulated modification [52]. Based on recent observations for recombinant glycoproteins produced in CHO cells, it can be assumed that O-glycosylation is more widespread than hitherto recognized [53].

12.3.3 Structure of Glycosylphosphatidylinositol Anchors

Glycosylphosphatidylinositol (GPI) anchors are widespread in eukaryotes, especially in parasitic protozoa. GPI anchors have been recognized as an important alternative mechanism for attaching various proteins to the cell membrane [54]. The GPI is anchored into the cell membrane by its lipid moiety and connected by its terminal ethanolamine phosphate, via a peptide linkage, to the C-terminus of a protein. Structural analysis of GPI anchors from different organisms has led to the proposal of an evolutionarily conserved core structure EtN-PO$_4$-6Manα1-2Manα1-6Manα1-4GlcNH$_2\alpha$1-6PI [55,56] (Fig. 12-4). This conserved core can be modified by a wide variety of carbohydrate side chains and ethanolamine phosphate residues. In some GPI structures, acylation (palmitic acid) of the inositol ring occurs. Also the lipid moiety exhibits extensive variety, for instance, acylglycerols in *Trypanosoma brucei*, alkylacylglycerols in *Leishmania*, and ceramides in *Dictiostelium, Saccharomyces* and *Paramecium* [55].

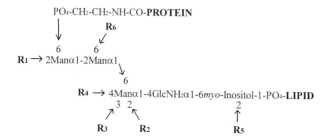

Fig. 12-4. Glycosylphosphatidylinositol anchor structure with evolutionarily conserved glycan backbone. Some frequently occurring substituents are: R$_1$, αMan, αGlc; R$_2$, ethanolamine phosphate; R$_3$, αGal$_{2-4}$; R$_4$, [αGlc]$_{+/-}$-βGalNAc, [βGal]$_{+/-}$-βGalNAc; R$_5$, palmitate; R$_6$, ethanolamine phosphate.

12.4 Biosynthesis of Glycoprotein Glycans

The biosynthesis of protein-bound oligosaccharides differs in several aspects from the biosynthesis of nucleic acids and proteins, which are synthesized as linear molecules with one type of linkage through a template-directed mechanism. Oligosaccharide chains can be branched as well as linear, and their monomeric units are connected to one another by many different linkage types. Oligosaccharides are not biosynthesized by a template-directed mechanism but indirectly by the concerted action of highly specific glycosyltransferases, enzymes which are under genetic control. A consequence of the lack of a template-directed mechanism for oligosaccharide synthesis is the existence of several different oligosaccharide chains at the same glycosylation position, designated as microheterogeneity. Therefore, glycosylation of a protein usually generates a set of glycoforms, all of which share an identical polypeptide backbone but are dissimilar in either the structure or disposition of their oligosaccharide units or both. The formation of glycoforms is not necessarily random but can be genetically regulated and is highly reproducible within one cell type under constant physiological conditions.

12.4.1 N-linked Carbohydrate Chains

The biosynthesis of N-glycans begins in the rER and is similar in lower and higher animal species as well as in the plant and fungal kingdoms. However, the ultimate structures of the mature carbohydrate chains vary enormously for a certain glycoprotein and are mostly species-, tissue-, organ-, and even cell-type-specific [43,57,58]. A lipid-linked oligosaccharide intermediate ($Glc_3Man_9GlcNAc_2$-PP-Dol) is constructed by sequential addition of individual sugar residues to dolichol pyrophosphate (dolichol phosphate cycle). Seven sugar residues (underlined in Fig. 12-5) are added directly from UDP-GlcNAc and GDP-Man on the cytoplasmic side of the ER, whereas the other residues are added from Man-P-Dol and Glc-P-Dol

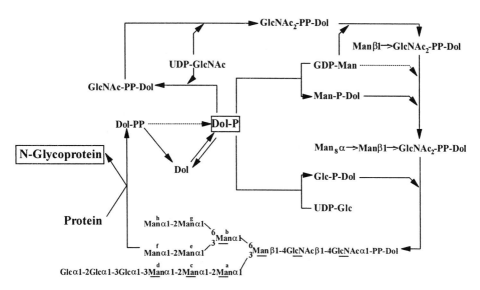

Fig. 12-5. The dolichol phosphate pathway leading to N-glycosylation of proteins. The superscript letters indicate the probable order (alphabetical) of addition of α-Man residues.

in the lumen of the ER [59–61]. The entire pre-assembled oligosaccharide is then transferred *en bloc* under the catalytic action of an oligosaccharyl transferase [62,63] to the acceptor amino acid (Asn) located in an Asn-X-Ser/Thr/(Cys) sequence within the polypeptide (X can be any amino acid except Pro) [64,65]. It has to be noted that not all consensus sequons are glycosylated and that probably additional secondary and tertiary structural elements of the protein are required to realize glycosylation [66–69]. Once transferred to the protein, the carbohydrate chain undergoes a series of trimming (within the lumen of ER and *cis*-Golgi) and elongation steps (within the *medial-* and *trans*-Golgi) before obtaining its final structure. This processing takes place through the interactive, stepwise action of glycosyl-transferases and exoglycosidases [61,70–72]. A schematic representation is shown in Fig. 12-6. Glycosyltransferases transfer monosaccharide residues from an activated donor, usually a nucleotide sugar, to the growing oligosaccharide chain. Exoglycosi-dases commonly remove monosaccharide units one by one. Processing of N-linked oligosaccharides is controlled by several factors such as the genetic control of

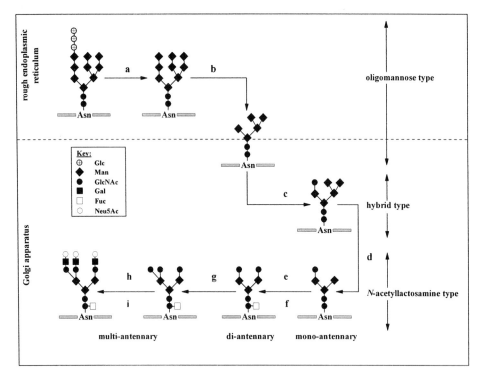

Fig. 12-6. Simplified scheme of the major pathway for the assembly of N-linked carbohydrate chains on newly synthesized glycoproteins. The figure indicates the topography of certain steps and the origin of oligomannose-, hybrid- and (mono-, di-, and multi-antennary) *N*-acetyllactosa-mine-type glycans. Involved enzymes are: a, α-glucosidases; b, α-mannosidases; c, *N*-acetylglu-cosaminyltransferase I; d, α-mannosidases; e, *N*-acetylglucosaminyltransferase II; f, fucosyltrans-ferase; g, *N*-acetylglucosaminyltransferase IV; h, β-galactosyltransferase; i, sialyltransferases.

enzyme expression, the intracellular localization of processing enzymes, the availability of substrates, and the substrate and acceptor specificity of the enzymes [73–75]. The exact control mechanism of all enzymes is not precisely understood and their action leads to different oligosaccharide structures synthesized at the same site on a protein, a phenomenon known as microheterogeneity. In some cases, sulfate, phosphate and/or Fuc residues may be added. Finally, the mature glycoprotein exits from the *trans*-Golgi network in membrane-bound vesicles. The contents of these vesicles are either secreted out of the cell, delivered to the plasma membranes as membrane glycoproteins (eventually linked to GPI anchors), or targetted to other organelles inside the cell [76].

12.4.2 O-linked Carbohydrate Chains

The biosynthesis of O-linked oligosaccharides is a post-translational process wherein carbohydrate is linked to hydroxyl groups of hydroxy amino acids within the polypeptide. Almost any amino acid bearing a hydroxyl group, including the less common hydroxyproline and 3-hydroxylysine, can be O-glycosylated. The biosynthetic data of O-glycans are mainly collected from studies of mucin-type structures [77–79]. O-glycan biosynthesis is initiated by GalNAc transfer to peptide, mainly in the *cis*-Golgi. Subsequently, the O-linked carbohydrate chains are mainly synthesized in the *medial-* and *trans*-Golgi by direct transfer of single monosaccharide residues from sugar nucleotides to the growing oligosaccharide chain, following pathways defined by the sequential action of (competing) glycosyltransferases [2,80]. Hence, the heterogeneity found in O-glycans is due to competition of glycosyltransferases for the same acceptor and the relative amounts of the various glycosyltransferases produced by a cell. Several core structures have been established which are substrates for further elongation [49] (Fig. 12-7). In contrast to the Asn-X-Ser/Thr

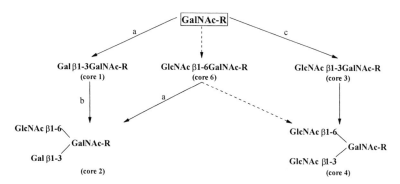

Fig. 12-7. Biosynthetic pathways of some O-glycan-core structures. Paths along solid lines are well established in mucin biosynthesis and paths along dotted lines can occur but are very slow. Some of the involved enzymes are: a, core 1 β3-Gal-transferase; b, core 2 β6-GlcNAc-transferase; c, core 3 β3-GlcNAc-transferase; d, core 4 β6-GlcNAc-transferase. R = Ser/Thr.

sequon for N-glycosylation, no specific amino acid sequence has been established for general O-glycosylation, although the substrate site is favored by the presence of relatively high concentrations of Pro, Ser, and Thr, and disfavored by strongly hydrophobic or hydrophylic residues. Notably, it has been found that Fuc can be linked to Ser/Thr within an apparently conserved sequence -Cys-X-X-Gly-Gly-Thr/Ser-Cys- (X = any amino acid), occuring in epidermal growth factor domains of several proteins [81]. The synthetic pathways for O-glycans are usually classified according to the types of linkage of carbohydrate to protein: GalNAc-O-Ser/Thr in mucins and mucin-type O-glycans, Xyl-O-Ser/Thr in proteoglycans, Gal-O-hydroxylysine in collagens, GlcNAc-O-Ser/Thr in nuclear and cytoplasmic proteins, and Man-O-Ser/Thr in yeast proteins [73]. The sugars commonly found in O-glycans are GalNAc, Gal, GlcNAc, NeuAc and Fuc. In mucins, the O-glycans are usually found in clusters on the polypeptide.

12.4.3 Glycosylphophatidylinositol Anchors

The biosynthesis of the GPI precursor mainly takes place in the lumenal face of the endoplasmic reticulum of the cell and is broadly similar in protozoan parasites and mammalian cells [82–84]. In general, it is assumed that the process involves the sequential transfer of monosaccharides. First, GlcNAc from UDP-GlcNAc is transferred by an α-GlcNAc transferase to phosphatidyl inositol (PI), followed by de-N-acetylation to glucosamine-PI. Subsequently, three α-Man residues coming from Man-P-Dol are added in single steps [85,86]. Ethanolamine phosphate from phosphatidyl ethanolamine is transferred to the terminal mannose. In mammalian cells, at least one extra ethanolamine phophate group is added and also, palmitoylation of inositol occurs early in the assembly of the GPI precursor [87]. Then, depending on the cell species, a complex series of fatty acid remodeling reactions can take place. The precise nature and composition of the lipid moiety (diacylglycerol, monoacylglycerol, alkylacylglycerol, ceramide) vary with protein and cell type. Additional α-Man residues, linked to the ethanolamine phosphate substituted terminal Man, are a common modification found in both lower and higher eukaryotes [88]. The final structure of the GPI anchor is depicted in Fig. 12.4. (Section 12.3.3). It is most probable that the glycosyltransferases involved in the GPI biosynthesis are different from those involved in N,O-glycoprotein synthesis [89,90]. The knowledge of the biosynthetic pathway still shows some major hiatus concerning the time and site of addition of extra sugar residues (e.g., Gal, Glc, Man, GalNAc, NeuAc) occurring in some protozoan species, and the precursors and enzymes involved. A (glyco)protein intended for anchoring to GPI must contain an N-terminal signal sequence for entry into the lumen of the endoplasmic reticulum and a GPI-signal sequence at the C-terminus. The GPI-signal peptide is cleaved and the newly exposed α-carboxyl group is directly attached to the pre-assembled GPI precursor via an amide bond involving the amine of ethanolamine [54,91]. However, several protozoa also synthesize free GPIs which are not covalently linked to protein and which appear to be metabolic end-products.

Due to the dramatic increase in studies of glycoconjugates in biological processes (glycobiology), a growing number of carbohydrates are now being reported which differ from the 'standard' N,O-glycans, in terms of their structure and protein linkage as well as their location in cell compartments [50]. Novel carbohydrate chains O-linked via Fuc or Glc found in coagulationfactor glycoproteins [92,93], O-linked GlcNAc found on a wide variety of cytoplasmic and nucleoplasmic proteins [51,52,94], and the discovery of mitochondrial glycoproteins [95,96], are just a few examples. Studies on the biosynthesis of these 'new' glycoconjugates are still in a preliminary stage.

12.5 Analysis of Glycoprotein Glycans

12.5.1 General Aspects

The primary structure analysis of glycoprotein glycans has still not reached the level of routine analysis and remains a highly specialized and laborious endeavour. No single technique is able to provide all information needed for the exact structural definition of a complex glycan. The structural analysis depends on the combined use of several physical, chemical, and biochemical techniques [97,98]. To date, several approaches based on ^{1}H-nuclear magnetic resonance (NMR) spectroscopy, mass spectrometry, enzymatic procedures, and profiling techniques are available but these current approaches have to be applied with great care. Since, in general, (recombinant) glycoproteins contain multiple-glycosylation sites and the carbohydrate chains usually exhibit considerable structural heterogeneity, it is (almost) not possible to analyze complete glycan structures on the intact glycoproteins. However, the rapid progress of multi-dimensional techniques in NMR spectroscopy and mass spectrometry implicate prospects for the future. Nevertheless, at the moment, a more practical approach is to release the attached carbohydrate chains from the protein, purify the oligosaccharide pool from the deglycosylated protein and reaction additives, and fractionate to individual, pure oligosaccharides, which can be identified using different chemical and spectroscopic methods. Methods for the structural analysis of (recombinant) glycoprotein glycans have been reviewed regularly [99–105]. For detailed laboratory protocols of all the main methods currently used to elucidate the structure and biosynthesis of glycoproteins, the reader is referred to the book *Glycobiology: A Practical Approach* [106].

A general strategy for characterizing and analyzing (recombinant) glycoprotein glycans applicable on most glycoproteins is summarized in Fig. 12-8. A good approach to obtain separately N- and O-glycans from N,O-glycoproteins is the enzymatic release of the N-glycans, followed by a chemical release of the O-glycans [107]. But first, in order to get an overview of the carbohydrate portion of a protein the intact glycoprotein is subjected to monosaccharide analysis. The determination of the monosaccharide composition is currently established after methanolysis and analysis of volatile derivatives by gas chromatography(-mass spectrometry) [108,109].

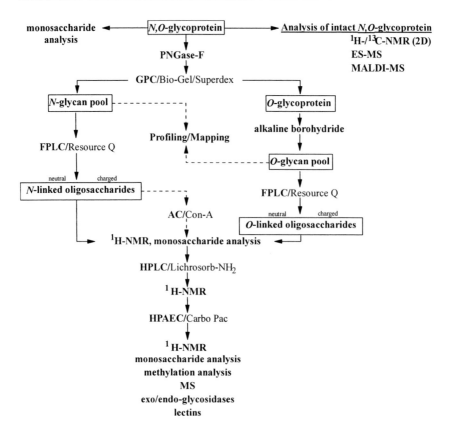

Fig. 12-8. A general strategy for the release, isolation, fractionation, and analysis of (recombinant) glycoprotein glycans.

This method allows simultaneous analysis of neutral monosaccharides, acetamido sugars, and sialic acids. A typical gas chromatogram is presented in Fig. 12-9. By using an internal standard, information about the carbohydrate content is also obtained. The occurrence of relatively large amounts of GalNAc and/or Man indicates the presence of O- and/or N-linked carbohydrate chains, respectively. It is obvious that monosaccharide analysis can also be performed on isolated pure glycans to determine the molar ratios of the constituent monosaccharide residues of an individual

Fig. 12-9. Monosaccharide composition analysis of glycoproteins by gas chromatography-mass spectrometry. (A) Gas chromatogram of a standard mixture of trimethylsilylated (methyl ester) methyl glycosides on a CP-Sil 5 column (25 m × 0.32 mm). Oven temperature program: 130° to 230 °C, at 4 °C min^{-1}. Each monosaccharide gives rise to multiple peaks due to α/β- and pyranose(p)/furanose(f) forms. (B) Gas chromatogram of the constituent monosaccharides of recombinant human erythropoietin (rhEPO), after methanolysis/re-N-acetylation/trimethylsilylation. (C, D, E, and F) (GC-)EI-mass spectra of the trimethylsilyl (methyl ester) methyl glycosides of αFuc$p$, αMan$p$, αGlcNAcp, and βNeuAc, respectively. IS = internal standard (mannitol).

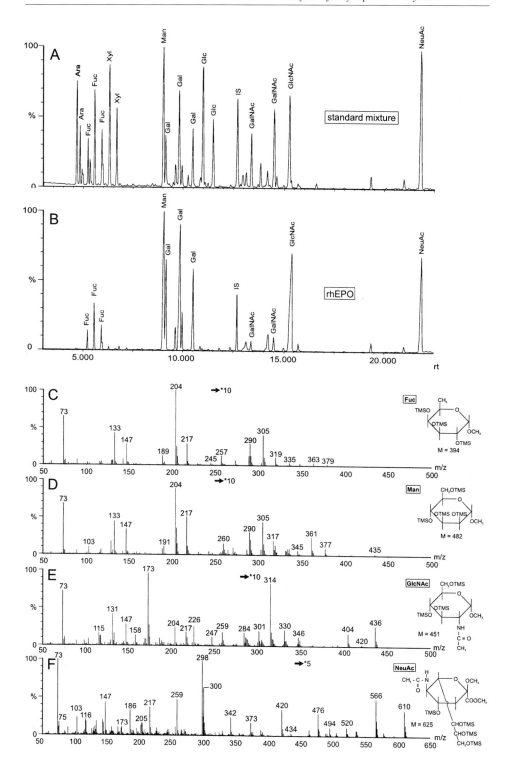

carbohydrate chain. The absolute configuration (D or L) of the monosaccharides are eventually determined by gaschromatography of their (-)2-butyl glycosides [110].

The presence of a GPI membrane anchor on a protein can be inferred by analysis for the components ethanolamine and myoinositol using amino acid analysis and GC-MS, respectively [111,112]. Another indication for the presence of GPI anchor is to analyze selectively for fatty acid components by releasing them with alkali, acidify to protonate the fatty acids, partition into organic solvent, and derivatize to the fatty acid methylesters for GC(-MS) analysis. For a detailed structural characterization of the GPI glycan moiety the protein must first be split off from the GPI.

12.5.2 Release of Carbohydrate Chains from Proteins

The liberation of the carbohydrate moieties from glycoproteins can be achieved in different ways depending on the type of linkage of the glycan to the protein. Commonly used procedures to generate carbohydrate chains are summarized in Table 12-2.

Alternatively, proteolytic digestion of the glycoprotein can be used to obtain glycopeptides which, after subsequent separation, can also give information about the glycosylation sites of the protein. The use of this method however, can be limited by the distribution of the sugar chains, since steric hindrance imposed by the carbohydrates may impair complete degradation of the peptide backbone, leaving adjacent carbohydrate moieties unseparated.

Table 12-2. Procedures to generate partial structures, representing one glycosylation position. Proteolytic and chemical cleavage of glycoproteins leads to the formation of glycopeptides, whereas the other procedures liberate carbohydrate chains from the protein. Conversion of reducing constituent monosaccharide residues into a fluorescent derivative or its corresponding alditol may follow the liberation of the carbohydrate chains.

Type of linkage	
N-glycosidic	O-glycosidic
Proteolytic digestion	Proteolytic digestion
Chemical protein degradation	Chemical protein degradation
Hydrazinolysis[a]	Hydrazinolysis[a]
Alkaline borohydride treatment[a]	Alkaline borohydride treatment[a]
Enzymatic hydrolysis	Enzymatic hydrolysis[b]

[a] Different conditions are applied for the release of N- and O-glycosidic carbohydrate chains.

[b] The enzyme generally used (endo-α-N-acetylgalatosaminidase), has a rather restricted specificity.

Chemical Methods

Hydrazinolysis [113] is a chemical method for releasing glycans. A fully automated hydrazinolysis procedure is achievable by the GlycoPrep$_{TM}$ 1000 (Oxford Glyco-Sciences) [114,115]. The chromatographic procedures necessary for recovery of the carbohydrate pool purified from amino acids, peptides (stemming from the degraded protein), and reactants, is integrated. Although different reaction conditions for N- and O-glycosidic linkage cleavages have been developed by Oxford Glyco-Sciences there is always mutual contamination of both types of carbohydrates. A further disadvantage of hydrazinolysis is the (limited) occurrence of side reactions leading to degradation or conversion of the carbohydrate chains [116]. This can complicate and compromise the structural analysis because of the introduction of additional, artificial heterogeneity. Notwithstanding the foregoing, a clear advantage of the automated hydrazinolysis procedure is its potential for relatively high throughput carbohydrate analysis, which makes it suitable for analysis of batch-to-batch consistency of (recombinant) glycoproteins. A chemical method, most often used for specific release of O-linked glycans is mild alkaline borohydride treatment (β-elimination reaction) [107,117]. To avoid peeling reaction at the reducing terminal, sodium borohydride is added to the reaction mixture to convert the oligosaccharides to the corresponding sugar alcohols (alditols) as soon as they are released from the polypeptide [118]. All chemical methods result however in various modifications of the released oligosaccharides.

For the removal of glycans from the glycoprotein/peptide with the aim of recovering the intact protein/peptide for sequence analysis, the use of anhydrous trifluoro-methanesulfonic acid (TFMS) has been found to be most successful [105]. N- and O-glycans are cleaved non-selectively leaving the primary structure of the protein/peptide intact, but the carbohydrate chains are destroyed.

Enzymatic Methods

Alternatively, carbohydrate chains can be released from the glycoprotein by enzymatic procedures. Two important types of N-glycan-releasing enzymes are known, namely the endo-β-*N*-acetylglucosaminidases (further referred to as endoglycosidases), which hydrolyze the GlcNAcβ1-4GlcNAc linkage in the *N,N'*-diacetylchitobiose unit of Asn-linked carbohydrate chains and the peptide-N^4-(*N*-acetyl-β-glucosaminyl) asparagine amidases (PNGases), which hydrolyze the β-aspartyl glycosylamine linkage. Specific studies [119–123] have shown that the glycan specificity of the endoglycosidases is rather rigid, which renders these enzymes unsuitable for obtaining a complete profile of the Asn-linked oligosaccharides. For instance, endoglycosidase H (Endo-H) from *Streptomyces plicatus* cleaves only oligomannose- and hybrid-type N-glycans. Two different amidohydrolases, namely PNGase-A (from almond emulsin) and PNGase-F (from *Flavobacterium meningosepticum*) are frequently used to split off intact N-linked oligosaccharides. Before using these enzymes, in some cases, the protein must be denatured to ensure that the N-glycosidic bond is accessible [124,125]. PNGases have however a unique resistance to denaturating substances like chaotropic agents, detergents, and disulfide reductants. The most important functional feature that PNGases-A and -F have in common, is the ability to release high-mannose, hybrid, as well as complex types of Asn-linked oligosaccharides provided that: (i) the glycosylation site is accessible to the enzyme; (ii) both the C- and N-terminal group of the Asn-residue bearing the carbohydrate chain are in peptide linkages; and, (iii) the carbohydrate chain attached to a glycoprotein comprises at least two residues [126]. Nonetheless, following digestion, it is always advisable to check (e.g., by SDS–PAGE or MALDI-MS) that no N-glycan remains associated with the protein since the enzyme frequently does not release all glycans with the same efficiency (time dependance). Furthermore, it is known that PNGase-F does not work when the Asn-linked GlcNAc bears a Fucα1-3 residue [127,128].

Until recently, the enzymatic release of O-linked glycans was restricted to only one type. Endo-α-N-acetylgalactosaminidase (from *Streptococcus pneumoniae*) releases only Galβ1-3GalNAc from Ser/Thr in the polypeptide. Substitution of the disaccharide by NeuAc, Fuc, or GalNAc residues or the absence of the Gal residue

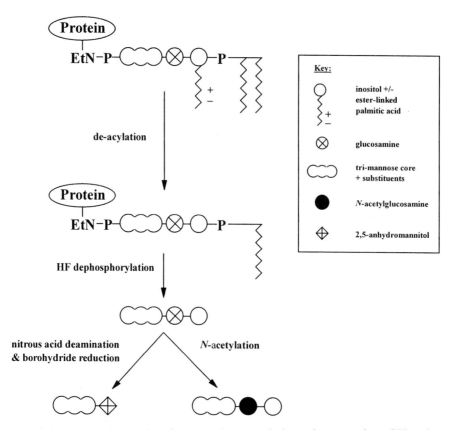

Fig. 12-10. A general procedure for generating neutral glycan fragments from GPI anchors.

abolishes hydrolysis. In addition, an endo-*N*-acetylgalactosaminidase from *Strepto-myces* has recently been shown to release more complex O-glycans [129].

In order to analyze the GPI anchor, a combination of chemical and enzymatic methods is used depending on the structural information desired [42,82]. In general, the first step in the characterization of GPIs is the enzymatic removal of the protein which can be linked via ethanolamine phosphate to the GPI glycan. Proteases (e.g., pronase, trypsin, papain) are used to this end. Phosphate groups are removed from the glycans by treatment with aquous HF yielding the GPI core glycan free from protein and lipid. The lipid portion can also specifically be released by enzymatic cleavage with phosphatidylinositol-specific phospholipase C from *Bacillus* (PIPLC) or GPI-specific phospholipase D (GPI-PLD). Palmitoylation of inositol renders the GPI structure resistant to cleavage by these enzymes. As an alternative for the last step, nitrous acid deamination can be applied which cleaves the linkage between myoinositol and glucosamine (converting glucosamine into 2,5-anhydromannose). A general procedure [54,111,130] for generating neutral glycan fragments from GPIs involves deacylation to iimprove solubility, dephosphorylation with ice-cold aquous HF and conversion to neutral species via *N*-acetylation or nitrous acid deamination/sodium borohydride reduction as depicted in Fig. 12-10. For isolation and fractionation of intermediates, several chromatographic procedures (described below for N,O-glycans) including HPAEC, are used. Subsequently, the resulting glycan fragments can be analyzed by ^{1}H-NMR spectroscopy as demonstrated in Fig. 12-11 for the glycan core derived from *Toxoplasma gondii* GPI [130].

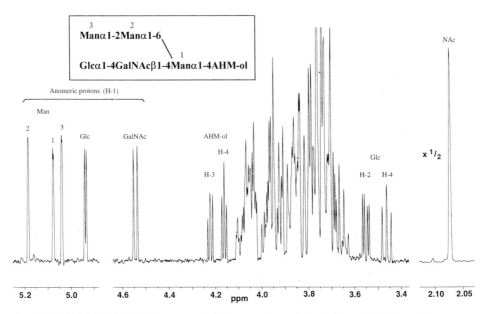

Fig. 12-11. 500-MHz ^{1}H-NMR spectrum of the core glycan derived from GPI of *Toxoplasma gondii*. The HOD signal (4.65–4.85 ppm) has been omitted. The relative scale of the NAc proton region differs from that of the rest of the spectrum. The spectrum was recorded in D$_2$O at 300 K.

12.5.3 Isolation and Fractionation of Oligosaccharides

Usually, after enzymatic release of the N-glycans from the protein, a gel filtration step (Bio-Gel P-100 or Superdex 75) is sufficient to separate the glycan pool from the residual high-molecular-mass protein. Direct structural analysis of the released oligosaccharides is mostly impeded by the heterogeneity of the sample and thus fractionation is essential. Separations are accomplished by exploiting differences in mass, charge, and hydrophilic properties of the oligosaccharides. Preparative and/or analytical separation of carbohydrate chains can be carried out using a variety of chromatographic or electrophoretic techniques (Table 12-3), including gel permeation chromatography, several types of HPLC (anion-exchange, amine-adsorption, and reversed-phase), high-pH anion-exchange chromatography (HPAEC), and affinity chromatography on immobilized-lectin columns. An important feature during fractionation is the detection method. Although, most glycans have (low) UV activity, for analytical purposes and especially when limited amounts of material are available, it is sometimes required to label the glycans. Commonly used methods are reductive amination with a fluorescent compound such as 2-aminobenzamide (2-AB) [131] or 2-aminopyridine (PA) [132], and reduction with alkaline sodium borotritide [133] for highly sensitive detection. Today, special developed glycan labeling and detection kits, eventually already in an automatical application, are on the market (Oxford GlycoSciences, Beckman, Bio-Rad, Takara Shuzo).

Table 12-3. Several fractionation methods and detection techniques used in the analysis of glycoproteins glycans.

Separation methods		Detection methods
Chromatography	Electrophoresis	
Affinity[a]	Capillary-zone	Fluorescence[b]
Amine-adsorption	Paper	Mass spectrometric
Anion-exchange	Polyacrylamide slab gel	Pulsed amperometric
Gas-liquid[c]		Radiochemical[d]
Gel-filtration		Refractive index
High-pH anion exchange		Ultra-violet absorbance
Paper		
Reversed-phase		
Thin-layer		

[a] Based on lectins or antibodies.
[b] Glycans are tagged with a fluorescent agent.
[c] Used in combination with mass spectrometric or flame-ionization detection; fractionated components are not recovered.
[d] A radioactive isotope is introduced into the glycans.

Separation According to Charge

Traditional methods like paper electrophoresis and low-pressure anion-exchange chromatography are more and more replaced by medium- and high-pressure ion-exchange chromatography, using FPLC and HPLC systems, respectively. Chromatography on a Mono Q or Resource Q anion-exchange column (Pharmacia FPLC system) offers a fast and reproducible separation method for sialyl-oligosaccharides, based on the sialic acid content [134,135]. However, the molecular mass of the compounds having the same number of sialic acid residues has a minor influence on the elution position. Separation according to the number of acidic (e.g., sulfate or carboxyl) groups can also be achieved by HPLC on MicroPak AX-5 and AX-10 [136] and Lichrosorb-NH$_2$ [137] columns. More recently, high-performance anion-exchange chromatography with pulsed amperometric detection (HPAEC-PAD) has been introduced for the separation and sensitive detection (10–100 pmol) of linkage and branch positional isomers of both neutral and acidic oligosaccharides (Dionex BioLC system with polymer-based pellicular column) [138,139]. In this application charge is introduced by ionization of hydroxyl groups of the oligosaccharide chains at alkaline pH (12–14). Correlation of the retention times with the structure of different oligosaccharides suggests that the accessibility of the readily ionizable hydroxyls to the quaternary amine stationary phase is a major determinant of the enhanced resolution [140,141].

Separation According to Mass

For fractionation of oligosaccharides according to their molecular mass, size exclusion chromatography (gel filtration) on Bio-Gel can be used [142]. Bio-Gel P-4 is most suitable for separation of (preferably neutral) oligosaccharides ranging in size equivalent to up to 24 glucose residues [133]. More recently, high-performance size exclusion chromatography using large pore polymeric column materials was introduced.

Separation According to Hydrophilic Properties and Structure

By HPLC on chemicaly modified silica, higher resolution can be obtained than by chromatography on soft gels. Reversed-phase HPLC, which depends on hydrophobic interaction between the sample and the C-18 stationary phase, has been used for the separation of a large array of neutral oligosaccharides [143,144]. Reversed-phase HPLC can also be applied to the fractionation of charged oligosaccharides by use of ion pairing reagents like triethylamine [145]. In normal-phase HPLC, using Lichrosorb-NH$_2$, the retention of carbohydrates on the amino-bonded silica is probably based on hydrogen bonding between the hydroxyl groups of the sugar and the amino group of the stationary phase. Alkylamino bonded silica columns have, compared with HPAEC, low-resolution properties for neutral and charged N- and O-glycans [142,146]. It has to be noted that the spatial structure of the carbohydrate chain determines the accessibility of the carbohydrate submolecular components to the column stationary phase, thereby influencing its chromatographic behavior in both normal- and reversed-phase HPLC.

Recently, some new semi-automated procedures based on chromatographic techniques are developed for (fast) structural analysis of glycoprotein glycans and these will be discussed in Section 12-6.

Separation According to Affinity Differences Towards Lectins

Numerous lectins have been purified and in some cases their properties are well defined. Their specific affinity for distinct sites in carbohydrate chains can be exploited for the isolation and purification of a wide range of oligosaccharides (and glycoconjugates) by affinity chromatography. Many examples can be found in the literature [147–151]. However, the binding specificity of lectins is in general complex, being determined in the first instance by the nature of the monosaccharides and their glycosidic linkages, but furthermore by steric factors, and in case of glycoconjugates even by non-specific interactions between the lectin and the carbohydrate-containing macromolecule. The lectin binding specificity is constantly redefined as the knowledge in this field increases and however positive this may be in itself, it may influence the benefits of lectin affinity chromatography for oligosaccharide purification procedures.

12.5.4 Structural Characterization of Oligosaccharides

The final goal in the analysis of carbohydrate chains is to determine the number, nature, order, and ringconformation (D/L and pyranose/furanose) of the constituent monosaccharide residues of the purified oligosaccharides. Furthermore, the anomericity (α or β) and the substitution pattern of the individual monosaccharides must be determined. Finally, the nature and localization of chemical substituents (e.g., *O*-methyl, acetate, lactate, sulfate, phosphate) potentially present on a given monosaccharide residue must be determined. Ultimately, this all leads to the description of the complete carbohydrate chain in a structural formula.

Since several biological functions are ascribed to carbohydrate chains as functional parts of glycoproteins (see Section 12.2), conformational analysis of these chains becomes more and more important [152]. In order to understand how carbohydrate chains are involved in biological processes, determination of the three-dimensional structures of these biomolecules is a crucial step. In the last decade an increasing effort has been made to establish three-dimensional structures of oligosaccharides as part of glycoproteins. Experimental data of carbohydrate chain conformations in solution are mainly obtained by advanced (multi-dimensional) NMR spectroscopy methods (e.g., NOESY, ROESY), which are compared with computer models obtained by theoretical methods (e.g., Molecular Dynamics simulations and Molecular Mechanics calculations). A detailed description is beyond the scope of this chapter, which is focused on primary structure determination.

Enzymatic and Chemical Analytical Methods

A powerful method for determining the monosaccharide sequence of carbohydrate chains is the sequential exoglycosidase digestion procedure [133,153,154]. Exoglycosidases are hydrolases that cleave monosaccharide residues from the non-reducing terminal of the carbohydrate chain. Most of the enzymes are highly specific toward their substrate, including the anomeric configuration. They are named according to their specificity, e.g., β-galactosidase cleaves the β-galactosyl linkage. Frequently used exoglycosidases are: α-mannosidase, β-galactosidase, β-N-acetylhexosaminidase; α-fucosidase and sialidase [101,153]. It has to be noted that aglycon specificity also plays a role for certain exoglycosidases. The principle of sequencing involves the stepwise degradation of the carbohydrate chain. The unknown, usually tritium-labeled, pure glycan is digested with an exoglycosidase and the remaining glycan is analyzed to determine if monosaccharides have been cleaved. After recovering of the digested glycan, it is subjected to another exoglycosidase. The amount of monosaccharide released by each exoglycosidase digestion can be estimated by analyzing the change in effective size of the glycan by gel permeation chromatography (Bio-Gel P-4 column) before and after the enzymatic digestion [155]. This procedure is repeated until the entire glycan sequence is revealed. Enzymatic analyses can be performed on small amounts (150 pmol) of labeled sample. As the method relies on the substrate specificity of each enzyme, special care must be taken to avoid contamination of each enzyme with other exoglycosidases.

Methylation analysis is the most reliable chemical method for the elucidation of the substitution pattern of the individual monosaccharide residues. It involves the methylation of all free hydroxyl groups of the oligosaccharide followed by the liberation of the methylated monosaccharides by hydrolysis and subsequent qualitative and quantitative analysis of volatile derivatives. The position of the free hydroxyl groups of the partially O-methylated monosaccharides indicate the positions in which the sugar residue was substituted. Effective and complete methylation is essential for obtaining reliable results. Many methylation methods have been reported (see review [156]), but Hakomori's procedure [157] using a mixture of methylsulfinylcarbanion and methyl iodide as the methylation reagent turned out to be most used. Usually, the hydrolysis products are converted into partially O-methylated alditol acetates by reduction with sodium borohydride, followed by acetylation with acetic anhydride [158]. Separation, identification, and quantitation of the various partially O-methylated alditol acetates can be performed by gas chromatography, in most cases combined with mass spectrometry [109]. An example of methylation analysis is depicted in Fig. 12-12.

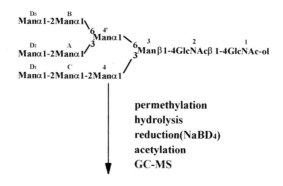

permethylation
hydrolysis
reduction(NaBD₄)
acetylation
GC-MS

PMAA	Mol. ratio	Structural feature
1,5-di-*O*-acetyl-2,3,4,6-tetra-*O*-methylhexitol	3	Man(1→
1,2,5-tri-*O*-acetyl-3,4,6-tri-*O*-methylhexitol	4	→2)Man(1→
1,3,5,6-tetra-*O*-acetyl-2,4-di-*O*-methylhexitol	2	→3,6)Man(1→
1,4-di-*O*-acetyl-3,6-di-*O*-methyl-2-*N*-methylacetamido-2-deoxyhexitol	1	→4)GlcNAc(1→
4-mono-*O*-acetyl-1,3,5,6-tetra-*O*-methyl-2-*N*-methyl-acetamido-2-deoxyhexitol	1	→4)GlcNAc-ol

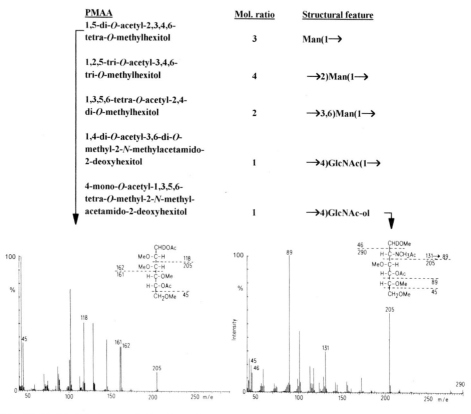

Fig. 12-12. Methylation analysis of an oligomannose-type oligosaccharide alditol. The EI-mass spectra of two partial methylated alditol acetates (PMAAs) are shown as example.

Mass Spectrometry

Different new techniques of mass spectrometry have been introduced in recent years for the structural study of carbohydrate chains. Methods using soft ionization techniques have become very important. The progress achieved in this field has been reviewed by several authors [159–161]. Already for a long time electron impact-mass spectrometry (EI-MS) in combination with gas chromatography, is used for the identification of monosaccharide derivatives. For instance, methanolysis of an oligosaccharide yields a mixture of methyl glycosides of neutral monosaccharides, amino sugars, and sialic acid (as methyl ester), which can be determined jointly by GC-MS after re-*N*-acetylation/trimethylsilylation [109,162] (Section 12.5.1, Fig. 12-9). GC-MS is also generally used for the identification of the partially methylated alditol acetates derived from neutral and *N*-acetylamino sugars in the methylation analysis of glycans. The positions of the *O*-methyl and *O*-acetyl groups can be deduced from the specific fragmentation observed in the highly characteristic EI-mass spectra (Section 12.4.1, Fig. 12-12). Ring size information can also be obtained.

Fast atom bombardment mass spectrometry (FAB-MS) has proven to be a valuable technique for the structural analysis of intact carbohydrate chains, especially with respect to molecular mass determination and sequence analysis. Positive- as well as negative-ion FAB-MS data have been reported for oligosaccharides, oligosaccharide-alditols, and glycopeptides as underivatized, permethylated, and peracetylated derivatives [163]. During FAB-MS, samples are ionized and desorbed from a liquid phase (matrix) using a beam of accelerated atoms (Ar, Xe, Cs). High-field and high-resolution magnetic sector mass spectrometers provide the possibility to investigate polar nonvolatile compounds or derivatives with relatively high molecular masses. In the positive mode the molecular mass can be deduced from the relatively abundant protonated molecular ion $(M+H)^+$, and from the frequently present cationized molecular ions $(M+Na)^+$ and $(M+K)^+$. In the negative-ion mode, $(M-H)^-$ acts as quasi-molecular ion. In Fig. 12-13, a di-antennary asialoglycopeptide is used as an example of sequence information deduced from a negative-ion FAB mass spectrum [109]. The most important fragmentation reaction involves cleavage of the glycosidic bond between the anomeric carbon atom and the interglycosidic oxygen, accompanied by a hydrogen migration. In the case of positive-ion FAB-MS of permethylated and peracetylated derivatives, cleavages have been detected on either site of the glycosidic oxygen. Charged fragments can result from the nonreducing as well as from the reducing end of the molecule.

Recently, another soft ionization technique, namely electrospray mass spectrometry (ESMS), has been introduced to the glycoprotein field [164]. ESMS is applicable to high-molecular mass molecules (~ 100 kDa), including intact glycoproteins. The method is particularly valuable to obtain information whether a protein has been post-translationally modified by the observation of any mass difference between the observed signal and that calculated as the sum of the amino acids present in the sequence. Having available some beforehand information about expected glycan structures, the probable composition of the glycoforms may also be observed. In principle, a stream of liquid containing the sample is directly injected into the ion

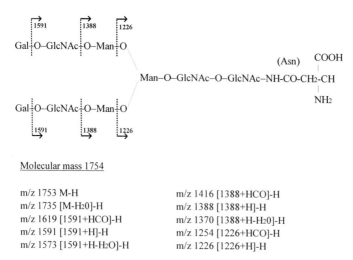

Fig. 12-13. Molecular mass and sequence information from the negative-ion FAB-mass spectrum of an asialo-di-antennary-*N*-acetyllactosamine-type N-glycan linked to asparagine.

source of a mass spetrometer. A spray of microdroplets is generated which, after passing a series of 'skimmers', encounters a drying gas. The net effect is the creation of charged molecular species, devoid of solvent, ready for analysis by (quatrapole) MS. Large molecules tend to carry multiple charges and the mass to charge (m/z) is measured. The data show a distribution of signals carrying varying numbers of net positive charges (working in the positive-ion mode) on protonatable basic sites (Lys, Arg) in the molecule. The peak top charge distribution data can be 'transformed' using a computer algorithm to produce a MS profile of intensity against mass.

MALDI-MS is the acronym for matrix-assisted laser desorption and ionization mass spectrometry. In principle, desorption and ionization of molecules (in a matrix) is induced by means of electromagnetic radiation from a laser and the mass of an ion is measured by determination of its m/z value using, in general, a time-of-flight (TOF) mass spectrometer [165,166]. This technique can be used to analyze intact glycoproteins as well as pure glycans and is therefore highly suitable to determine the number of glycans and their individual mass. Basically, a particular molecular species (e.g., a carbohydrate chain or a particular glycoform of a protein) yields only one MS signal (in contrast to ESMS) and hence by correlating mass to primary structures data are obtained to support an assigned structure. The method can be used for (preferably deacidified) glycan pools to get a mass profile or on pure glycans for exact mass determination. Small sample amounts (0.1 μg pure glycan or 10 μg glycoprotein) are required. The application of MALDI-MS in the glycoprotein field is rapidly increasing and numerous studies have appeared in literature recently [97,167–169]. In Fig. 12-14, the MALDI mass spectrum of FSH is depicted as an example of the potential of the technique for the identification of intact glycoproteins.

For batch-control procedures, several approaches, including HPLC in an on-or offline configuration with MS, have been described [170–173]. In these approaches

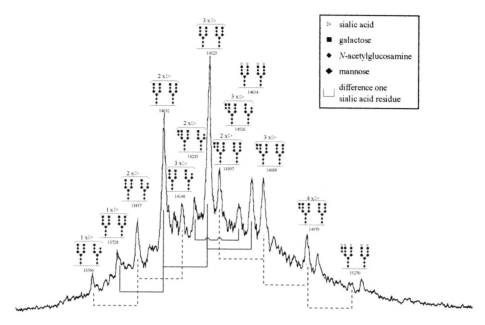

Fig. 12-14. MALDI-mass spectrum of follicle stimulating hormone (FSH) glycoprotein.

either free or derivatized oligosaccharides and glycopeptides were separated and subsequently analyzed by ES-, MALDI- or FAB (tandem)-MS.

[1]H-nuclear Magnetic Resonance Spectroscopy

It is but 20 years ago that high-resolution ^{1}H-NMR spectroscopy was introduced for
the elucidation of the primary structure of glycans derived from glycoproteins
[174,175]. Nowadays, it is one of the most commonly used methods because, in
addition to determining the primary structure, NMR can provide information on
the conformation and molecular dynamics of the molecule in solution.

For interpretation of the ^{1}H-NMR spectrum of a carbohydrate chain in terms of
primary structural assignment use is made of structural-reporter group signals as
listed in Table 12-4. In carbohydrates, most of the sugar-skeleton protons give resonance signals in a crowded region between δ 3.5 and δ 3.9 ppm. The chemical shifts
of specific types of protons resonating at clearly distinguishable positions outside of
this bulk region in the spectrum, together with their coupling constants and the signal
line-widths bear information on the primary structure. As example, a ^{1}H-NMR spectrum is depicted in Fig. 12-15. The conversion of the chemical shift values of the
structural-reporter group signals to carbohydrate structures makes use of extensive
libraries of reference compounds [176,177]. Recently, database-related computer
programs have become available which contain tables of ^{1}H-NMR chemical shift
values and corresponding carbohydrate structures, and literature references [178,–

Table 12-4. ¹H-NMR structural-reporter group signals for carbohydrate chains of glycoproteins.

- Anomeric protons H-1

- Amide protons (in H_2O)

- Man H-2 and H-3

- GalNAc-ol H-2, H-3, H-4 and H-5

- Sialic acid H-3 protons (equatorial and axial)

- Fuc H-5 and H-6 protons (CH_3)

- Gal H-3 and H-4

- Protons shifted out of the bulk region due to glycosylation shifts

- Protons shifted out of the bulk region due to the presence of non-carbohydrate substituents like acyl, sulfate, and phosphate groups

- Protons belonging to substituents on carbohydrate residues like *O*-methyl, *N,O*-acetyl and *N*-glycolyl groups

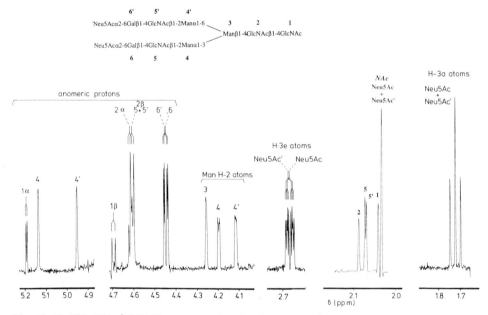

Fig. 12-15. 500-MHz ¹H-NMR spectrum showing the structural-reporter-group regions of a di-antennary-*N*-acetyllactosamine-type N-glycan. The relative scale of the NAc proton region differs from that of the rest of the spectrum. The spectrum was recorded in D_2O at 300 K.

181]. Significant improvement in the interpretation of NMR spectra of complex carbohydrates derives from the application of multi-dimensional (2D and 3D) NMR methods stimulated by dramatic advances in computer technology and in the construction of high-field, superconducting magnets over the past decade. A limitation of the NMR spectroscopic analysis (although non-destructive) is still the amount of sample (at least 20 µg of pure glycan) required. However, using a micro-probe this may be lowered to about 5 µg. Several excellent books [182–184] and reviews [185–189] dealing with the application of NMR techniques for macromolecular structural analysis have appeared. As mentioned before, due to the multiple glycosylation sites and high heterogeneity, causing a multitude of (overlapping) protein- and carbohydrate-derived NMR signals, it was not possible to analyze carbohydrate structures directly on the intact glycoproteins by NMR spectroscopy. Recently, however, structural NMR studies of both carbohydrate and protein moieties of intact complex glycoproteins appeared in literature, showing that considerable progress has been made in this field [190–192]. The structure assessment by NMR spectroscopy of the α-subunit of human chorionic gonadotropin (hCG), an intact glycoprotein subunit, was probed by ^{1}H and gradient-enhanced ^{1}H-^{13}C heteronuclear correlation spectroscopy [193,194]. Also, the glycosylated amino-terminal adhesion domain of the human T-cell surface glycoprotein CD2 was solved in solution by NMR methods [195]. Other intact glycoproteins investigated by ^{1}H and/or ^{13}C NMR spectroscopy are: hen phosvitin [196], IgG [197], submaxillary mucins [198], glycophorin A [199] and pineapple stem bromelain [200]. Another exciting development is the

Table 12-5. Methods to obtain information about specific carbohydrate features.

Information	Methods
Carbohydrate content, composition, D/L-configuration	Colorimetric determinations; GC-monosaccharide analysis; GC-absolute configuration determination; NMR-spectroscopy
Molecular mass of glycoprotein/glycan (presence of glycosylation)	Gel filtration chromatography; Mass profile FAB/ES/ MALDI-mass spectrometry; SDS–PAGE (before/after enzyme treatment)
Nature of carbohydrate–peptide linkage (N/O)	Proteolytic digestion; Amino acid analysis; Examination of alkali lability; hydrazinolysis
Type of glycans (oligomannose, complex, hybrid), glycoforms	GC-monosaccharide analysis; Size/charge profile analysis; Capillary electrophoresis
Number/proportions of glycans present	Size/charge profile analysis; Mapping by HPLC, HPAEC, FACE, MALDI-MS
Sequence of monosaccharide residues	Digestion by exoglycosidases; Partial hydrolysis; NMR-spectroscopy; mass spectrometry
Positions of glycosidic linkages	Methylation analysis/GC-MS; FAB-MS; NMR-spectroscopy
Anomeric configuration	Digestion by exoglycosidases; NMR-spectroscopy
Certain structural determinants	Antibody responses; Endo/exo-glycosidases; Affinity chromatography (lectins)
Type of charged substituents	Size/charge profile analysis; HPLC; HPAEC; NMR-spectroscopy
Spatial structure of glycoprotein/glycan	X-ray analysis; (2D/3D) NMR-spectroscopy; Molecular dynamics, mechanics and modeling

progress that has been made in expression and labeling of glycoproteins (e.g., hCG) in CHO cells with stable NMR isotopes like ^{13}C and ^{15}N [201]. This 'friendly' labeling procedure for mammalian post-translationally-modified proteins holds promise to solve the structure of intact glycoproteins in solution more routinely. Moreover, it opens the way to determine the conformation of (^{13}C-, ^{15}N-labeled) carbohydrate chains on the surface of glycoproteins, and furthermore, the assessment of interactions of the carbohydrate chains with the peptide backbone come into reach.

Table 12-5 summarizes the various methods that can be applied to obtain information about specific carbohydrate features.

12.6 Oligosaccharide Profiling/Mapping

Because the biological and physico-chemical properties of (recombinant) glycoproteins frequently are significantly influenced by the type of glycosylation, knowledge about the glycosylation pattern is of utmost importance. Although in many cases the exact function of the carbohydrate chains has not been established as yet, much attention should be paid to obtain the 'right' glycosylation pattern. Therefore, means of assessing batch-to-batch consistency during production of glycoprotein pharmaceuticals, in terms of protein glycosylation, is urgently required. The necessary information can be obtained by the procedures described earlier in this chapter, but for routine analysis of production batches a simpler, less labor-intensive procedure is needed. Currently, methods are being developed, based on oligosaccharide-profiling procedures, facilitating reliable and fast batch control of glycoproteins. Next to the analysis of the monosaccharide composition (discussed in Section 12.1; Fig. 12-9), the glycans are released from the protein backbone and directly scanned by a combination of different separation techniques, such as more-dimensional HPLC [202,203] and HPAEC [140,141]. Also, gel and capillary electrophoresis seem to be highly promising to obtain oligosaccharide profiles [204–208]. Identification of the glycans is commonly based upon the co-elution (chromatography) or co-migration (electrophoresis) with standard compounds. Comparison of the profiles (oligosaccharide-fingerprints) provides direct information on the (structural features of the) intact carbohydrate chains present on the (recombinant) glycoproteins of different batches. During profile/mapping analysis, fractionation of the different glycans in the pool of released carbohydrates is normally performed by either of three separation procedures using the chromatographic techniques described earlier in this chapter:

1. *Mono-Q in combination with Lichrosorb-NH$_2$*. Released carbohydrates are first fractionated by FPLC on a Mono-Q anion-exchange column on the basis of the sialic acid (or other acidic substituents) content, and then subfractionated by HPLC using a Lichrosorb-NH$_2$ colum [135]. Detection in both procedures is usually performed by UV.
2. *High-pH anion-exchange chromatography with pulsed amperometric detection*. HPAEC-PAD has been introduced for the separation and sensitve detection of car-

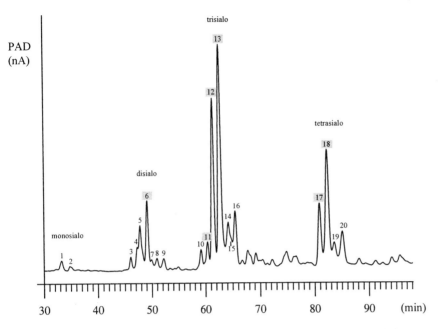

Fig. 12-16. Glycosylation profile of a glycoprotein. HPAEC elution pattern of the enzymatically released N-glycans of bovine fetuin. The column (CarboPac PA-1, 25 × 0.4 cm) was eluted with 0.1 M NaOH, using a NaOAc gradient (0–230 mM), at a flow rate of 1 ml min^{-1}. Detection was performed by pulsed amperometry.

bohydrates [141]. Fractionation is mainly on basis of charge, leading to the formation of carbohydrate charge-clusters. Linkage and branch positional isomers of carbohydrates are separated within each cluster (Fig. 12-16).

3. *GlycoSep C in combination with GlycoSep H.* This relatively new procedure can be compared with the Mono-Q/Lichrosorb-NH$_2$ method. The main differences are that the carbohydrates are labeled prior to separation with 2-aminobenzamide (2-AB) at the reducing terminus [131] and the use of volatile elution buffers, thus circumventing various desalting steps. The first separation step is performed on the GlycoSep C column (Oxford GlycoSciences) on the basis of the sialic acid content (or the overall negative charge). Thereupon subfractionation takes place on the GlycoSep H column (Oxford GlycoSciences) on basis of hydrophobicity/size of the 2-AB labeled glycans. Detection in both procedures is performed by fluorescence detection.

Currently, attempts are made to develop reproducible and (partly) automated analysis instruments, of which some are already commercially available (Table 12-6). Although, both the monosaccharide compositional data and the profiling data allow the fast comparison of different batches of a (recombinant) glycoprotein, for each specific purpose, criteria must be set to define, on the basis of the obtained chromatograms, whether two batches are identical with respect to glycosylation or

Table 12-6. Several commercially available semi-automatic carbohydrate analysis instruments.

System	Company	Performance
GlycoPrep™ 1000	Oxford GlycoSciences[a]	Release (hydrazinolysis) and recovery of N- and O-linked carbohydrate chains from glycoproteins[b]
GlycoMap™ 1000	Oxford GlycoSciences	Size exclusion chromatography of released (radiolabeled) carbohydrate chains with radiochemical or refractive index detection
RAAM™ 2000 (Reagent Array Analysis Method)	Oxford GlycoSciences	Degradation of fractionated and released glycans in an array of defined multiple exoglycosidase mixtures[c,d]
GlycoTAG™	TaKaRa[e]	Two-dimensional high performance liquid chromatography of released glycans which are labeled with 2-aminopyridine[f,g]
FACE (Fluorophore-assisted carbohydrate electrophoresis)	Glyko[h]	Polyacrylamide slab gel electrophoresis of released glycans which are labeled with 2-aminoacridone[i]
Glyco Doc™ Imaging System	Bio-Rad[j]	Polyacrylamide slab gel electrophoresis of released fluorescent labeled glycans

[a] Oxford GlycoSciences, Abingdon, U.K.
[b] Often used in combination with HPAEC.
[c] Performed with neutral carbohydrate chains only.
[d] Used in combination with GlycoMap™ 1000 and Glycosequencer: a gel filtration profile of degradation products is indicative for a structure.
[e] TaKaRa, Kyoto, Japan.
[f] Best performed with neutral carbohydrate chains.
[g] Carbohydrate samples are automatically derivatized on the Palstation 4000.
[h] Glyko, Novato, U.S.A.
[i] Analysis is performed with a Glycoscan Fluorescence Electrophoresis System.
[j] Bio-Rad Laboratories, Hercules, California, U.S.A.

not. In situations like the occurrence of undefined carbohydrate chains in an oligosaccharide-map, or the setting-up of a fingerprinting procedure for routine batch control, as a first step an approach involving a full delineation of the structure of each oligosaccharide component is still required.

12.7 Structural Analysis of N- and O-linked Glycans of Recombinant Human Erythropoietin (rhEPO)

To illustrate the general strategy for the characterization of the carbohydrate chains from a recombinant glycoprotein, the analytical procedure for recombinant human erythropoietin (rhEPO) expressed in CHO cells is described here [106,209]. The molecular mass of EPO is 34–39 kDa, and the polypeptide chain has one O-glycosylation site at Ser126 and three N-glycosylation sites at Asn24, Asn38, and Asn83, respectively [210]. The total carbohydrate content is 40 % of the molecular mass. EPO has been expressed in various heterologous cell systems, e.g., CHO [211], BHK [212], Y2 cells derived from NIH-3T3 [213], insect cells [214], and cultured tobacco cells [215]. Several studies on the structure and function of the carbohydrate chains of (rh)EPO have been carried out [216,217]. The recombinant human glycoprotein is an important therapeutic agent for the treatment of anemia associated with renal failure, since it stimulates red cell proliferation and differentiation in the bone marrow [218,219]. Glycosylation (both the extent and the precise oligosaccharide structures attached) significantly influences the biological and physico-chemical properties of (rh)EPO [220–223].

12.7.1 Liberation of the N-linked Carbohydate Chains

rhEPO (50 mg) was dissolved in 5 ml 50 mM Tris–HCl, pH 8.4, containing 50 mM EDTA. Subsequently, 1 % (by volume) 2-mercaptoethanol and 1 % (mass/volume) SDS were added, and the mixture was kept for 3 minutes at 100 °C. After cooling to room temperature, the sample was diluted twice with incubation buffer, and PNGase-F (from *F. meningosepticum*, Boehringer Mannheim) was added (0.2 U mg^{-1} EPO). The mixture was incubated for 16 h in an 'end-over-end' mixer at ambient temperature, then the solution was heated for 3 minutes at 100 °C, and after cooling, another aliquot of PNGase-F (0.2 U mg^{-1} EPO) was added. The incubation was continued for 24 h. The deglycosylation was checked by SDS–PAGE to be essentially complete as the molecular mass of rhEPO shifted from 35 kDa (native) to 22 kDa (N-deglycosylated). The N-glycan pool was isolated by gel filtration chromatography on a Bio-Gel P-100 column (47 × 2 cm), eluted with 25 mM NH$_4$HCO$_3$, pH 7.0, at a flow rate of 22 ml h^{-1}. The eluent was monitored at 206 nm. After desalting on a Bio-Gel P-2 column (45 × 1 cm) using water as eluent, the N-glycan pool was ready for fractionation.

12.7.2 Liberation of the O-linked Carbohydrate Chains

The P-100 void volume fraction (containing the protein possessing the O-linked glycan) was lyophilized and suspended (approx. 10 mg ml^{-1}) in 0.1 M NaOH, contain-

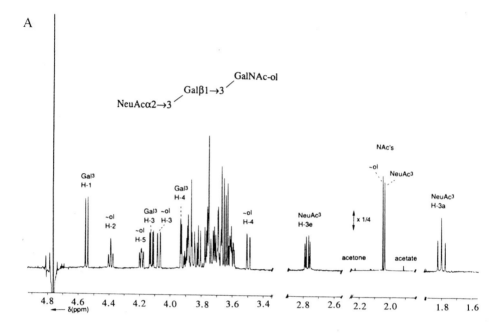

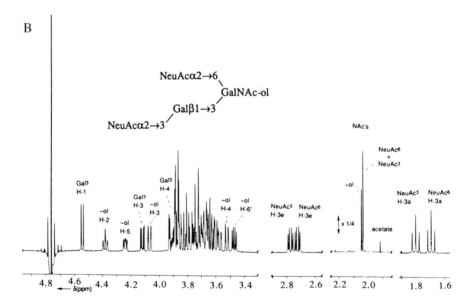

Fig. 12-17. 500-MHz ^{1}H-NMR spectra of the released O-glycans of recombinant human erythropoietin (rhEPO) from CHO cells. The relative scale of the NAc proton region differs from that of the rest of the spectrum. The spectra were recorded in D_2O at 300 K.

ing 1 M NaBH$_4$. The solution was kept for 24 h at 40 °C, then cooled on ice and neutralized with 4 M HOAc. Boric acid was removed by repetitive co-evaporation with MeOH, containing 1 % HOAc. Finally, the material was resuspended in water and centrifuged at 12000 g. After desalting of the supernatant on a Bio-Gel P-2 column (45 × 1 cm) using water as eluent, the O-glycan pool was ready for fractionation.

12.7.3 Fractionation and Structural Determination of the O-Glycans

Separation according to charge on a Mono Q HR 5/5 anion-exchange column (Pharmacia) yielded two carbohydrate-containing fractions, denoted O1 and O2, which were purified by HPLC on Lichrosorb-NH$_2$, yielding the fractions O1.2 and O2.1, respectively. Both fractions were investigated by 500-MHz ^{1}H-NMR spectroscopy and revealed the ^{1}H-NMR spectra of a monosialylated trisaccharide alditol Neu5Acα2-3Galβ1-3GalNAc-ol and a disialylated tetrasaccharide alditol Neu5Acα2-3Galβ1-3(Neu5Acα2-6)GalNAc-ol, respectively, as depicted in Fig. 12-17. In a recent study comparing the O-glycans from urinary EPO and rhEPO from CHO cells, in rhEPO also Galβ1-3(Neu5Acα2-6)GalNAc-ol was detected [211,224]. It has to be noted that the O-glycans of rhEPO are completely different from those of urinary EPO containing only GalNAc and Neu5Acα2-6GalNAc [225].

12.7.4 Fractionation and Structural Determination of the N-Glycans

The enzymatically released N-glycan pool was first fractionated according to charge on a Mono Q HR 5/5 anion-exchange column (Pharmacia), yielding four carbohydrate-containing fractions, denoted N1, N2, N3, and N4, having elution positions corresponding to mono-, di-, tri-, and tetrasialylated complex type N-glycans, respectively (Fig. 12-18). A further subfractionation of each Mono Q fraction was obtained by HPLC on Lichrosorb-NH$_2$. Because a detailed discussion of the structure elucidation of all rhEPO-derived N-glycans is beyond the scope of this chapter, only the tetrasialylated fraction N4, which was separated into eight fractions (N4.1–N4.8) by HPLC (Fig. 12-19), will be discussed. For the same reason, we will only focus on the major fractions N4.4, N4.6, and N4.8. At this stage an investigation by ^{1}H-NMR spectroscopy of the fractions is already beneficial to get an impression of further heterogeneity. Fraction N4.4 contained a single compound which could be identified by ^{1}H-NMR spectroscopy as being a tetrasialylated tetra-antennary oligosaccharide:

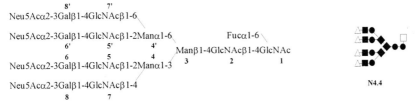

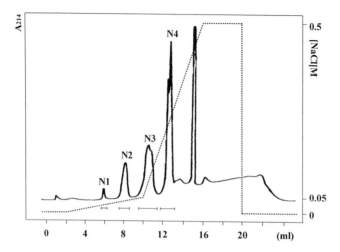

Fig. 12-18. FPLC elution pattern of the enzymatically released N-glycans of recombinant human erythropoietin (rhEPO) from CHO cells. The column (HR 5/5 Mono Q) was eluted with a NaCl gradient in H_2O as indicated, at a flow rate of 2 ml min^{-1}. Detection was performed at 214 nm.

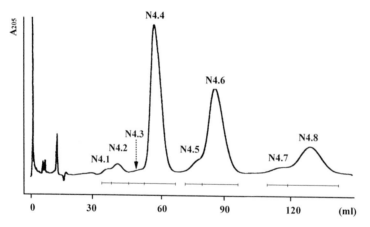

Fig. 12-19. HPLC elution pattern of FPLC fraction N4 derived from rhEPO. The column (Lichrosorb-NH₂) was eluted with 30 mM K_2HPO_4/KH_2PO_4 pH 7/acetonitrile (37.5:62.5, v/v), at a flow rate of 2 ml min^{-1}. Detection was performed at 205 nm.

This is the major constituent of all N-linked carbohydrate chains of rhEPO. Fraction N4.8 also contained a single compound and its structure was determined by ¹H-NMR spectroscopy as being:

The location of two extra *N*-acetyllactosamine units in the upper branches was proved by enzymatic digestion experiments as described below. From the ^{1}H-NMR spectrum of fraction N4.6 the presence of a mixture of tetrasialylated tetra-antennary oligosaccharides, each containing one extra *N*-acetyllactosamine unit, was deduced. The components in fraction N4.6 were separated by HPAEC, affording the subfractions N4.6.1 and N4.6.2 (Fig. 12-20). ^{1}H-NMR spectroscopy in conjunction with enzymatic digestion experiments identified the structures of the compounds to be as follows:

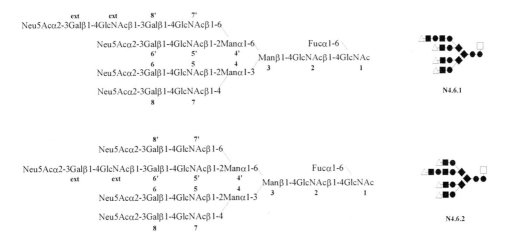

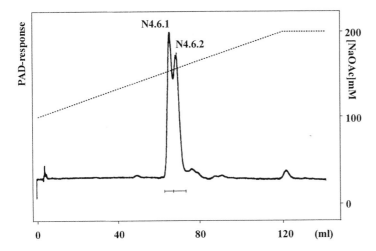

Fig. 12-20. HPAEC elution pattern of HPLC fraction N4.6 derived from rhEPO. The column (CarboPac PA-1, 25 × 0.9 cm) was eluted with 0.1 M NaOH, using a NaOAc gradient as indicated, at a flow rate of 4 ml min[-1]. Detection was performed by pulsed amperometry.

To determine the branch location of the *N*-acetyllactosamine repeating units in the tetra-antennary oligosaccharides, the compounds were degraded with endo-β-galactosidase, followed by digestion with *N*-acetyl-β-glucosaminidase [226] (Fig. 12-21). Endo-β-galactosidase (from *Bacteroides fragilis*, Boehringer Mannheim) hydrolyzes the β-galactosidic linkage in the Galβ1-4GlcNAc element in a poly-(*N*-acetyllactosamine) sequence when the galactose residue is not terminal [227]. After digestion, the trisialo compound (in case of one extra *N*-acetyllactosamine unit) and the

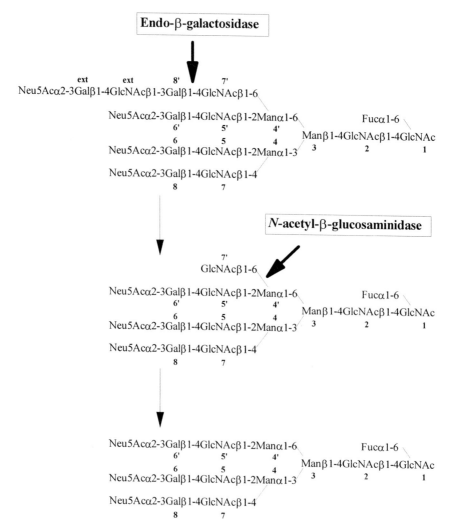

Fig. 12-21. Determination of the branch location of an *N*-acetyllactosamine repeating unit in a tetra-antennary oligosaccharide derived from rhEPO. Succesive incubation with endo-β-galactosidase and *N*-acetyl-β-glucosaminidase yielded a tri-antennary oligosaccharide which could be identified by ^{1}H-NMR spectroscopy.

disialo compound (in case of two extra *N*-acetyllactosamine units) were isolated by Mono Q anion-exchange chromatography. These compounds were then treated with *N*-acetyl-β-glucosaminidase (from jack beans, Sigma) to release the terminal Glc-NAc residue. In this way oligosaccharides (N4.6.1D, N4.6.2D, and N4.8D, see Table 12-7) were obtained that could unambiguously be identified by high-resolution ¹H-NMR spectroscopy, providing conclusive evidence for the structures of the intact compounds. The results obtained by ¹H-NMR spectroscopy in terms of ¹H-chemical shifts of the structural-reporter group protons are usually presented in the form of a table (see Table 12-7).

Table 12-7. ¹H-chemical shifts of the structural-reporter-group protons of the constituent monosaccharides of some N-linked oligosaccharides derived from recombinant human erythropoietin. For the short-hand symbolic notation, see text.

Reporter group	Residue	Chemical shifts (ppm) in						
		N4.4	N4.6.1	N4.6.2	N4.8	N4.6.1D	N4.6.2D	N4.8D
H-1	GlcNAc-1α[a]	5.182	5.182	5.182	5.182	5.182	5.181	5.181
	GlcNAc-1β	4.688	4.690	4.689	4.689	4.692	4.690	4.691
	GlcNAc-2α[a]	4.659	4.659	4.659	4.659	4.664	4.663	4.663
	GlcNAc-2β	4.664	4.663	4.664	4.664	4.667	4.668	4.668
	Man-**4**	5.131	5.130	5.130	5.129	5.114	5.115	5.114
	Man-**4'**	4.857	4.854	4.858	4.855	4.905	4.884	4.915
	GlcNAc-**5**	4.563	4.564	4.563	4.561	4.560	4.557	4.559
	GlcNAc-**5'**	4.593	4.589	4.602	4.60	4.574	–	–
	Gal-**6**	4.542	4.542	4.542	4.541	4.547	4.542	4.543
	Gal-**6'**	4.545	4.547	4.455	4.454	4.547	–	–
	GlcNAc-**7**	4.542	4.542	4.542	4.541	4.547	4.542	4.543
	GlcNAc-**7'**	4.545	4.547	4.545	4.546	–	4.546	–
	Gal-**8**	4.542	4.542	4.542	4.541	4.547	4.542	4.543
	Gal-**8'**	4.559	4.467	4.557	4.468	–	4.566	–
	GlcNAc$_{ext}$	–	4.701	4.699	4.700[b]	–	–	–
	Gal$_{ext}$	–	4.556	4.557	4.556[b]	–	–	–
H-2	Man-**3**	4.203	4.206	4.205	4.21[c]	4.212	4.21	4.213
	Man-**4**	4.220	4.219	4.219	4.22	4.212	4.21	4.213
	Man-**4'**	4.090	4.091	4.081	4.080	4.11	3.96	3.96
H-3	Gal**6**	4.117	4.117	4.116	4.116	4.117	4.117	4.115
	Gal**6'**	4.117	4.117	n.d.	n.d.	4.117	–	–
	Gal**8**	4.117	4.117	4.116	4.116	4.117	4.117	4.115
	Gal**8'**	4.117	n.d.[d]	4.116	n.d.	–	4.117	–
	Gal$_{ext}$	–	4.117	4.116	4.116	–	–	–
H-4	Gal**6'**	n.d.	n.d.	4.161	4.161	n.d.	–	–
	Gal**8'**	n.d.	4.159	n.d.	4.161	–	n.d.	–

Table 12-7. (Continued).

Reporter group	Residue	Chemical shifts (ppm) in						
		N4.4	N4.6.1	N4.6.2	N4.8	N4.6.1D	N4.6.2D	N4.8D
H-3a	Neu5Ac	1.804[e]	1.802[e]	1.803[e]	1.801[e]	1.802[f]	1.802[f]	1.800[b]
H-3e	Neu5Ac	2.756[e]	2.757[e]	2.757[e]	2.757[e]	2.756[f]	2.756[f]	2.756[b]
NAc	GlcNAc-1	2.038	2.037	2.038	2.037	2.040	2.037	2.038
	GlcNAc-2α[a]	2.094	2.094	2.092	2.091	2.095	2.091	2.092
	GlcNAc-2β	2.090	2.090	2.088	2.088	2.095	2.087	2.092
	GlcNAc-5	2.047	2.047	2.047	2.047	2.044	2.044	2.043
	GlcNAc-5′	2.038	2.037	2.038	2.037	2.044	–	–
	GlcNAc-7	2.075	2.075	2.074	2.075	2.074	2.073	2.073
	GlcNAc-7′	2.038	2.037	2.038	2.037	–	2.052	–
	GlcNAc$_{ext}$	–	2.037	2.038	2.037[g]	–	–	–
	Neu5Ac	2.031[h]	2.031[h]	2.031[h]	2.031[h]	2.031[i]	2.031[i]	2.031[g]
H-1	Fucα[a]	4.901	4.900	4.898	4.897	4.893	4.888	4.886
	Fucβ	4.909	4.908	4.905	4.905	4.901	4.897	4.893
CH$_3$	Fucα[a]	1.211	1.211	1.210	1.210	1.211	1.210	1.209
	Fucβ	1.222	1.223	1.222	1.222	1.223	1.222	1.221

[a] α and β stand for the α and β anomers of GlcNAc-1; [b] Signal stemming from two protons; [c] Values given with only two decimals because of spectral overlap; [d] n.d.; not determined; [e] Signal stemming from four protons; [f] Signal stemming from three protons; [g] Signal stemming from two NAc groups; [h] Signal stemming from four NAc groups; [i] Signal stemming from three NAc groups.

12.7.5 Discussion

The glycosylation of rhEPO is of great importance for its biological functioning. It has been demonstrated that removal or modification of the glycan chains, or prevention of glycosylation at specific sites, results in altered *in vivo* and *in vitro* activity [225,228]. Although rhEPO posesses only one O- and three N-glycosylation sites, over 35 different N-linked oligosaccharides and two O-linked oligosaccharides could be identified. The established N-glycan structures ranged from mono-sialylated di-antennary structures to fully-sialylated (containing extra *N*-acetyllactosamine units) tetra-antennary structures. Between these two extremes a great variety of di-, tri/tri'- and tetra-antennary structures, containing 1–4 sialic acid residues and/or 1–2 extra *N*-acetyllactosamine units was found [209]. These results illustrate nicely the formidable heterogeneity that can occur in a (recombinant) glycoprotein. Compared with natural EPO isolated from the urine of patients with aplastic anemia, the N-glycan heterogeniety was rather similar [228,229]. However, some differences in composition and relative amounts could be noticed. Recombinant EPO seems to have a somewhat higher content of tetra-antennary oligosaccharides and a tendency to contain higher amounts of N-linked carbohydrate chains with Galβ1-4GlcNAcβ1-3 repeats. Furthermore, the carbohydrate chains of natural EPO contain Neu5Ac in α2-3 and α2-6 linkage to Gal, while rhEPO from CHO cells contains three types of sialic acid, namely Neu5Ac (95%), Neu5Gc (2%) and Neu5,9Ac$_2$ (3%), only in α2-3 linkage. The presence of Neu5Gc and Neu5,9Ac$_2$ has also been observed in other recombinant glycoproteins expressed in CHO cells [230,231]. With respect to the α1-6 fucosylation of the Asn-bound GlcNAc residue in CHO and BHK EPOs, amounts between at least 80% and nearly 100% have been reported [216].

Glycosylation of recombinant glycoproteins is influenced by numerous factors as outlined in Chapter 5. With the above-described example, the extreme importance of an accurate and reliable analysis of recombinant glycoprotein therapeutics with respect to their carbohydrate content in terms of composition and structure is clearly demonstrated. In the context of therapeutic administration of recombinant-DNA glycoproteins, the glycosylation patterns of the engineered proteins demand thorough consideration with regard to applicability, tolerance, and patent position.

Acknowledgements

The authors gratefully acknowledge Drs J.J.M. van Rooijen, R. Gutiérrez Gallego, and C.H. Hokke for their contributions. The rhEPO study was supported by the Netherlands Foundation for Chemical Research (SON/NWO) and Organon International BV (Oss, The Netherlands).

Abbreviations

Man	D-mannose
Fuc	L-fucose
Gal	D-galactose
Glc	D-glucose
Xyl	D-xylose
Ara	L-arabinose
Rha	L-rhamnose
Rib	D-ribose
GlcNH$_2$	D-glucosamine
GlcNAc	*N*-acetyl- D-glucosamine
GlcNAc-ol	*N*-acetyl- D-glucosaminitol
GalNAc	*N*-acetyl- D-galatosamine
GalNAc-ol	*N*-acetyl- D-galatosaminitol
NeuAc/Neu5Ac	*N*-acetylneuraminic acid (sialic acid)
NeuGc/Neu5Gc	*N*-glycolylneuraminic acid

GlcA	glucuronic acid
GalA	galacturonic acid
MurNAc	*N*-acetylmuramic acid
rER	rough Endoplasmic Reticulum
Ala	L-alanine
Arg	L-arginine
Asn	L-asparagine
Cys	L-cysteine
Glu	L-glutamine
Gly	L-glycine
His	L-histidine
HyLys	5-hydroxy-L-lysine
HyPro	4-hydroxy-L-proline
Lys	L-lysine
Pro	L-proline
Ser	L-serine
Thr	L-threonine
ADP	adenosine diphosphate
CMP	cytidine 5'-monophosphate
GDP	guanosine diphosphate
UDP	uridine diphosphate
Dol-P	dolichol phosphate
GPI	glycosyl phosphatidyl inositol
AC	Affinity chromatography
FPLC	Fast protein liquid chromatography
GPC	Gel permeation chromatography
HPLC	High performance liquid chromatography
HPAEC	High pH anion exchange chromatography
PAD	Pulsed amperometric detection
GC	Gas chromatography
MS	Mass spectrometry
FAB	Fast atom bombardment
EI	Electron impact
ES	Electrospray
MALDI	Matrix assisted laser desorption ionization
TOF	Time of flight
NMR	Nuclear magnetic resonance
δ	chemical shift
SDS–PAGE	Sodium dodecyl sulfate polyacrylamide gel electrophoresis
PNGase-F	peptide-N^4-(*N*-acetyl-β-glucosaminyl) asparagine amidase-F
ConA	Concanavalin A
FSH	Follicle stimulating hormone (follitropin)
hCG	human chorionic gonadotropin
hCG-α	α-subunit of human chorionic gonadotropin
hCG-β	β-subunit of human chorionic gonadotropin
rhEPO	recombinant human erythropoietin

cDNA	complementary DNA
CHO	Chinese hamster ovary
BHK	Baby hamster kidney
EtN	Ethanolamine
2-AB	2-aminobenzamide
HF	Hydrofluoric acid
PA	Pyridylamino
PIPLC	Phosphatidyl inositol-phospholipase C
TFMS	Trifluoromethane sulfonic acid
TMS	Trimethylsilyl

References

[1] Paulson, J. C., *Trends Biochem Sci*, 1989, *14*, 272–276.
[2] Jentoft, N., *Trends Biochem Sci*, 1990, *15*, 291–294.
[3] Ginsberg, V., Robbins, P. (Eds.), *Biology of Carbohydrates*, Vol 3. New York: J. Wiley, 1991.
[4] Cumming, D. A., *Glycobiology*, 1991, *1*, 115–130.
[5] Elbein, A. D., *Trends Biotechnol*, 1991, *9*, 346–352.
[6] Kobata, A., *Eur J Biochem*, 1992, *209*, 483–501.
[7] Rasmussen, J. R., *Curr Opin Struct Biol*, 1992, *2*, 682–686.
[8] Hart, G. W., *Curr Opin Cell Biol*, 1992, *4*, 1017–1023.
[9] Kornfeld, S., Mellman, I., *Annu Rev Cell Biol*, 1989, *5*, 483–525.
[10] Varki, A., in: *Cell surface carbohydrates and cell development*, Fukuda, M. (Ed.), Ann Arbor: CRC Press, 1992, pp. 25–69.
[11] Kretz, K. A., Carson, G. S., Morimoto, S., Kishimoto, Y., Fluharty, A. L., O'Brien, J. S., *Proc Natl Acad Sci USA*, 1990, *87*, 2541–2544.
[12] Wetzler, L. M., Barry, K., Blake, M. S., Gotschlich, E. C., *Infect Immunol*, 1992, *60*, 39–43.
[13] Kjellen, L., Lindahl, U., *Annu Rev Biochem*, 1991, *60*, 443–475.
[14] Hardingham, T. E., Fosang, J., *FASEB J*, 1992, *6*, 861–870.
[15] Ruoslahti, E., *Annu Rev Cell Biol*, 1988, *4*, 229–255.
[16] Ruoslahti, E., Yamaguchi, Y., *Cell*, 1991, *64*, 867–869.
[17] Meri, S., Pangburn, M. K., *Proc Natl Acad Sci USA*, 1990, *87*, 3982–3986.
[18] Bremer, E., Hakomori, S., *Biochem Biophys Res Commun*, 1982, *106*, 711–718.
[19] Zaleska, M. M., Erecinska, M., *Proc Natl Acad Sci USA*, 1987, *84*, 1709–1712.
[20] Bhaskar, K. R., Garik P., Turner, B. S., Bradley, J. D., Bansil, R., Stanley, H. E., LaMont, P., *Nature*, 1992, *360*, 458–461.
[21] Varki, A., *Glycobiology*, 1992, *2*, 25–40.
[22] Rose, M. C., *Am J Physiol Lung Cell*, 1992, *263*, L413–L429.
[23] Yolken, R. H., Peterson, J. A., Vonderfecht, S. L., Fouts, E. T., Midthun, K., Newburg, J., *J Clin Ivest*, 1992, *90*, 1984–1991.
[24] Inoue, M., Ton, S., Ogawa, H., Tanizawa, O., *Am J Clin Pathol*, 1991, *96*, 711–716.
[25] Schwarz, A. L., *Annu Rev Immunol*, 1990, *8*, 195–229.
[26] Karlsson, K. A., *Curr Opin Struct Biol*, 1995, *5*, 622–635.
[27] Schenkman, S., Jiang, M. S., Hart, G. W., Nussenzweig, V., *Cell*, 1991, *65*, 1117–1125.
[28] Ofek, I., Sharon, N., *Curr Top Microbiol Immunol*, 1990, *151*, 91–114.
[29] Goldkorn, I., Gleeson, P. A., Toh, B.-H., *J Biol Chem*, 1989, *264*, 18768–18774.
[30] Dennis, J. W., Laferte, S., Waghorne, C., Breitman, M. L., Kerbel, R. S., *Science*, 1987, *236*, 582–585.

[31] Lowe, J. B., Stoolman, L. M., Nair, R. P., Larsen, R. D., Berhend, T. L., Marks, R. M., *Cell,* 1990, *63,* 475–484.

[32] Walz, G., Aruffo, A., Kolanus, W., Bevilacqua, M., Seed, B., *Science,* 1990, *250,* 1132–1135.

[33] Foxall, C., Watson, S. R., Dowbenko, D., Cennie, C., Lasky, L. A., Kiso, M., Hasegawa, A., Asa, D., Brandley, B. K., *J Cell Biol,* 1992, *117,* 895–902.

[34] Stamencovic, I., Sgroi, D., Aruffo, A., *Cell,* 1992, *68,* 1003–1004.

[35] Carey, D. J., Crumbling, D. M., Stahl, R. C., Evans, D. M., *J Biol Chem,* 1990, *265,* 20627–20633.

[36] Drake, S. L., Klein, D. J., Mickelson, D. J., Oegema, T. R., Furcht, L. T., McCarthy, J. B., *J Cell Biol,* 1992, *117,* 1331–1341.

[37] Li, M., Kurata, H., Itoh, N., Yamashina, I., Kawasaki, T., *J Biol Chem,* 1990, *265,* 11295–11298.

[38] Ashwell, G., Harford, J., *Annu Rev Biochem,* 1982, *51,* 531–554.

[39] Drickamer, K., *Cell,* 1991, *67,* 1029–1032.

[40] Allen, H. J., Kisailus, E. C. (Eds.), *Glycoconjugates: Composition, Structure and Functions,* New York: Marcel Dekker Inc, 1992.

[41] Varki, A., *Glycobiology,* 1993, *3,* 97–130.

[42] Ferguson, M. A. J., Williams, A. F., *Annu Rev Biochem,* 1988, *57,* 285–320.

[43] Stevens, V. L., *Biochem J,* 1995, *310,* 361–370.

[44] Bailey, A. J., Sims, T. J., Avery, N. C., Halligan, E. P., *Biochem J,* 1995, *308,* 385–390.

[45] Rosa, J. L., Perez, J. X., Ventura, F., Tauler, A., Gil, J., Shimoyama, M., Pilkis, S. J., Bartrons, R., *Biochem J,* 1995, *309,* 119–125.

[46] Hoshino, S., Kikkawa, S., Takahashi, K., Itoh, H., Kaziro, Y., Kawasaki, H., Suzuki, K., Katada, T., Ui, M., *FEBS Lett,* 1990, *276,* 227–231.

[47] Montreuil, J., *Adv Carbohydr Chem Biochem,* 1980, *37,* 157–223.

[48] Hanisch, F. G., Chai, W., Rosankiewicz, J. R., Lawson, A. M., Stoll, M. S., Feizi, T., *Eur J Biochem,* 1993, *217,* 645–655.

[49] Brockhausen, I., in: *Glycoproteins. New Comprehensive Biochemistry;* Montreuil, J., Vliegenhart, J. F. G., Schachter, H. (Eds.) Amsterdam: Elsevier, 1995, Vol. 29, pp. 201–259.

[50] Hayes, B. K., Hart, G. W., *Curr Opin Struct Biol,* 1994, *4,* 692–696.

[51] Roquemore, E. P., Chou, T.-Y., Hart, G. W., *Methods Enzymol,* 1994, *230,* 443–460.

[52] Haltiwanger, R. S., Kelly, W. G., Roquemore, E. P., Blomberg, M. A., Dong, L. Y., Kreppel, L., Chou, T. Y., Hart, G. W., *Biochem Soc Trans,* 1992, *20,* 264–269.

[53] Stults, N. L., Cummings, R. D., *Glycobiology,* 1993, *3,* 589–596.

[54] Mayor, S., Menon, A. K., Cross, G. A. M., *J Cell Biol,* 1991, *114,* 61–71.

[55] Frankhauser, C., Homans, S. W., Thomas Oates, J. E., McConville, M. J., Desponds, C., Conzelmann, A., Ferguson, M. A. J., *J Biol Chem,* 1993, *268,* 26365–26374.

[56] Azzouz, N., Striepen, B., Gerold, P., Capdeville, Y., Schwarz, R. T., *EMBO J,* 1995, *14,* 4422–4433.

[57] Goochee, C. F., Gramer, M. J., Andersen, D. C., Bahr, J. B., Rasmussen, J. R., *Biotechnology,* 1991, *9,* 1347–1355.

[58] Lis, H., Sharon, N., *Eur J Biochem,* 1993, *218,* 1–27.

[59] Kornfeld, K., Kornfeld, S., *Annu Rev Biochem,* 1985, *54,* 631–664.

[60] Elbein, A. D., *FASEB J,* 1991, *5,* 3055–3063.

[61] Moremen, K. W., Trimble, R. B., Herscovics, A., *Glycobiology,* 1994, *4,* 113–125.

[62] Kean, E. L., *J Biol Chem,* 1991, *266,* 942–946.

[63] Gilmore, R., *Cell,* 1993, *75,* 589–592.

[64] Jenkins, N., Parekh, R. B., James, D. C., *Nature Biotechnol,* 1996, *14,* 975–981.

[65] Grinnell, B. W., Walls, J. D., Gerlitz, B., *J Biol Chem,* 1991, *266,* 9778–9785.

[66] Kaplan, H. A., Welply, J. K., Lennarz, W. J., *Biochim Biophys Acta,* 1987, *906,* 161–173.

[67] Abeijon, C., Hirschberg, C. B., *Trends Biochem Sci,* 1992, *17,* 32–36.

[68] Cummings, R. D., in: *Glycoconjugates: Composition, Structure and Function:* Allen, H. J., Kisailus, E. C. (Eds.), New York: Marcel Dekker Inc. 1992, pp. 333–360.

[69] Shakin-Eshleman, S. H., Spitelnik, S. L., Lakshni, K., *J Biol Chem,* 1996, *271,* 6363–6366.

[70] Hirschberg, C. B., Snider, M. D., *Annu Rev Biochem,* 1987, *56,* 63–87.

[71] Paulson, J. C., Colley, K. J., *J Biol Chem,* 1989, *264,* 17615–17618.

[72] Kleene, R., Berger, E. G., *Biochim Biophys Acta,* 1993, *1154,* 283–325.

[73] Schachter H., *Curr Opin Struct Biol,* 1991, *1,* 755–765.

[74] Sharper, J. H., Sharper, N. L., *Curr Opin Struct Biol,* 1992, *2,* 701–709.

[75] Van den Eijnden, D. H., Joziasse, D. H., *Curr Opin Struct Biol,* 1993, *3,* 711–721.

[76] Hughes, R. C., in: *Protein Glycosylation: Cellular, Biotechnological and Analytical Aspects:* Conradt, H. S. (Ed.), Weinheim: GRF Monographs VCH, 1991, pp. 1–11.

[77] Carraway, K. L., Hull, S. R., *BioEssays,* 1989, *10,* 117–121.

[78] Carraway, K. L., Hull, S. R., *Glycobiology,* 1991, *1,* 131–138.

[79] Schachter, H., Brockhausen, I., in: *Glycoconjugates: Composition, Structure and Function;* Allen, H. J., Kisailus, E. C. (Eds.), New York: Marcel Dekker Inc., 1992, pp. 263–332.

[80] Strous, G. J., Dekker, J., *Crit Rev Biochem Molec Biol,* 1992, *27,* 57–92.

[81] Harris, R. J., Spellman, M. W., *Glycobiology,* 1993, *3,* 219–224.

[82] Tomas, J. R., Dwek, R. A., Rademacher, T. W., *Biochemistry,* 1990, *29,* 5413–5422.

[83] Ferguson, M. A. J., *Curr Opin Struct Biol,* 1991, *1,* 522–529.

[84] Tartakoff, A. M., Singh, N., *Trends Biochem Sci,* 1992, *17,* 470–473.

[85] Nenon, A. K., Mayor, S., Schwarz, R. T., *EMBO J,* 1991, *9,* 4249–4258.

[86] Brewis, I. A., Ferguson, M. A. J., Mehlert, A., Turner, A. J., Hooper, N. M., *J Biol Chem,* 1995, *270,* 22946–22952.

[87] Costello, L. C., Orlean, P., *J Biol Chem,* 1992, *267,* 8599–8605.

[88] McConville, M. J., Ferguson, M. A. J., *Biochem J,* 1993, *294,* 305–324.

[89] Ferguson, M. A. J., *Biochem Soc Trans,* 1992, *20,* 243–256.

[90] Englund, P. T., *Annu Rev Biochem,* 1993, *62,* 121–138.

[91] Amthauer, R., Kodukula, K., Brink, L., Udenfriend, S., *Proc Natl Acad Sci USA;* 1992, *89,* 6124–6128.

[92] Nishimura, H., Takao, T., Hase, S., Shimonishi, Y., Iwanaga, S., *J Biol Chem,* 1992, *67,* 17520–17525.

[93] Kuraya, N., Omichi, K., Nishimura, H., Iwanaga, S., Hase, S., *J Biochem,* 1993, *114,* 763–765.

[94] Chou, C.-F., Omary, M. B., *J Biol Chem,* 1993, *268,* 4465–4472.

[95] Levrat, C., Ardail, D., Louisot, P., *Int J Biochem,* 1990, *22,* 287–293.

[96] Gasnier, F., Rousson, R., Lerme, F., Vaganay, E., Louisot, P., Gateau-Roesch, O., *Eur J Biochem,* 1992, *206,* 853–858.

[97] Dwek, R. A., Edge, C. J., Harvey, D. J., Wormald, M. R., *Annu Rev Biochem,* 1993, *62,* 65–100.

[98] Lennarz, W. J., Hart, G. W. (Eds.), *Methods Enzymol,* 1994, *230.*

[99] Kamerling, J. P., Hård, K., Vliegenhart, J. F. G., in: *From Clone to Clinic;* Crommelin, D. J. A., Schellekens, H. (Eds.), Dordrecht: Kluwer Acad. Publ., 1990, pp. 295–304.

[100] Parekh, R. B., Patel, T. P., *Trends Biotechnol,* 1992, *10,* 276–280.

[101] Hounsell, E. F., in: *Methods in Molecular Biology: Glycoprotein Analysis in Biomedicine:* Hounsell, E. F. (Ed.), Totowa, NJ: Humana Press Inc., 1993, Vol. 14.

[102] Spellmann, M., *Anal Chem,* 1990, *62,* 1714–1722.

[103] Kamerling, J. P., *Pure Appl Chem,* 1994, *66,* 2235–2238.

[104] Verbert, A. (Ed.), *Methods in Glycoconjugates.* Reading: Harwood Academic Publishers, 1995.

[105] Kamerling, J. P., *Biotechnologia Aplicada,* 1996, *13,* 167–180.

[106] Fuduka, M., Kobata, A. (Eds.), *Glycobiology, A Practical Approach,* Oxford: Oxford University Press IRL, 1993.

[107] Damm, J. B. L., Kamerling, J. P., Van Dedem, G. W. K., Vliegenthart, J. F. G., *Glycoconjugate J*, 1987, *4*, 129–144.
[108] Kamerling, J. P., Gerwig, G. J., Vliegenthart, J. F. G., Clamp, J. R., *Biochem J*, 1975, *151*, 491–495.
[109] Kamerling, J. P., Vliegenhart, J. F. G., in: *Clinical Biochemistry: Principles, Methods, Applications*. Lawson, A. M. (Ed.), Berlin: Walter de Gruyter, 1989, pp. 176–263.
[110] Gerwig, G. J., Kamerling, J. P., Vliegenthart, J. F. G., *Carbohydr Res*, 1979, *77*, 1–7.
[111] Menon, A. K., *Methods Enzymol*, 1994, *230*, 418–442.
[112] Schneider, P., Ferguson, M. A. J., *Methods Enzymol*, 1995, *250*, 614–640.
[113] Takasaki, S., Mizuochi, T., Kobata, A., *Methods Enzymol*, 1982, *83*, 263–268.
[114] Fox, S., *Genetic Engineering News*, 1991, Nov/Dec.
[115] Patel, T. P., Bruce, J. A., Merry, A., Bigge, C., Wormald, M., Jaques, A., Parekh, R. B., *Biochemistry*, 1993, *32*, 679–693.
[116] Bendiak, B., Cummings D. A., *Carbohydr Res*, 1986, *151*, 89–103.
[117] Zinn, A. B., Plantner, J. J., Carlson, D. M., in: *The Clycoconjugates*. Horowitz, M. I., Pigman, W. (Eds.), New York: Academic Press, 1977, pp. 69–85.
[118] Whistler, R. L., BeMiller, J. N., *Adv Carbohydr Chem*, 1985, *13*, 289–329.
[119] Elder, J. H., Alexander, S., *Proc Natl Acad Sci USA*, 1982, *79*, 4540–4544.
[120] Plummer, T. H., Jr, Elder, J. H., Alexander, S., Phelan, A. V., Tarentino, A. L., *J Biol Chem*, 1984, *259*, 10700–10704.
[121] Donald, A. S. R., Feeney, J., *Biochem J*, 1986, *236*, 821–828.
[122] Maley, F., Trimble, R. B., Tarentino, A. L., Plummer, T. H., Jr., *Anal Biochem*, 1989, *180*, 195–204.
[123] Stenbe, K., Gross, V., Hosel, W., Thran-thi, T. A., Decker, K., Heinrich, P. C., *Glycoconjugate J*, 1986, *3*, 247–254.
[124] Chu, F. K., *J Biol Chem*, 1986, *261*, 172–177.
[125] Langer, B. G., Hong, S. K., Schmelzer, C. H., Bell, W. R., *Anal Biochem*, 1987, *166*, 212–1257.
[126] Tarentino, A. L., Plummer, T. H., Jr, *Methods Enzymol*, 1994, *230*, 44–57.
[127] Treffer, V., Altmann, F., Marz, L., *Eur J Biochem*, 1991, *199*, 647–652.
[128] Staudacher, E., Altman, F., Marz, L., Hård, K., Kamerling, J. P., Vliegenthart, J. F. G., *Glycoconjugate J*, 1993, *9*, 82–85.
[129] Ishii-Karahasa, I., Iwasw, H., Hotta, K., Tanaka, Y., Omura, S., *Biochem J*, 1992, *288*, 475–482.
[130] Striepen, B., Zinecker, C. F., Damm, J. L. B., Melgers, P. A. T., Gerwig, G. J., Koolen, M., Vliegenthart, J. F. G., Dubremetz, J.-F., Schwarz, R. T., *J Mol Biol*, 1997, *266*, 797–813.
[131] Bigge, J. C., Patel, T. P., Bruce, J. A., Goulding, P. N., Charles, S. M., Parekh, R. B., *Anal Biochem*, 1995, *230*, 229–238.
[132] Hase, S., in: *Methods in Molecular Biology: Glycoprotein Analysis in Biomedicine;* Hounsell, E. F. (Ed.), Totowa: Humana Press Inc., 1993, Vol. 14, pp. 69–80.
[133] Yamashita, K., Mizuochi, T., Kobata, A., *Methods Enzymol*, 1982, *83*, 105–126.
[134] Van Pelt, J., Damm, J. B. L., Kamerling, J. P., Vliegenthart, J. F. G., *Carbohydr Res*, 1987, *169*, 43–51.
[135] Damm, J. B. L., Voshol, H., Hård, K., Kamerling, J. P., Van Dedem, G. W. K., Vliegenthart, J. F. G., *Glycoconjugate J*, 1988, *5*, 221–233.
[136] Green, E. B., Baenziger, J. U., *Anal Biochem*, 1986, *158*, 42–49.
[137] Bergh, M. L., Koppen, P., Van den Eijnden, D. H., *Carbohydr Res*, 1981, *94*, 225–229.
[138] Townsend, R. R., Hardy, M., Oledino, J. D., Carter, S. R., *Nature*, 1988, *335*, 379–380.
[139] Manzi, A. E., Diaz, S., Varki, A., *Anal Biochem*, 1990, *188*, 20–32.
[140] Townsend, R. R., Hardy, M. R., *Glycobiology*, 1991, *1*, 139–147.
[141] Hermentin, P., Witzel, R., Doenges, R., Bauer, R., Haupt, H., Patel, T., Parekh, R. B., Brazel, D., *Anal Biochem*, 1992, *206*, 419–429.

[142] Van Pelt, J., Kamerling, J. P., Vliegenthart, J. F. G., Verheijen, F. W., Galjaard, H., *Biochim Biophys Acta,* 1988, *965,* 36–45.
[143] Dua, V. K., Dube, V. E., Li Y.-T., Bush, C. A., *Glycoconjugate J,* 1985, *2,* 17–30.
[144] Bendiak, B., Orr, J., Brockhausen, I., Velh, G., Phoebe, C., *Anal Biochem,* 1988, *175,* 96–105.
[145] Henderson, S. K., Henderson, D. E., *J Chromatogr Sci,* 1986, *24,* 198–203.
[146] Mutsaers, J. H. G. M., Van Halbeek, H., Vliegenthart, J. F. G., Tager, J. M., Reuser, A. J. J., Kroos, M., Galjaard, H., *Biochim Biophys Acta,* 1987, *911,* 244–251.
[147] Debray, H., Montreuil, J., Lis, H., Sharon, N., *Carbohydr Res,* 1986, *151,* 359–370.
[148] Narasimhan, S., Freed, J. C., Schachter, H., *Carbohydr Res,* 1986, *149,* 65–83.
[149] Liener, I. E., Sharon, N., Goldstein, I. J. (Eds.), *The Lectins: Properties, Function and Applications in Biology and Medicine.* Orlando: Academic Press, 1986.
[150] Sharon, N., Lis, H., *Lectins.* London: Chapman & Hall, 1989.
[151] Kobata, A., Endo, T., *J Chromatogr,* 1992, *597,* 111–122.
[152] Lis, H., Sharon, N., *Annu Rev Biochem,* 1986, *53,* 35–67.
[153] Kobata, A., *Anal Biochem,* 1979, *100,* 1–14.
[154] Kobata, A., *Eur J Biochem,* 1992, *209,* 483–501.
[155] Mizuochi, T., in: *Methods in Molecular Biology: Glycoprotein Analysis in Biomedicine:* Hounsell, E. F. (Ed.), Totowa: Humana Press Inc., 1993, Vol. 14, pp. 55–68.
[156] Jay, A., *J Carbohydr Chem,* 1996, *15,* 897–923.
[157] Hakomori, S. I., *J Biochem Tokyo,* 1964, *55,* 205–208.
[158] Stellner, K., Saito, H., Hakomori, S. I., *Arch Biochem Biophys,* 1973, *155,* 464–469.
[159] Harrison, A. G., Cotter, R. J., *Methods Enzymol,* 1990, *193,* 3–37.
[160] McCloskey, J. A., *Methods Enzymol,* 1991, *193,* 539–768.
[161] Settineri, C. A., Burlingame, A. L., in: *Carbohydrate Analysis-HPLC and Capillary Electrophoresis.* El Razzi, Z. (Ed.), Elsevier Amsterdam, Sciences, 1995, pp. 447–514.
[162] Dell, A., Oates, J. E., in: *Analysis of Carbohydrates by GLC and MS,* Bierman, G. J., Mc Ginnis, G. (Eds.), Boca Raton, Fl.: CRC Press, 1988.
[163] Puffing, K. L., Welpy, J. K., Huang, E., Henoin, J. D., *Anal Chem,* 1992, *64,* 1440–1448.
[164] Linsley, K. B., Cham, S -Y., Chan, S., Reinhold, B. D., Lisi, P. J., Reinhold, V. N., *Anal Biochem,* 1994, *219,* 207–217.
[165] Hillenkamp, F., Karas, M., *Methods Enzymol,* 1991, *193,* 283–298.
[166] Cotter, R. J., *Anal Chem,* 1992, , 1027–1031.
[167] Billeci, T. M., Stults, J. T., *Anal Chem,* 1993, *65,* 1709–1716.
[168] Stahl, B., Klabunde, T., Witzel, H., Krebs, B., Steup, M., Karas, M., Hillenkamp, F., *Eur J Biochem,* 1994, *220,* 321–330.
[169] Sutton, C. W., O'Neill, J. A., Cottrell, J. S., *Anal Biochem,* 1994, *218,* 34–36.
[170] Smith, K. D., Harbin, A. M., Carruthers, R. A., Lawson, A. M., Hounsell, E. F., *Biomed Chromatogr,* 1990, *4,* 261–266.
[171] Carr, S. A., Barr, J. R., Roberts, G. D., Anumalu, K. R., Taylor, P. B., *Methods Enzymol,* 1990, *193,* 501–518.
[172] Huberty, M. C., Vath, J. E., Yu, W., Martin, S. A., *Anal Chem,* 1993, *65,* 2791–2800.
[173] Liu, J., Volk, K. J., Kerns, E. H., Klohr, S. E., Lee, M. S., Rosenberg, I. E., *J Chromatogr,* 1993, *632,* 45–56.
[174] Dorland, L., Haverkamp, J., Schut, B. L., Vliegenthart, J. F. G., Spik, G., Strecker, G., Fournet, B., Montreuil, J., *FEBS Lett,* 1977, *77,* 15–20.
[175] Vliegenthart, J. F. G., Van Halbeek, H., Dorland, L., *Pure Appl Chem,* 1981, *53,* 45–77.
[176] Vliegenthart, J. F. G., Dorland, L., Van Halbeek, H., *Adv Carbohydr Chem Biochem,* 1983, *41,* 209–374.
[177] Kamerling, J. P., Vliegenthart, J. F. G., *Biol Magn Reson,* 1992, *10,* 1–194.
[178] Doubet, S., Bock, K., Smith, D., Darvill, A., Albersheim, P., *Trends Biochem Sci,* 1989, *14,* 475–477.
[179] Van Kuik, J. A., Hård, K., Vliegenthart, J. F. G., *Carbohydr Res,* 1992, *235,* 53–68.

[180] Feizi, T., Bundle, D., *Curr Opin Struct Biol,* 1996, *6,* 659–662.
[181] Van Kuik, J. A., Vliegenthart, J. F. G., *Trends Food Sci Technol,* 1993, *4,* 73–77.
[182] Dabrowski, J., in: *Two-Dimensional NMR Spectroscopy: Application for Chemists and Biochemists;* Croasmun, W. R., Carlson, R. M., (Eds.) New York: VCH, 1987, pp. 349–386.
[183] Friebolin, H., *Basic One- and Two-Dimensional NMR Spectroscopy.* Weinheim: VCH, 1993.
[184] Van der Ven, F. J. M., *Multidimensional NMR in Liquids.* New York: VCH, 1995.
[185] Hård, K., Vliegenthart, J. F. G., in: *Glycobiology: A Practical Approach.* Fukuda, M., Kobata, A. (Eds.), Oxford: IRL Press, 1993, pp. 223–242.
[186] Serianni, A. S., in: *Glycoconjugates: Composition, Structure and Function;* Allen, H. J., Kisailus, E. C. (Eds.), New York: Marcel Dekker, 1992, pp. 71–102.
[187] Van Halbeek, H., *Curr Opin Struct Biol,* 1994, *4,* 697–709.
[188] Van Halbeek, H., *Methods Enzymol,* 1994, *230,* 132–168.
[189] Peters, T., Pinto, B. M., *Curr Opin Struct Biol,* 1996, *6,* 710–720.
[190] Mer, G., Hietter, H., Lefevre, J.-F., *Nature Struct Biol,* 1996, *3,* 45–53.
[191] Wyss, D. F., Wagner, G., *Curr Opin Biotech,* 1996, *7,* 409–416.
[192] De Beer, T., Van Zuylen, C. W. E. M., Leeflang, B. R., Hård, K., Boelens, R., Kaptein, R., Kamerling, J. P., Vliegenthart, J. F. G., *Eur J Biochem,* 1996, *241,* 229–242.
[193] De Beer, T., Van Zuylen, C. W. E. M., Hård, K., Boelens, R., Kaptein, R., Kamerling, J. P., Vliegenthart, J. F. G., *FEBS Lett,* 1994, *348,* 1–6.
[194] Van Zuylen, C. W. E. M., De Beer, T., Rademacher, G. J., Haverkamp, J., Thomas-Oates, J. E., Hård, K., Kamerling, J. P., Vliegenthart, J. F. G., *Eur J Biochem,* 1995, *231,* 754–760.
[195] Wyss, D. F., Choi, J. S., Wagner, G., *Biochemistry,* 1995, *34,* 1622–1634.
[196] Brockbank, R. L., Vogel, H. J., *Biochemistry,* 1990, *29,* 5574–5583.
[197] Gilhespy-Muskett, A. M., Partridge, J., Jefferis, R., Homans, S. W., *Glycobiology,* 1994, *4,* 485–489.
[198] Gerken, T. A., Butenhof, X. J., Shogren, R., *Biochemstry,* 1989, *28,* 5536–5543.
[199] Dill, K., Hu, S., Berman, E., Paola, A. A., Lacombe, J. M., *J Prot Chem,* 1990, *9,* 129–136.
[200] Lommerse, J. P. M., Kroon-Batenburg, L. M. J., Kroon, J., Kamerling, J. P., Vliegenthart, J. F. G., *J Biomol NMR,* 1995, *5,* 79–94.
[201] Lustbader, J. W., Birken, S., Pilak, S., Pound, A., Chait, B. T., Mirza, U. A., Ramnarain, S., Canfield, R. E., Brown, J. M., *J Biol NMR,* 1996, *7,* 295–304.
[202] Fomiya, N., Awaya, J., Kuromo, M., Endo, S., Arata, Y., Takahashi, N., *Anal Biochem,* 1988, *171,* 73–90.
[203] Rice, K. G., Takahashi, N., Namiski, Y., Tran, A. D., Lisi, P. J., Lee, Y. C., *Anal Biochem,* 1992, *206,* 278–287.
[204] Edge, C. J., Rademacher, T. W., Wormald, M. R., Parekh, R. B., Butters, T. D., Wing, D. R., Dwek, R. A., *Proc Natl Acad Sci USA,* 1992, *89,* 6338–6342.
[205] Starr, C. M., Masada, I., Hagka, C., Seid, B., *Glycobiology,* 1993, *3,* 511–520.
[206] Hermentin, P., Doenges, R., Witzel, R., Hokke, R. C., Vliegenthart, J. F. G., Kamerling, J. P., Conradt, H. S., Nimtz, M., Brazel, B., *Anal Biochem,* 1994, *221,* 29–41.
[207] Susuki, S., Honda, S., *TRAC,* 1995, *14,* 279–288.
[208] Hu, G. F., *J Chromatogr,* 1995, *705,* 89–103.
[209] Hokke, C. H., Bergwerff, A. A., Van Dedem, G. W. K., Kamerling, J. P., Vligenthart, J. F. G., *Eur J Biochem,* 1995, *228,* 981–1008.
[210] Krantz, S. B., *Blood,* 1991, *77,* 419–434.
[211] Inoue, N., Takeuchi, M., Asano, K., Shimizu, R., Takasaki, S., Kobata, A., *Arch Biochem Biophys,* 1993, *301,* 375–378.
[212] Nimtz, M., Wray, V., Rüdiger, A., Conradt, H. S., *FEBS Lett,* 1995, *365,* 203–208.
[213] Goto, M., Akai, K., Murakami, A., Hashimoto, C., Kawanishi, G., Takahashi, J., Ishimoto, A., Chioba, H., Sasaki, R., *Bio/Technology,* 1988, *6,* 67–71.

[214] Wojchowski, D. M., Orkin, S. H., Sytkowski, A. J., *Biochim Biophys Acta,* 1987, *910,* 224–232.

[215] Matsumoto, S., Ikura, K., Ueda, M., Sasaki, R., *Plant Mol Biol,* 1995, *27,* 1163–1172.

[216] Watson, E., Bhide, A., Van Halbeek, H., *Glycobiology,* 1994, *4,* 227–237.

[217] Storring, P. L., Gaines Das R. E., *J Endocrinol,* 1992, *134,* 459–484.

[218] Takeuchi, M., Kobata, A., *Glycobiology,* 1991, *1,* 337–346.

[219] Koury, M. J., Bondurant, M. C., *Eur J Biochem,* 1992, *210,* 649–663.

[220] Dube, S., Fischer, J. W., Powell, J. S., *J Biol Chem,* 1988, *263,* 17516–17521.

[221] Fukuda, M. N., Sasaki, H., Lopez, L., Fukuda, M., *Blood,* 1989, *73,* 84–89.

[222] Delorme, E., Lorenzini, T., Griffin, J., Martin, F., Jacobsen, F., Boone, T., Elliott, S., *Biochemistry,* 1992, *31,* 9871–9876.

[223] Higuchi, M., Oh-Eda, M., Kuboniwa, H., Tomonoh, K., Shimonaka, Y., Ochi, N., *J Biol Chem,* 1992, *267,* 7703–7709.

[224] Takeuchi, M., Takasaki, S., Miyazaki, H., Kato, T., Hoshi, S., Kochibe, N., Kobata, A., *J Biol Chem,* 1988, *263,* 3657–3663.

[225] Tsuda, E., Kawanishi, G., Ueda, M., Masuda, S., Sasaki, R., *Eur J Biochem,* 1990, *188,* 405–411.

[226] Hokke, C. H., Kamerling, J. P., Van Dedem, G. W. K., Vliegenthart, J. F. G., *FEBS Lett,* 1991, *286,* 18–24.

[227] Scudder, P., *J Biol Chem,* 1984, *259,* 6586–6591.

[228] Takeuchi, M., Takasaki, S., Shimada, M., Kobata, A., *J Biol Chem,* 1990, *265,* 12127–12130.

[229] Sasaki, H., Bothner, B., Dell, A., Fukuda, M., *J Biol Chem,* 1987, *262,* 12059–12076.

[230] Hokke, C. H., Bergwerff, A. A., Van Dedem, G. W. K., Van Oostrum, J., Kamerling, J. P., Vliegenthart, J. F. G., *FEBS Lett,* 1990, *275,* 9–14.

[231] Bergwerff, A. A., Van Oostrum, J., Asselberg, F. A. M., Burgi, R., Hokke, C. H., Kamerling, J. P., Vliegenthart, J. F. G., *Eur J Biochem,* 1993, *212,* 639–656.

Part Three
Economics, Safety and Hygiene

13 Biosafety

Yusuf Chisti

13.1 Introduction

The bioprocessing industry has an undeniable record of safe operation. Yet, equally undeniable is the continuing public concern regarding the safety of biotechnology [1], including bioprocessing. Many hazards are associated with industrial bioprocessing: genetically modified organisms with real or perceived risks may have to be handled; highly pathogenic bacteria, viruses, and potentially contaminated substances such as blood may need to be processed; or the bioproduct may be so active that minute amounts may cause allergenic, toxic, or other activity-associated reactions in personnel exposed to it [1]. To assure safe processing, bioprocess engineers, operators, and managers must be intimately aware of the nature of the biohazard, the containment and regulatory issues, and how design and operation must satisfy the biosafety demands. This chapter examines risk assessment, biohazard containment and inactivation practices, and other biosafety issues relevant to industrial bioprocessing. Considerations relating to deliberate release of genetically modified organisms (GMOs) are not discussed as that subject is outside the scope of this chapter. Deliberate release has been reviewed elsewhere [2].

13.2 Risk Assessment

Most micro-organisms and animal cells used in industrial processes pose little or no risk to human health and the environment; nevertheless, some high-risk human, animal, and plant pathogens are used (Table 13-1). By definition, a pathogen is any

Table 13-1. Some commercially used hazardous microorganisms.

Bordetella pertussis	Yellow fever virus
Clostridium tetani	Poliovirus
Corynebacterium diphtheriae	Rabies virus
Mycobacterium tuberculosis	Rubella
Salmonella typhi	Foot-and-mouth disease virus

micro-organism or virus that can cause disease in any other living organism, including other micro-organisms. Whenever a pathogen is used, it must be contained. In the long run, the trend is to move to safer processes by replacing harmful microbes with non-pathogenic recombinant producers.

The nature of the viable agent – microorganisms, viruses, animal and plant cells – has the greatest impact on the containment needs. Whenever possible, 'generally recognized as safe' or GRAS species should be used in commercial processes. Recombinant variants of GRAS organisms are preferable to non-GRAS microbes. When non-GRAS species must be used, known pathogens should be avoided, or variants not capable of producing disease should be preferred. For example, pathologically incompetent *Escherichia coli* K-12 strains are used in producing recombinant proteins. Many micro-organisms have a long history of safe use in food [3], and at least one previously unused species has been successfully commercialized as human food after exhaustive safety testing [4].

Use of new species or those with unknown risks must be preceded by assessment of risk [5,6] including pathogenicity testing [7]. The internal institutional biosafety committee, in consultation with the published guidelines and the Recombinant Advisory Committee of the U.S. National Institutes of Health (NIH), establishes the appropriate containment level for a strain.

Understanding of how viable and bioactive substances spread and invade the body is essential to assessing risk of processing. Microbial entry into the body occurs through inhalation of aerosols and particles; transfer to mouth via contaminated hands; damaged skin; and eye–hand contact or splashes. In addition, bioactive substances may be absorbed through intact skin. Aerosols and airborne particles are particularly troublesome sources of contamination. Aerosols spread easily and widely. Particles smaller than 5 μm dry instantly and remain suspended in the environment for long periods while circulating with air currents [8]. Many operations generate aerosols, including centrifugation, homogenization, mixing, blending, aeration of liquids, leakage of liquids under pressure, and handling of solids. Laboratory procedures can contribute. For example, aerosols are generated by bursting bubbles, breakage of liquid film as in pipeting, drops falling on surfaces, splashes, ultrasonic vibrations, sampling with syringe and needle, and during pouring and siphoning [9,10]. Contaminated apparels, hands, equipment and process streams, and circulating air spread micro-organisms, as do insects, rodents, and other pests.

Although relatively few infection episodes have been associated with industrial activity – the majority having occurred in research and diagnostic laboratories – only about 20 % of the laboratory-acquired infections have been ascribed to specific causes [8]. Unknown or unrecognized causes for most of the events suggest a continuing insufficiency of knowledge on the links between operational practices and infection. Thus, continuing vigilance is advised. Potential sources of contamination include direct accidental inoculation (needles, sharps, cuts or abrasions, animal bites), inhalation of aerosols, ingestion, and contact of contaminated material (hands, spills, contaminated surfaces) with membranes [8].

In addition to microbes, most physiologically active fermentation products – antibiotics, mycotoxins, enzymes, steroids, hormones, vaccines, deactivated microorganisms, antibodies, and other proteins – can be disruptive to health, and certain products are highly toxic. Aflatoxins are potent carcinogens. The fermentation conditions – temperature, pH, type of substrate, agitation, metabolic energy source, dissolved oxygen and carbon dioxide, the nature and concentration of micronutrients, metal ions, and other chemicals – influence the spectrum of biochemicals synthesized by an organism [3]. Under certain environmental conditions organisms such as *Aspergillus flavus* and *Aspergillus oryzae* are known to produce lethal toxins [11]. Species of the genus *Claviceps* and some members of other genera produce toxic ergot alkaloids. Mycotoxin production is widespread among fungi [12].

Several types of micro-organisms are known to cause allergenic reactions when inhaled in large amounts. Implicated organisms include Actinomycetes, *Aspergillus* sp., *Aspergillus niger, Aureobasidium pullulans, Bacillus subtilis,* Baculoviruses, *Candida tropicalis, Penicillium* sp., and *Penicillium citrinum* [11]. Allergenic reactions may be rapid, or the response may not occur until several hours after exposure, making connecting to the allergen difficult. Reactions may be extremely serious and, occasionally, fatal. Severe allergenic reactions to *Bacillus subtilis* proteases are well known [11].

The hazard posed by nonviable bioactive material such as cytotoxic agents or endotoxins may not be eliminated by sterilization. In such cases, additional chemical decontamination of work areas, equipment and waste streams would be necessary using validated processes [13]. For example, solutions of sodium hydroxide (0.1 M) readily inactivate the botulinum toxin and are recommended for surface decontamination [14].

Gram-negative bacteria produce thermostable endotoxins. Endotoxin-containing aerosols may be generated, for example, during cell disruption. Inhaled endotoxins are implicated in allergenic response; parenteral administration causes a pyrogenic reaction and other symptoms. Adverse reactions to endotoxins have been observed with *Enterobacter agglomerans, Flavobacterium* sp., *Methylophilus methylotrophus, Methylomonas methanolica, Pseudomonas aeruginosa, Serratia marcescens* and *E. coli* [11]. Up to 4 % of the dry weight of *E. coli* K-12 has been estimated to be endotoxin [11]. An action threshold value of 30×10^{-12} kg m^{-3} for airborne endotoxin has been recommended [11].

Increasingly, industrial processes use recombinant micro-organisms and animal cells. Some specific issues relating to such use are discussed in Sections 13.2.1 and 13.2.2. Biosafety considerations for solid-state fermentation processes have been discussed by Chisti [3], and issues relating to composting have been treated by Stentiford and Dodds [15]. The commonly used bioprocessing schemes and individual unit operations have been detailed elsewhere [16–20].

13.2.1 Recombinant Microorganisms

Development of containment requirements for recombinant microbes must consider environmental and ecological consequences of inadvertent release. Important considerations include survivability and colonization potential of the organism in the environment, and the organism's ability to transfer any part of it's genome to indigenous populations [21]. Survival and persistence studies are carried out in ecosystems such as activated sludge, mammalian gastro-intestinal tract, soil, and river water [21]. Gene transfer studies may be combined with those of persistence. Such work should be 'designed to show that the recombinant construct behaves similarly to the host in a representative ecosystem where the organism could be introduced inadvertently' [21]. Assessments of potential biohazard should take into account characteristics of the unaltered parent, the unaltered plasmid vector, and the transposable elements. The U.S. Food and Drug Administration (FDA) has discussed these aspects in some detail [21]. As a general guide, a recombinant production strain should not have any known combination of pathogenicity, high colonization ability, and high genetic transfer competency [21]. In addition to the genes of interest, selectable marker genes are introduced into the host during transformation. Safety of such markers and the proteins they encode remains a subject of debate [22]. Specifically, the effects of any antibiotic resistance markers should be considered: such markers may facilitate colonization of gastro-intestinal tracts of fermentation process workers receiving antibiotic therapy [21]. Indeed, evidence is emerging that antibiotics in animal feeds alter the intestinal microflora in farm animals, and similar altered ecosystems become established in farm-workers that routinely contact the animals.

Recombinant *E. coli* K-12 strains and their plasmidless hosts are unable to establish in environments consistent with various deliberate release scenarios [23]. Moreover, those strains are non conjugating and apparently incapable of transferring genes [23] to other organisms.

13.2.2 Animal Cells

Cell cultures may be contaminated with pathogenic viruses (e.g., HIV, hepatitis B) and mycoplasma (e.g., *Mycoplasma pneumoniae*). Again, immunosuppressed individuals are especially susceptible even to otherwise harmless viruses. Human cell lines are particularly high-risk and so are those derived from nonhuman primates; other mammalian cells are somewhat less risky, but may harbor agents capable of producing disease in humans (e.g., rabies, bovine spongiform encephalopathy agent, Hantaan virus in rodents). Avian and fish cells, and those from invertebrates may be lower risk. Previously uncontaminated cells can become infected during processing through human contact, or use of contaminated sera.

Other than being contaminated, established animal cell lines are potentially tumorigenic. Immunocompromised individuals are particularly susceptible, but a healthy immune system may be effectively circumvented if the transformed cell culture is compatible with an individual [24]. In one case, an accidentally inoculated (needle puncture, human tumor cells) laboratory worker developed a tumor [24]. Clearly, as with any microbiological process, workers with temporarily damaged skin should not handle viable material.

Because cell lines can harbor undetected viruses, and the lability of therapeutic proteins rules out the use of severe treatments that are capable of destroying viruses, a contamination-free product cannot be guaranteed; however, the risk of contamination can be reduced to extremely low levels using a multifaceted approach including use of exhaustively characterized cells and in-process controls.

Cells used in production originate in a manufacturer's working cell bank (MWCB) that is derived from a master cell bank (MCB). Only well-characterized cell lines are used in production. Characterization must assess identity of the cell line, it's microbial and viral contamination, genetic stability, and, for genetically modified cells, the genetic construct must be verified [24]. Guidelines for characterization of cell banks have been established by the FDA, and common practices have been described by Lubiniecki [25].

Use of a well-characterized cell in production is insufficient assurance of a safe product; additional in-process controls are necessary. Controls that need implementing include assessment of pre-harvest culture broth for relevant viruses and establishment of acceptance/rejection criteria for viral loads taking into account the validated capabilities of downstream virus removal/inactivation steps. Combinations of those approaches reduce risk to extremely low levels. In fact, so far not a single case of viral infection has been associated with the use of cell culture-derived biopharmaceuticals (see also [12]). Product purity and viral safety issues for cell culture-derived therapeutic proteins have been discussed further by others [12,25–29].

13.3 Containment Levels

Containment requirements are based on assessment of potential biohazard of an agent as reflected in it's risk classification [5]. Conventional pathogens have been categorized into four risk groups with increasingly stringent containment needs. The U. S. Centers for Disease Control and Prevention/National Institutes of Health recognize biosafety containment levels BL1–4 for laboratory operations. A different biosafety level assignment is used for large-scale processing: GLSP, BL1-LS, BL2-LS, and BL3-LS. The BL1-LS (LS = large scale) corresponds to BL1 of the laboratory scale, and so forth. The GLSP is a lower level than the BL1 designation. The BL4 has no equivalent at the large scale, and agents requiring BL4 containment are not used in commercial production. BL4 organisms require high-level containment; they are pathogenic and hazardous to laboratory personnel; and they produce transmissible diseases for which no prophylaxis or treatment exists. The biosafety level assignment considers whether an agent is pathogenic, poses a hazard to laboratory and plant personnel, is transmissible to the community, and availability of prophylaxis or treatment. Pathogenicity, or the ability to produce disease, depends on factors such as virulence, invasiveness, and infectivity. The specific risk categories for various organisms have been noted by Richardson and Barkley [14]. Table 13-2 lists the BLx containment requirements for several pathogens.

In addition to the United States Public Health Service CDC-NIH biohazard classification system [14], other guidelines have been established by the World Health Organization (WHO), the European Federation of Biotechnology (EFB), and national agencies in the United Kingdom [18], Canada [30], as well as other countries. Here, the focus is on the U.S. practices that are similar to those of the other developed

Table 13-2. Biohazard level classification of some pathogens.

Microorganism	BL2	BL3	BL4
Bacteria			
Bordetella pertussis	●		
Clostridium tetani	●		
Corynebacterium diphtheriae	●		
Coxiella burnettii			●
Mycobacterium tuberculosis		●	
Neisseria meningitidies		●	
Salmonella typhi	●		
Yersinia pestis		●	
Fungi			
Aspergillus flavus	●		
Viruses			
Ebola			●
Hantaan			●
Hepatitis B		●	
HIV		●	
Influenza	●		
Lassa			●
Poliovirus	●		
Rabies virus		●	
Rubella	●		
Yellow fever		●	

The containment levels shown are for relatively small-scale operations. For large-scale work, use of the next higher containment level is recommended (except BL3). Consult Richardson and Barkley [14] for further guidance.

countries. Practices and regulatory frameworks for other jurisdictions have been noted by Collins and Beale [20] and by Hambleton et al. [19]. Table 13-3 lists the four risk categories used by the WHO in classifying hazardous micro-organisms.

Of the four biosafety levels that are relevant to large-scale work, the GLSP level is suitable for non-pathogenic and non-toxigenic strains (including genetically modified variants) that have an extended history of safe industrial use. The U.S. GLSP derives from the 'good industrial large-scale practice', or GILSP guidelines originally established by the Organization for Economic Cooperation and Development (OECD) in 1986 for use with suitable recombinant strains. Ideally, all industrial processes should comply with those minimal requirements. GLSP is appropriate to organisms satisfying the following criteria [31]:

1. The host organism is non-pathogenic and free of adventitious agents. The host has an extended record of safe industrial use, or it has built-in incompetencies that limit it's survival in the environment, and it has no adverse environmental consequences.
2. The genetically modified version is non-pathogenic and safe in an industrial setting, and without adverse environmental consequences.

Table 13-3. Biohazard risk classification used by the World Health Organization.

Risk group 1	(no or very low individual and community risk) A microorganism that is unlikely to cause human or animal disease.
Risk group 2	(moderate individual risk, low community risk) A pathogen that can cause human or animal disease but is unlikely to be a serious hazard to laboratory workers, the community, livestock, or the environment. Laboratory exposures may cause serious infection, but effective treatment and preventive measures are available and the risk of spread of infection is limited.
Risk group 3	(high individual risk, low community risk) A pathogen that usually causes serious human or animal disease but does not ordinarily spread from one individual to another, directly or indirectly. Effective treatment and preventive measures are available.
Risk group 4	(high individual and community risk) A pathogen that causes serious human or animal disease and that can be readily transmitted from one individual to another, directly or indirectly. Effective treatment and preventive measures are not usually available.

3. The DNA vector is well-characterized and free from known harmful sequences. To the extent possible, the size of the insert is limited to that necessary for the intended function, and the insert does not increase the stability of the construct in the environment unless required by the intended function; the insert is poorly mobilizable, and it does not transfer resistance markers to micro-organisms not known to naturally acquire resistance if such acquisition would compromise use of a drug for controlling disease in man, veterinary medicine, or agriculture.

GLSP requirements further include: (i) implementation of a health and safety program; (ii) suitably trained personnel and written operational procedures; (iii) facilities, equipment, protective clothing and practices appropriate to risk; (iv) regulatory compliance with regard to discharges to the environment; (v) minimization of aerosol generation to prevent adverse risk to employee health; and (vi) a spill control plan within the emergency response plan [31]. Often, bioprocesses must comply also with the Good Manufacturing Practices (GMP) regulations; hence, in certain areas an otherwise GLSP process may actually meet much higher standards. For example, GLSP does not require any special containment, but to meet product protection requirements, use of enclosed equipment is advisable for processing of biopharmaceuticals [32].

Genetically modified *Saccharomyces cerevisiae, E. coli* K-12, *B. subtilis, Aspergillus oryzae,* and Chinese hamster ovary (CHO) cells, for example, can be used under GLSP classification. However, it should be noted that the bioactivity and the nature of the specific recombinant product being produced could strongly affect the acceptable containment level. Thus, an otherwise GLSP species being used to make a toxic or unusually bioactive substance, may have to be contained not because of an intrinsic 'biohazard', but because of the nature of the product. GLSP-compliant processing aims to attain the lowest practicable exposure of workplace and the environment to any physical, chemical, or biological agent [18].

Processes requiring greater containment than GLSP should preferably be designed to one level higher than the minimum acceptable biosafety level. Typically, the cost of building to BL2-LS level is a minor increment over BL1-LS [33]. Indeed, many U.S. companies have designed production facilities to BL2-LS containment requirements where the BL1-LS measures would have sufficed [34]. Occasionally, the containment requirements for a given micro-organism differ slightly between jurisdictions. Again, the preferred practice is to err toward caution. The GLSP and BLx-LS containment requirements are summarized in Table 13-4.

As noted earlier, depending on the risk, process support laboratories may have to comply with BL1-3 laboratory standards. Even a minimum containment GLSP process support laboratory should comply with good microbiological practices (Table 13-5) that are intended to protect both the operator and the product. Table 13-5 lists minimal requirements. Detailed guidelines on laboratory practices appear elsewhere [10,14,30].

In view of the different guidelines for laboratory- and large-scale operations, the question of demarcation between the two scales is important. In the U.S., processes larger than 10 L are assessed as 'large scale'. In the U.K., there is no specific volume guideline to distinguish between large and small scale [18]. For otherwise identical circumstances, the risk can be reasonably assumed to increase with the scale of operation, although the contrary has been argued [18].

The EFB Working Party on Safety in Biotechnology continues to produce 'reports' [9,24,35–39] that provide a useful perspective on biosafety issues. These reports have discussed plant pathogens [37,38], handling of micro-organisms of various risk classifications [35], assessing the impact on human health [9], hazard-based classification of micro-organisms [39], work with human and animal cell cultures [24], and general biosafety issues [36]. The regulatory approaches to biosafety issues in the European Community and the United States have been compared [31]. Because the published guidelines specify only general requirements – not methods of compliance – ongoing consultations with experts and the internal biosafety committee [40] are essential. Future European standards are likely to be quite specific on 'technical specifications, codes, methods of analysis and lists of organisms' than the current U.S. practices.

In most countries, responsibility for biosafety issues is spread over several regulatory agencies. In the United States, biotechnology products and production facilities may come under the jurisdiction of the FDA, the U.S. Department of Agriculture (USDA), the Environmental Protection Agency (EPA), and the Occupational Safety and Health Administration (OSHA). The roles of the relevant agencies and the specific Acts under which they are empowered have been summarized elsewhere [34].

---➤

[a] Complete change. [b] Minimize release using procedural controls. [c] Minimize release using engineered controls. [d] Prevent release. [e] Comply with local environmental codes. [f] Inactivate by validated means. [g] Required in all GMP-compliant processing.
Also consult Richardson and Barkley [14] and the Canadian Medical Research Council laboratory biosafety guidelines [30].

Table 13-4. The biohazard containment requirements.

Specification	GLSP	BL1-LS	BL2-LS	BL3-LS
1. Only authorized personnel allowed	●	●	●	●
2. Written procedures and training for good housekeeping and safety	●	●	●	●
3. Implementation and enforcement of institutional codes for hygiene and safety	●	●	●	●
4. Protective work wear and changing facilities	●	●	●	●[a]
5. Hand washing facilities	●	●	●	●
6. No eating, drinking, smoking, mouth pipeting, or cosmetics application in work area	●	●	●	●
7. Institutional accident reporting system		●	●	●
8. Biosafety manual			●	●
9. Medical surveillance			●	●
10. Closed equipment or other primary containment for processing viable agent	●[b]	●	●	●
11. Inactivation of culture by validated procedures before removal from closed systems		●[c]	●[d]	●[d]
12. Enclosed sampling, material additions, and transfers to/from closed systems to prevent/minimize aerosols, surface contamination, etc.		●	●	●
13. Treatement of exhaust gases from closed equipment		●[c]	●[d]	●[d]
14. Inactivation of viable agent by validated means before opening of closed systems	●	●	●	●
15. Emergency plan, systems and procedures for handling large accidental spills		●	●	●
16. No leakage of viable agent from rotating seals or other penetrations and mechanical devices		●[c]	●	●
17. Evaluation and monitoring of integrity of containment in closed systems			●	●
18. Evaluation/validation of containment with host organism prior to using recombinant organism	●[e]	●[f]	●[f]	●[f]
19. Containment and treatment of effluent before discharge			●	●
20. Permanent identification of closed process equipment and use of identification on batch records			●	●
21. Display of universal biohazard sign on contained equipment when processing viable agent			●	●
22. Display of universal biohazard sign on doors to contained areas during operation				●
23. Low-pressure operation of process systems				●
24. Operations to be in a controlled area:				●
Separate specified entry				●
Air-locks at all entrances (including emergency exits)				●
Readily cleaned and decontaminated finishes[g]				●
Protection of utilities, services, process piping and wiring against contamination				●
Separate gowning and washing facilities at each entrance; shower facilities in close proximity				●
Personnel should shower before leaving controlled area				●
Areas sealable for fumigation				●
Sealed penetrations into area				●
Ventilation (controlled negative pressure in area; HEPA filtered exhausts, once through ventilation)				●

Table 13-5. Good microbiological practices for safe handling of potentially risky microorganisms [35].

1. Operators should have a basic knowledge of microbiology. All personnel should be aware of the risks of cultivated pathogens. Only essential personnel with the necessary training should be allowed into the work area. Practices that prevent spread of pathogens should be routinely followed. Suitable full-front laboratory apparels should be used. Work wear should remain within the work area. Hands must be washed with suitable disinfectant soap after removing latex gloves.

2. No eating, drinking, smoking, mouth pipeting, and application of cosmetics in the work area. No contact between work area materials or tools and the mouth of operators. Use of good aseptic technique.

3. Aerosol generating activities (e.g., filling of bottles and tubes, centrifugation) should be confined to biological safety cabinets. Any washing activities require special care.

4. Infected waste is sealed in containers the outside of which is disinfected prior to transfer to autoclave or incinerator.

5. Use of validated thermal or chemical sterilization processes that assure the requisite kill.

6. Use of reliable equipment.

7. Disinfection of all work surfaces and hands after normal work.

8. Disinfection of work surfaces, floors and hands after spill of infectious material.

9. An emergency action plan with details of first aid, cleaning and disinfection. Staff trained to deal with emergencies.

10. Decontamination of laboratory clothing.

13.4 Risk Management

Design and operation of a bioprocessing facility must assure safety of personnel within the facility and those in the surrounding community [41]. Protection is achieved through a combination of engineered facilities, processes, and equipment; worker training and education; use of personnel protective equipment; operational practices; validation of machinery and methodologies; controlled access to facilities; biosafety committee or subcommittee; and medical and environmental surveillance [8]. Typically, enclosed process equipment or biological safety cabinets provide primary containment to the viable material. In the event of inadvertent release from process equipment, further secondary containment is provided by the building. Environmental monitoring seeks to ascertain the satisfactory functioning of primary and secondary containment [42].

   This section details the various aspects of risk management, including selection and use of biological safety cabinets which are invariably encountered at numerous stages of bioprocessing. Also included is a section on handling of biohazardous spills.

13.4.1 Biological Safety Cabinets

Biological safety cabinets are the primary means of containment in process support laboratories and during early stages of culture development. Based on design and the protection afforded, biological safety cabinets are designated as Class I, II, and III. Capabilities of the various classes are summarized in Table 13-6 [43]. Note that laminar flow 'clean benches' are not biological safety cabinets and should not be used to handle potentially hazardous material.

Class I cabinets (Fig. 13-1) do not protect the work area against microbial or particulate contamination. The operator is protected so long as a minimum linear air velocity of 0.4 m s^{-1} is maintained through the front opening [43]. The cabinet is hard-ducted to the building exhaust system (Fig. 13-1).

Class II biosafety cabinets (Fig. 13-2) protect the operator, the product, and the environment. The work area is bathed in downward laminar flow of particle-free, recirculated air. In addition, air from the room is drawn in through the front opening to prevent leakage of aerosols and contaminated air. The linear air flow rate at the opening should be 0.4 m s^{-1} or greater. HEPA filtered air may be exhausted into

Table 13-6. Protection capability and biohazard suitability of various classes of biological safety cabinets.

Biohazard level	Protection provided			Cabinet class
	Personnel	Product	Environment	
BL 1–3	Yes	No	Yes	I
BL 1–3	Yes	Yes	Yes	II (A, B1–3)
BL 4	Yes	Yes	Yes	III (B1, B2)

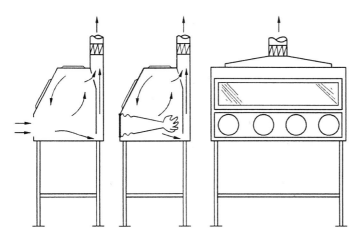

Fig. 13-1. Class I biological safety cabinet.

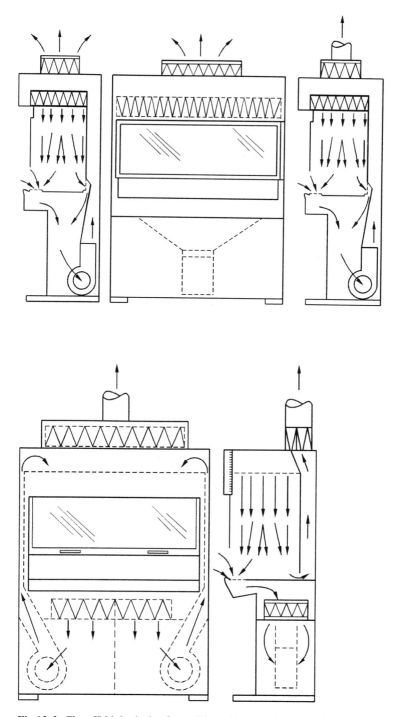

Fig.13-2. Class II biological safety cabinets Types A (top) and B.

the laboratory as in Class II Type A cabinets (Fig. 13-2), or the discharge may be hard-piped to the building exhaust system as in Class II Types B1–3. Type B1 cabinets recirculate part of the air over the work area, hence they may be used to process only minute amounts of volatiles. Type B2 cabinets are total exhaust devices that may be used also for some chemical containment, so long as the fumes are not susceptible to electrical ignition. As a general rule, no class of biological safety cabinets is suited to handling volatile toxic substances, but nonvolatile toxic chemicals can be handled in all classes of cabinets.

Class II Type B3 cabinets are ducted Type A devices that like other Type B cabinets provide a minimum linear air velocity of 0.51 m s^{-1} at the opening. All positive-pressure contaminated plenums within a Type B3 cabinet are surrounded by negative-pressure chambers to prevent leakage to the environment.

Class III biosafety cabinets (Fig. 13-3) are designed for handling BL4 biohazard agents. The cabinet is a fully sealed chamber with HEPA filtered air inlet and exhaust. The front end is provided with a sealed window and ports with heavy-duty, arm-length rubber gloves. Access to the chamber is through a side-mounted, disinfectant-filled dunk tank or through a sterilizable double-door pass-through such as an autoclave. The operator, the environment and the work area are protected. Air from Class III cabinets must be exhausted through two HEPA filters in series, or one HEPA filter and an incinerator. A dedicated, independent exhaust system exterior to the cabinet is used to maintain air flow [43]. The enclosed work chamber is kept at a lower pressure than the laboratory. Usually a 0.5 inch (1.3 cm) water gage pressure differential is maintained [43]. Class III cabinets are usually installed only in maximum containment laboratories having other suitable safeguards.

Biological safety cabinets rely on HEPA filters at air exhaust and/or intake to provide requisite protection to personnel, product and the environment [43,44]. Gener-

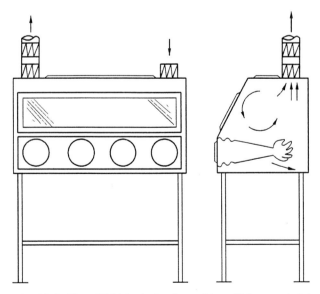

Fig. 13-3. Class III biological safety cabinet. Side-mounted dunk tank not shown.

ally, HEPA filters are rated to remove particles down to 0.3 μm with an efficiency of 99.97 %, but more expensive higher efficiency (99.99 % or higher) filters are available. The 0.3 μm particles are least easily filtered compared with larger or smaller particles; hence that size is used for HEPA filter performance specifications [44]. Filters are susceptible to shock-induced mechanical damage; therefore, performance and integrity of a biosafety cabinet must be certified after initial installation, after relocation, after repair, and at yearly intervals. The specific certification tests depend on the type of cabinet. Some essential tests include the downflow velocity and volume testing (Class I and II); the inflow velocity test (Class I and II); the negative-pressure testing (Class II and III); air flow smoke patterns tests (Class I and II); the HEPA filters leak tests (Class I–III); the cabinet leak test (Class II and III); and testing of alarms and interlocks (Class III). HEPA filters should be decontaminated prior to replacement. Decontamination is usually done with formaldehyde or hydrogen peroxide vapor, and decontamination provisions should be provided during installation. Class II cabinets usually have sensors for monitoring the pressure drop across the HEPA filter, and a low exhaust flow alarm is provided.

Proper technique is essential for safe use of biological safety cabinets. The containment air curtain at the opening of Class II cabinets is easily disrupted by rapid sweeping arm movements into and out of the cabinet [43]. Other activities such as rapid movement of personnel around the cabinet and opening/closing of room doors also disrupt the air barrier [43]. Arm movements into/out of the cabinet should be slow and perpendicular to the front face of the cabinet [43]. The number of arm entries should be minimized by preparing a checklist of the required materials and placing them in the cabinet. The seating height should be adjusted so that the face of the operator is above the front opening [43]. Manipulations should be delayed for about 1 minute after arms/hands are placed inside the cabinet [43]. Arms, hands or other objects should not rest across the front grill or the room air may flow into the sterile work area. All manipulations inside the cabinet should be at least 10 cm (4 inches) from the front grill [43]. Any aerosol-generating equipment should be operated in the rear of the cabinet [43]. Used pipets should be discarded into a horizontal, disinfectant-containing discard tray kept within the cabinet [43]. Potentially contaminated material should not be brought out until after surface decontamination with a suitable disinfectant [43]. The cabinet air blower should be switched on at least 3–5 minutes before commencing work [43]. The work surfaces and the interior walls should be disinfected before and after use by wiping with 70 % ethanol or other suitable disinfectant. If chlorine bleach is used, a second wiping with sterile water is needed to remove residual chlorine, which is corrosive to stainless steel [43]. Also, any material and containers placed in the cabinet should be wiped with 70 % ethanol to reduce risk of contamination. Consult Richmond and McKinney [43] for additional guidance on selection, installation, and proper use of biological safety cabinets. Management of spills within a biosafety cabinet is discussed in Section 13.4.2.

13.4.2 Spill Management

Primary containment is provided by fully closed process equipment such as fermenters, centrifuges, heat exchangers, and pumps. However, even the best designed primary containment can fail; therefore, emergency response procedures must be inplace. Accidental release from process equipment poses a potential risk to employees and the local environment [45]. The extent of risk depends on the pathogenicity, virulence, invasiveness, and infectivity of the agent, and the volume being handled [45]. Release or spill may occur within the confines of a biological safety cabinet, in unconfined areas within a facility, or more broadly due to catastrophic failure of a process equipment [45].

For spills within a biosafety cabinet, immediate decontamination should be effected by a trained, suitably equipped (latex gloves, laboratory coat, safety glasses) technician while the cabinet air circulation system continues to operate [45]. Decontamination requires flooding the work tray with a disinfectant while minimizing aerosol generation. Thorough contact of the spill and the disinfectant is necessary for a preferred minimum of 30 minutes. The spill is then absorbed into disposable cloth or paper towels and discarded into autoclavable bags. The work surface, the cabinet walls, and any equipment inside is wiped with a disinfectant-soaked cloth. If the spill extends to the exhaust grills, the catch basin should be flooded (30 minutes) with disinfectant which is then drained into an autoclavable bag. A disinfectant-soaked cloth is used to wipe the grill and the catch basin [45]. The outside of the autoclavable container and bag should be wiped with a disinfectant-soaked cloth. Upon completion of cleanup, all solid material (including gloves, wiping cloth, lab coat, and any contaminated garments) that came into contact with the viable agent should be placed into an autoclavable bag. This material should be autoclaved at 121 °C for a minimum of 1 h or other suitable period that has been previously established and validated for a particular load size and distribution [45]. Once contaminated gloves and clothing have been removed, germicidal soap should be used to wash arms, hands, and face.

Disinfectants suitable for most purposes are chlorine bleach (500 ppm available chlorine), iodine solution (25–1600 ppm available iodine), formaldehyde (0.2–8.0 %), and 2 % glutaraldehyde. Only chlorine bleach is satisfactory for treating liquids; other agents noted are suited for wiping surfaces, glassware, etc. Disinfectants such as quaternary ammonium compounds, ethyl alcohol, and isopropyl alcohol are not broadly effective. Chlorine bleach should not be used on stainless steel process equipment, and care should be taken to ascertain that bleach is compatible with the fluid being disinfected. The disinfection capability of chlorine bleaches declines with increasing pH.

Readily accessible spill carts should be provided to deal with small unconfined spills [45]. A spill cart should have supplies of chlorine- and iodine-based disinfectants (e.g., 5 % Wescodyne, and 5 % Clorox), spill control supplies (autoclavable squeegee, autoclavable dust pan, autoclavable forceps, autoclavable biohazard bags, bucket, disposable wipes, and spill pillows), as well as protective clothing and equipment (disposable lab coats and jump suits, disposable latex gloves, disposable safety glasses, autoclavable boots, and half-face or full-face respirator with HEPA filter cartridges). The following treatment procedure, adapted from Van Houten [45] is recommended: Warn others of spill and leave the area holding your breath to avoid inhaling potentially hazardous aerosols; remove contaminated clothing, folding contaminated areas inwards and discard into an autoclavable bag; wash potentially contaminated body areas as well as face, arms, and hands with germicidal soap; shower if necessary; wear protective clothing (disposable lab coat, latex gloves, safety glasses, autoclavable boots) and, if necessary, HEPA filter-equipped half or full-face mask; enter the area with the spill cart; use spill pillows to isolate floor drains if present and connected directly to sewer (i.e., not connected to a biokill treatment facility); encircle the spill with disinfectant, ensuring adequate spill–disinfectant contact while minimizing aerosolization; allow a 30-minute contact time; pick up broken glass and other sharp objects with the forceps, dust pan, squeegee, and place them in a leak-proof autoclavable container; use disposable wipes or spill pillows to mop up the liquid and discard it into an autoclavable bag; wipe the outside of the autoclavable bags with a disinfectant-soaked cloth. Use an uncontaminated, disinfectant-soaked cloth to wipe the area of the spill. Upon completing clean-up, all solid material that came in contact with the viable agent should be placed into autoclavable bags. Bags, containers, and contaminated clothing should be decontaminated in an autoclave at 121 °C for 1 hour or other prevalidated period. Arms, hands, and face should be washed with germicidal soap. Shower if necessary. The spill cart should also be disinfected. Exposed personnel may have to undergo prophylactic or other treatments and medical surveillance in accordance with pre-established policies. Larger, unconfined spills are usually handled by especially trained spill response personnel using procedures similar to the ones noted for smaller spills.

13.4.3 Buildings and Facilities

The design of a facility determines it's ability to provide secondary containment. Facilities processing especially hazardous material should preferably be located away from heavily built-up areas. In extreme cases, dispersal patterns for any inadvertently released material should be considered for all the meteorological scenarios relevant to the location. Access to the facility should be restricted to authorized personnel, with certain areas being 'off limit' to all but the relevant personnel [41]. Access control may require security fencing, electronic card controlled entry, vandal-proof exterior windows, etc. In addition to containment, the design of the facility needs to assure protection of the product. Products need to be protected against contamination, particularly microbial contamination to which they are highly susceptible [32]. Protection must be provided throughout manufacturing and storage [32]. All facilities need an effective insect and rodent control program.

Certain bioprocess facilities may require holding or handling of infected or potentially infected or suspect animals. Design requirements for such facilities are beyond the scope of this chapter; consult Richardson and Barkley [14] for guidance.

Table 13-7. Design concepts for biohazard containment[a].

1. Controlled access
2. Work areas at negative pressure relative to surroundings
3. HEPA filtered air exhaust
4. Additional containment of aerosol-generating activities
5. Personnel training
6. Personnel protective equipment
7. Decontamination of bioactive process wastes
8. Medical surveillance of 'at-risk' personnel
9. Environmental monitoring

[a] Adapted from Flickinger and Sansone [13].

The principal concepts for design and operation of biohazard containment facilities are summarized in Table 13-7. Specific features are discussed in the following sections. The guidelines given comply with the U.S. FDA recommendations [32].

13.4.3.1 Layout

The layout of the facility affects efficiency of operations, the potential for containment, and prevention of cross-contamination. Attention to layout is required by the GMPs [46]. Because the flow of personnel, equipment, materials and air in the facility must be controlled, the building and the process must be closely integrated by design [47]. 'Contained', 'clean', and 'dirty' areas should be identified on the process flow sheets and the building layout drawings. Movement of personnel, equipment, process streams, and air across containment boundary must assure integrity of containment through a combination of engineered systems and operational protocols. In general, flows must be unidirectional [32], from clean to dirty areas, and not vice versa. The dirty and clean paths should not cross [32], and there should be no back-tracking. The biohazard containment areas and the aseptic product filling areas should be located in different wings of the facility, with no sharing of common hallways or direct access [32]. Adequate space must be provided for various uses. There should be no overcrowding in work areas, especially the contained areas.

13.4.3.2 Air Handling

Quality of the ventilation air and it's flow in a facility are crucial to containment, biosafety, as well as protection of the product. Air in a facility is handled by the heating, ventilation and air conditioning (HVAC) system. Designing the HVAC system requires classification of process areas as 'clean', 'dirty', or 'contained'. Air-locks are used to isolate the contained areas and those with critical cleanliness requirements, from other zones.

Effective containment depends critically on management of air flow and pressure differentials [32]. Area pressure differentials help to prevent airborne contaminants from intruding into contamination-free parts of a facility. Pressure differentials of 1.3 mm (0.05 inches) of water are typically used between adjacent areas. Areas containing infectious agents (e.g., viral vaccine) must be maintained under negative pressure relative to the surroundings [41,48]. Otherwise, the air flow is generally from 'clean' to 'dirty' areas. Access to a contained process area should be through an air-lock that is maintained at a lower pressure than the contained area and the outer access corridor [41]. Access should be restricted (e.g., card-controlled entry).

The HVAC system design should assure that unwanted air pressure differentials do not develop in the event of mechanical failure. Visual and audible indication of ventilation failure should be provided. When feasible, once-through ventilation is preferred to prevent spread of contamination or the likelihood of cross-contamination [41]. Air from the facility is exhausted usually on the roof, away from any air intake. All air from contained areas in a BL3-LS facility should be exhausted through HEPA filters, and the area should be under negative pressure with respect to the surroundings (Table 13-4).

The BL2-LS facility design guidelines do not specify secondary containment through air flow management, or HEPA filtering of the area's air supply and exhaust. Primary containment must be achieved by using closed systems or appropriate biosafety cabinets (Table 13-4). The requirements notwithstanding, HEPA filtered air supply is recommended for minimizing the potential for contaminating the product [33]. Pipework should be minimized in contained areas, and wall penetrations and light fixtures should be sealed.

Aerosol build-up in work areas can be reduced by HEPA filtered ventilation with a sufficient number of air changes per hour – 20–30 are not unusual in BL2-LS processing areas [33]. The mandated minimum number of air changes in various areas may have to be exceeded to account for factors such as humidity and heat rejection in the area, it's typical function, personnel capacity, production of vapors and fumes, and generation of aerosols. The relative humidity in most processing areas is controlled at $40-50 \pm 5\%$. Relative humidities exceeding 50% promote corrosion, whereas values lower than 40% lead to problems with static electricity. Ventilation is discussed further by First [44], del Valle [48], and Lee [49].

Air quality is of particular concern in the processing environment of the sterile final dosage forms of pharmaceuticals. Areas where the sterile product, containers, and closures are exposed to the environment are designated as 'critical' areas. Examples include 'fill rooms' and other aseptic processing areas. For a long time, the air quality in critical areas was required to be at least Class 100, that is, no more than 100 particles of $\geq 0.5\ \mu m$ per cubic foot of air. Higher standards are now in demand. In critical areas, the number of colony forming units (CFU) should not exceed one per cubic foot ($0.03\ m^3$) of space, and a 0.05-inch (1.3 mm) water gage positive-pressure differential must be maintained relative to adjacent areas. Class 100 areas are typically contained within Class 100,000 areas in which the particle count of $\geq 0.5\ \mu m$ particles does not exceed 10^5 per cubic foot ($0.03\ m^3$). In addition, the CFUs do not exceed 25 per 10 cubic feet ($0.3\ m^3$) of air. Moreover, the Class 100,000 surrounding space must equal or exceed 20 air changes per hour. A higher number, generally 60–75 air changes per hour, must be provided in critical aseptic fill rooms [48]. The Class 100 area itself should be designed for at least 600 air changes per hour. HEPA filtered air is almost always supplied at the ceiling, and low wall-mounted returns are preferred. Aseptic filling areas need to be maintained at positive pressure [32]. Additional issues relating to HVAC system design for protection of the product have been discussed by del Valle [48] and Dobie [50].

13.4.3.3 Construction, Finishes and Practices

How easily and well a facility may be cleaned, sanitized or decontaminated depends on it's construction, and finish. The building and room finishes are subject to GMP and containment guidelines [46,47,30,51]. Processing areas should have non-shedding, smooth, impervious, splash-resistant finishes that are capable of being cleaned and disinfected. Even a GLSP-level fermentation area should be capable of being hosed down. Contained areas should be capable of being sealed and decontaminated by disinfectant spray and by fumigation. Decontamination procedures should be established beforehand [41] and validated. Formaldehyde vapor is a commonly used fumigant, but it is carcinogenic. In addition, mixing of formaldehyde with chlorine-containing disinfectants can produce potent lung carcinogens. Paraformaldehyde gas may be used to surface-sterilize heat-sensitive equipment.

Floors must slope to drains. Drains should have a slope of at least 2 cm per linear meter to assure complete drainage. Floor drains in areas that are susceptible to spills should not be connected directly to the municipal sewer [33]; instead, the drains should be piped to the biokill system. Drains should be provided with 20 cm water trap seals to prevent back-escape of vapor, gases, and aerosols from the containment sump into the work area. Sometimes, in addition to water traps, the contained drains are provided with check valves that prevent back-flow in case of a pressure build-up in the containment sump [33]; but usually suitable venting of sumps and tanks is sufficient to preclude pressure build-up. All traps should be decontaminated after a spill, and on a regular basis by pouring several trap volumes of a disinfectant solution down the drain.

Catastrophic failure of pressure vessel fermenters is unlikely, but large leakages of fluid may occur from failed valves, ports, or gaskets. All material released up to the full volume of the largest fermenter should be contained within a diked area emptying into a sump, and treated through the biokill system. After the spill has been treated, the diked area should also be disinfected (see Section 13.4.2).

The spill containment and treatment system should be capable of functioning on demand [32], and it should have sufficient capacity to handle the entire process volume. A minimum containment capacity of twice the production capacity has been recommended [32] which is quite reasonable considering that the process fluid as well as the subsequent wash effluent must be contained.

Generally, concrete floors with troweled-on epoxy finish are preferred in processing areas, but in some low-traffic laboratory areas welded PVC flooring (not tiles) may be used. The epoxy finish (or vinyl) should extend at least 10 cm up the walls, and the edges should be coved. Walls are generally concrete masonry units with epoxy paint. Ceilings are suspended dry wall, painted, or epoxy finished. In aseptic areas and those processing BL1-LS or higher agents, all internal corners (including wall-to-ceiling) should be preferably coved and all finishes should be flush (base flush with wall, door and window frames flush with walls). Sealed windows are the norm. In addition, any exterior windows should be break-resistant whenever high-level containment is required. Light fixtures should be flush mounted, and, in high-level containment areas, they should be of a type that can be serviced from outside the contained area.

The area furnishings should be of a sanitary design, resistant to water, process chemicals, and disinfectants. Usually, work surfaces are stainless steel or epoxy tops with baked epoxy painted casework [33]. Placement of equipment and furnishings should not interfere with cleaning and disinfection [33]. Equipment may be placed on housekeeping pads with radiused edges, or raised on legs that comply with hygienic design standards. Floor penetrations should be minimized and penetrations should be fully sealed to prevent seepage [33]. Supporting equipment from wall-mounted brackets, or from overhead supports is preferable [33].

Hand-washing facilities should be provided near exits in the contained areas. Sink faucets (taps) should be automatic, or elbow- or foot-operated. Suitable disinfectant soap should be provided. Although hand air dryers have been recommended [33], paper towels are preferable. Electric dryers recirculate particles and aerosols that are deposited on hands. The BL2-LS facility design has been discussed by Miller and Bergmann [33] and the essential requirements are noted in Table 13-4.

A BL3-LS contained area should have separate facilities for gowning and washing at each entrance and showering facilities should be provided in close proximity (Table 13-4). Air-locks should be provided at all entrances and exits (including emergency exits) in a BL3-LS contained area. Only a minimal number of essential personnel should be allowed into contained areas. Monitoring should be from outside the contained area, through sealed windows, intercoms, and closed-circuit television.

Containment features for pilot-scale fermentation facilities for producing cytotoxic agents and oncogenic viruses have been described [13]. Some of the methods noted are no longer state-of-the-art, but are effective nonetheless. For example, more elegant and reliable contained sampling methods are now available [52]. Specific construction details of animal cell culture facilities have been noted by Donnelly [53] and Lubiniecki [25].

The GMP regulations require provision of separate facilities for handling of spore-forming microorganisms. In addition, dedicated, segregated facilities are needed for processing penicillins (and other β-lactams) because cross-contamination with penicillins and penicillin-containing substances cannot be reasonably prevented in a multiproduct facility [12]. Separate air handling systems are necessary if a building processes penicillins as well as non-penicillin products. Similarly, in facilities producing several viral vaccines, Damm [41] recommends complete isolation of areas dealing with different viruses.

Containment and decontamination capabilities must function under normal conditions as well as during emergencies [54]. Provisions and practices must be in-place for evacuation during emergencies such as fire. Consequences of loss of power on containment should be evaluated during design, and emergency power should be provided to prevent loss of containment. In addition, some essential process equipment may require standby power and steam supply.

13.4.4 Process Equipment

The design of process machinery determines it's primary containment capability. Moreover, specific equipment may have specific hazards associated with it's use. For example, high-pressure equipment such as cell disruptors may generate sprays of contaminated fluid in the event that a gasket fails [55]. Similarly, aerosols are produced during centrifugation and submerged aeration of culture broth in fermenters.

Design and evaluation of process equipment require identifying areas where potential leakage of contained material could occur (Table 13-8). Points of possible leakage should be minimized, better contained, or, when feasible, eliminated. Containment capabilities should be assessed for all equipment, including fermenters, centrifuges, filters, solvent extraction units, cell disruption devices, spray and freeze dryers, sterilizers and autoclaves, pumps, valves, pipes, and heat exchangers, bottling and vialing machinery, downstream purification units, and waste treatment systems. In addition, the HVAC system and the utilities should be assessed for potential of becoming contaminated. Potentially contaminated lubricating fluids, steam condensate, wash fluids, etc., should be treated through the contaminated waste system.

Process equipment, control cabinets, electrical housings, switches, etc., should be splash-resistant. Alternatively, control cabinets and instrumentation may be located outside the contained area and serviced through sealed cables [40]. Utility lines (e.g., steam, air, water, vacuum) that are connected directly to process equipment are at risk of becoming contaminated [40]. Protective measures include maintaining a positive pressure in the delivery lines relative to contaminated equipment, use of microbial-grade filters and back-flow preventers.

To the extent feasible, equipment entering and leaving the contained area should be decontaminated by thermal sterilization in double-door autoclaves connecting the inside and the outside of the work area. Waste should be sterilized in a separate autoclave that is not used for process sterilizations [33]. Autoclaves should have interlocked doors, and fail-safe devices that prevent opening the autoclave until a complete sterilization cycle has been implemented. Pre-vacuum type autoclaves are preferred for sterilization of solid waste and other general process uses. Sterilization cycles and load configurations should be pre-validated. Principles of thermal inactivation have been described elsewhere [16].

Table 13-8. Points of potential release of contained material.

– Shaft seals (vessels, pumps)	– Perforated pipes, vessels, equipment
– Flanges (pipes, valves, vessels)	– Leakage into cooling water
– Pumps	– Tubings and hoses
– Points of entry of probes and sensors	– Sanitary connections
– Valve packings	– Rupture discs
– Ruptured diaphragms in valves	– Pressure relief valves
– Sample points	– Exhaust gases
– Perforated metal bellows	– HVAC ducts, seals, filters
– Leaking seals and gaskets anywhere	– Waste collection and treatment system

13.4.4.1 Fermentation Plant

Fermenters usually contain large amounts of potentially hazardous viable material. Frequently, in addition to containing the viable agent used in production, entry of any other viable material into the fermenter must be prevented. Leakages from fermenters are not uncommon and considerations relating to containment and treatment of spills have already been discussed (Sections 13.4.2 and 13.4.3). In addition, all through operation, most fermenters need to be aerated, and, the fermenter exhaust gases must also be sterilized. Usually, two 0.2 μm rated absolute hydrophobic filters in series are used to filter sterilize the exhaust air. Those filters are often heated to above dew point to prevent condensation, and, on larger or highly aerated vessels, filters are preceded by condensers [40]. In addition, cyclonic separators installed before the filters may be used to protect the filter against contamination by foam and spray. Mechanical foam breakers may also be used [56]. In some cases, one

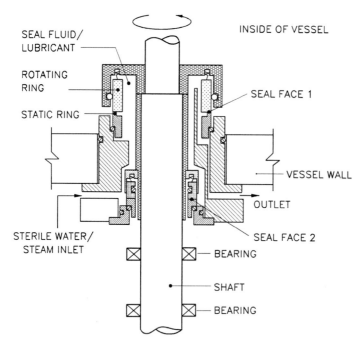

Fig. 13-4. Double mechanical seal with water lubrication. A mechanical seal consists of two seal rings made of materials such as silicon carbide, tungsten carbide, or carbon. The flat faces of the rings press against each other to form the seal. The small gap between the faces is lubricated by a film of liquid (either culture broth or the sterile sealing fluid) or a film of gas (in dry-running seals). One of the seal rings forms a static seal with the shaft and rotates with the shaft. The other ring makes a static seal with the vessel or the stationary seal housing, and remains stationary. Two such seals in close proximity on the same shaft constitute a double mechanical seal. A film of fluid between the running and the static faces is essential to the sealing action, and any damage (e.g., scratching) to the seal faces would produce leakage. Seals must be periodically replaced as a part of the preventive maintenance program.

exhaust gas filter may be followed by an incinerator. Integrity of the exhaust filters should be checked in-place using the forward flow diffusion method after the filter is sterilized, but before use.

Fermenters are protected against overpressurization with a rupture disc [57] that should be piped to a HEPA-filter-vented containment/treatment system. Sometimes, the rupture disc is followed by a pressure relief valve so that the vessel returns to a contained state once the pressure is released. This arrangement should have an attained pressure indicator for detecting disc rupture. In addition, use of overpressure sensors to shut off the sparger air supply in case of overpressurization has been recommended [13,40], but this practice is unusual: normally, the air supply regulator is set to a pressure significantly lower than the value required to open the rupture disc; thus, in case of a blocked exhaust filter, the vessel would attain a pressure equal to the air supply pressure that is still well within the vessel's capabilities.

Double mechanical seals (Fig. 13-4) with sterile lubricating water between them are used to seal agitator shafts in fermenters requiring high-level containment. The pipework required for sterilizing and lubricating the seal with sterile clean steam condensate is shown in Fig. 13-5. The pressure in the lubricating water chamber

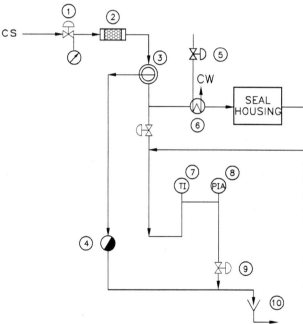

Fig. 13-5. Pipework for sterilization and sterile lubrication of mechanical seals. The inlet and outlet shown on the seal housing correspond to those shown on the seal assembly in Fig. 13-4. For sterilization, the clean steam (CS) regulating valve ① is set to the desired pressure. The valve below the sight glass ③ and valve ⑨ are opened to drain the system. Steam enters the system through strainer ② and the sterilization setpoint

temperature is attained on temperature indicator ⑦. Valve ⑨ is now pulsed to allow drainage of condensate. Steam trap ④ also serves to drain condensate. After predetermined sterilization time, valve ⑨ is closed and the cooling water supply valve ⑤ is opened. Clean steam is condensed and the condensate is cooled to ~ 25 °C in the condenser ⑥. The pipe section above valve ⑨, the seal housing and the sight glass assembly fill up with the cooled sterile water needed to lubricate the seal. Steam pressure on top of the water assures that the seal housing remains under positive pressure with respect to the fermenter. Pressure is monitored at indicator ⑧ that is equipped with a low-pressure alarm.

(Fig. 13-4) is kept higher than that in the fermenter, hence any leakage is into the vessel [40,57]. Low-pressure alarms are recommended for the sealing fluid chamber. Miller and Bergmann [33] recommend using a 'collection tube' in the seal fluid chamber drain for detecting seal failure. Presumably, debris or colored matter would accumulate in the tube if the seal on the culture side failed. A conductivity sensor in the water chamber should provide a better method of detecting leakage of culture fluid into the chamber. Most fermentation media are relatively conductive, whereas uncontaminated sterile water produced by condensing clean steam is a poor conductor of electricity. Rotating seals can be eliminated altogether by using magnetically coupled agitators, but torque consideration limit the vessel size to about 800 L with animal cell culture bioreactors [57], and only to about 80 L with microbial fermenters. Large airlift bioreactors that do not require mechanical agitation are particularly suited to containment [16,58]. In addition, airlift reactors are more reliable than the mechanically stirred ones.

Added protection against microleaks of viable agents can be provided by specifying double O-ring seals on all fermenter entry ports. *In situ* sterilizable probes that can be removed from a fermenter during cultivation, and decontaminated in-place before exposing the surroundings are available. The fermenter flanges may have double O-ring gaskets with a zone of live steam in between. Similarly, a live steam bar-

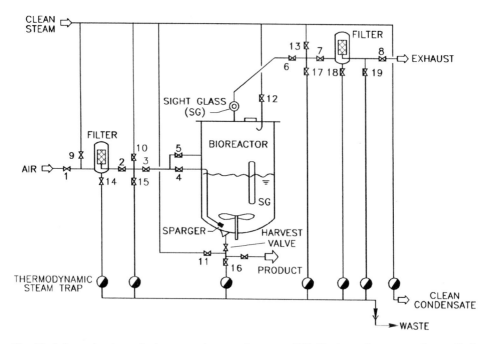

Fig. 13-6. Steam barrier at the harvest valve on a fermenter [52]. During culture, steam is supplied to the outlet side of the closed harvest valve through valve 11. The condensate drains through valve 16 and the steam trap to the biokill system. Inlet and exhaust air is sterilized through submicron absolute filters. A double mechanical seal at the point of entry of the agitator shaft (Fig. 13-4) assures leak-free operation.

rier can be maintained on the outside of valves that connect directly to the fermenter (Fig. 13-6) [52]. All contaminated or potentially contaminated condensate should drain to the biokill system.

Closed systems should be sampled such that exposed surfaces are not contaminated, and no aerosols are released [40]. This requirement must be met at the BL2-LS and higher. Needle-and-syringe sampling through rubber septa does not meet containment requirements. Moreover, hypodermic needles have been frequently implicated in accidental inoculation of handling personnel. Contained sampling that releases no viable aerosols is detailed in Fig. 13-7 [57]. In contrast, an uncontained sampling system is depicted in Fig. 13-8. The contained sampling principles detailed in Fig. 13-7 can be incorporated in automated sampling devices one of which, available from Bioengineering AG, is shown in Fig. 13-9. The sample container should be opened within a suitable biosafety cabinet.

The acid and alkali reservoirs in BL3-LS fermentation plants should be stainless steel pressure vessels that are hardpiped to the fermenter [40]. For lesser level containment, glass reservoirs connected using silicone rubber or other similar tubing and peristaltic pumps are acceptable for pH control. Hardpiping, as opposed to using rapid connection couplings, is preferred for minimizing aerosol generation when high-level containment is necessary [13]. Culture transfer between fermenters should

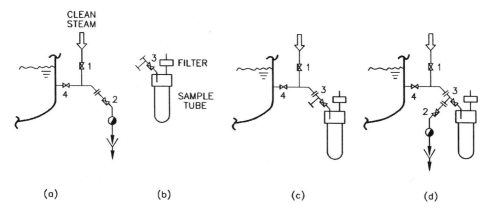

Fig. 13-7. Sterile, contained, aerosol-free sampling [52]. (a) Prior to sampling, the fermenter sampling valve 4 is closed and a clean steam barrier is maintained on the outside of the valve (steam supply valve 1) and condensate is removed (valve 2) to the contained drain. For sampling, valves 2 and 1 are closed, and the system is allowed to cool. The steam trap and valve 2 assembly is disconnected at the sanitary quick coupling. A presterilized (autoclave) sample container (b) having a 0.2 µm breathing filter is attached to the fermenter as in (c). Valve 3 of the sampling device remains closed. The steam trap assembly is reconnected as shown in (d). Valves 1 and 2 are opened in sequence to sterilize the connection using steam at 121 °C for 25 minutes. Valves 2 and 1 are now closed in sequence. After the assembly has cooled, the sample is withdrawn by opening valves 3 and 4. After the sample has been collected, valves 4 and 3 are closed, and the connection is resterilized by opening valves 1 and 2. Upon cooling, the connection is returned to state (a), and the contained sample container is opened in a biological safety cabinet. The system shown is suitable for biohazard containment as well as sampling of bioactive substances that are inactivated by thermal sterilization.

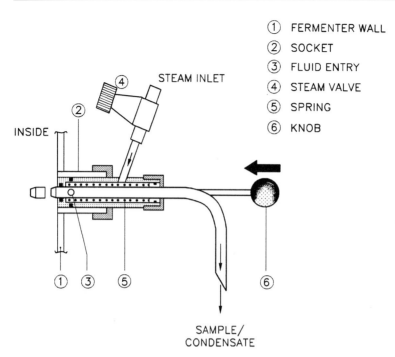

①	FERMENTER WALL
②	SOCKET
③	FLUID ENTRY
④	STEAM VALVE
⑤	SPRING
⑥	KNOB

Fig. 13-8. Assembly for uncontained sampling (shown closed). The sampling assembly is mounted in port ② on the wall ① of the fermenter. For sterilization, steam supplied through valve ④ surrounds and enters the sample pipe at ③; condensate issues from the sample outlet. Once the assembly has sterilized, the steam supply is shut off and the system is allowed to cool. The sample is withdrawn by pushing the knob ⑥ to move the fluid inlet ③ into the fermenter. Releasing the knob causes the spring ⑤ to push the valve into closed position. (Diagram courtesy of Bioengineering AG.)

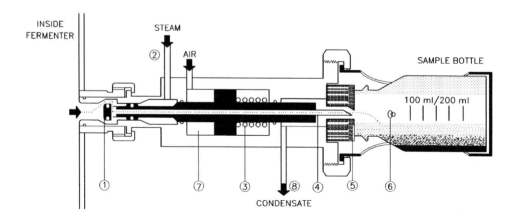

be through permanent hardpiped transfer lines in BL3-LS facilities [40]. Pertinent transfer practices have been described by Chisti [52]. A transfer system meeting the BL2-LS criteria is shown in Fig. 13-10 [52]. Although the needle-and-diaphragm type of connections are still frequently used, particularly during inoculation of fermenters, latest designs of safety connection devices (Fig. 13-11) have virtually eliminated the need for needle-type connectors.

Fully contained processing within closed equipment is feasible and has indeed been implemented, for example, in production of recombinant human interferon using *E. coli* K-12 [59]. Facilities manufacturing vaccines such as mumps, measles,

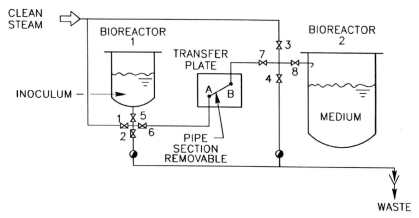

Fig. 13-10. A transfer system meeting the BL2-LS standards [52]. For transferring inoculum from bioreactor 1 to bioreactor 2, a pipe section is connected (sanitary couplings) between points A and B on the transfer plate. The entire transfer pipe between the two fermenters is now sterilized by supplying steam through valves 1 and 3 while valves 6 and 7 are open. The condensate drains through valves 2 and 4. Valves 5 and 8 remain closed. Once the system has sterilized and cooled, valves 5 and 8 are opened, and contents of fermenter 1 are transferred to fermenter 2 by pressurizing (sterile air) vessel 1 relative to vessel 2. Upon completion of the transfer, valves 5 and 8 are closed. The entire transfer line is steam sterilized and cooled prior to removing pipe section A–B. As noted by Chisti [52], the correct sequencing of the various valves is important during sterilization and transfer. All contaminated condensate is piped to the biokill system.

Fig. 13-9. Contained sampling. Normally, valve ① is closed by the action of spring ③ and the needle ④ is retracted into the housing. For sampling, a closed sampling bottle with a rubber diaphragm cap ⑤ and a breathing filter ⑥ is attached to the sampling device. Steam is now run through ② to sterilize the sample path, including the outside of the rubber diaphragm on the sampling bottle. The condensate is withdrawn at ⑧. Once sterilization is complete and the assembly has cooled, valve ① is opened by pneumatic action of compressed air supplied to chamber ⑦. The spring is compressed, the needle moves through the rubber diaphragm into the bottle, and sample flows in. Releasing the air pressure in chamber ⑦ closes the valve by the action of spring ③. The assembly must be re-sterilized and cooled before the sample bottle is removed. The entire sampling operation, including installation and removal of bottles can be automated, and sampling can be done at pre-programmed intervals. (Diagram courtesy of Bioengineering AG.)

CLOSED   OPEN

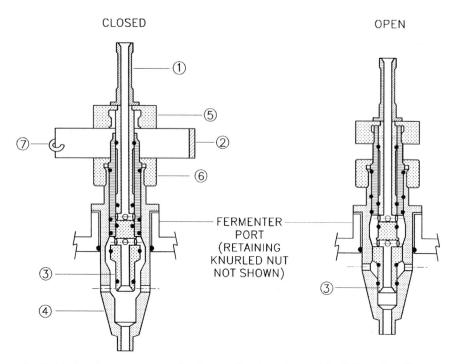

Fig. 13-11. A safety connection valve for coupling inoculum, acid, alkali, and antifoam reservoirs to the fermenter. The nipple ① on the closed valve assembly is connected to transfer tubing attached to an empty reservoir equipped with a breathing filter. The entire set-up (safety connection valve, transfer tube, and reservoir with filter) is autoclaved and cooled. The reservoir is filled with culture fluid inside a biological safety cabinet. The valve body ④ is now installed in a port on top of the fermenter. The fermenter is sterilized after installation of all reservoirs (e.g., inoculum bottle, acid and alkali containers), and cooled. For inoculation, the protective clamp ② is removed and the valve ③ is pushed in to allow pumping of the fluid into the fermenter. The system shown is satisfactory for GLSP processing, but not for higher-level containment.

rubella, varicella, or hepatitis need particular attention to containment [41]. Fermenter design practices noted by Chisti [52,57] are generally sufficient for GLSP and BL1-LS operations, and can be easily extended to BL3-LS level (e.g., use of double mechanical seals; two exhaust filters in series). More general design issues relating to equipment for submerged culture [16,52,57] and solid-state culture [3] have been addressed elsewhere.

13.4.4.2 Downstream Processing

During processing, the viable agent should be removed or inactivated as early as feasible in the process sequence. This usually means inactivation upon completion of fermentation, prior to downstream processing. Inactivation of *E. coli* K-12 using sulfuric acid in commercial processing has been noted [59]. Alternatively, cells may be removed by microfiltration or ultrafiltration; however, operations such as centrifugation and macroporous filtration (e.g., filter presses) do not remove all viable particles. Once the viable agent has been deactivated or removed, and the bioproduct poses no special risk, subsequent processing may proceed in open systems [18] with due regard to the GMP-dictated requirements for protecting the product. When inactivation or removal are not feasible (e.g., live vaccines), downstream processing must be contained. Containment of some downstream processing machinery – particularly centrifuges and cell-disruption devices – has been discussed by Deans and Stewart [60], and other commercially relevant mechanical cell-disruption equipment has been described [16,55].

In addition to the viable agent, the physiological and toxicity profiles of products and by-products constrain the choice of process equipment. Certain process schemes may be ruled out by the extent of containment needs and the amount and types of waste streams that would need to handled. In one case, a difficult-to-contain rotary drum filter that also generated difficult-to-dispose filter aid-mixed solids was replaced with fully contained ultrafiltration for recovery of cefoxitin [61].

Sometimes, primary containment within the process equipment may not be feasible, or contaminated machinery may have to be dismantled – for example, during harvesting of solids from tubular bowl batch centrifuges – and processing within enclosures would be necessary for primary containment.

Isolation of aerosol-generating process equipment in HEPA-filter-exhausted, negative-pressure enclosures is a suitable means of containment [13,41]. The containment room doors should be interlocked with the equipment so that the doors can be opened only when the equipment has stopped running, and sufficient time has elapsed for several space volumes of air to be exhausted from the contained area. Ideally, the equipment itself should be designed for primary containment, and isolation within enclosures should be an added safety measure to contain accidental release from high-pressure devices such as centrifuges and cell disruptors. Performance of the primary and the secondary containment should be validated and monitored regularly.

Multiproduct facilities may require product-dedicated process equipment to eliminate the likelihood of cross-contamination. This is especially so for equipment that cannot be reasonably freed of all traces of a product. For example, chromatography media and membrane filters may have to be dedicated to specific products. The use status – clean, in use, dirty, washed, sterile, etc. – of equipment should be clearly identified at all times. Downstream purification and formulation areas should process only one product at a time. A label control program must be in place.

Handling of a concentrated, bioactive product can be the most dangerous part of bioprocessing [11]. Fine powders are easily aerosolized; hence, handling of freeze-dried cultures or toxins can be particularly hazardous [62]. Because of their aerosol-generating potential, spray-dryers and freeze-dryers require special attention to containment [11]. Freeze-drying of biohazardous substances has been discussed by Adams [62]. As a general rule, automation and fully enclosed mechanized operation can significantly enhance containment while reducing the need for human–process contact.

13.4.4.3 Other Systems

Heat exchange devices are frequently encountered in bioprocessing plants. Corroded heat exchangers (vessel jackets, plate heat exchangers, condensers, shell-and-tube exchangers, etc.) or those with leaking gaskets can contaminate the cooling fluid with the viable agent. Heat exchange equipment should be selected with regard to containment requirements, and cooling water may have to be monitored for contamination.

Clean-in-place (CIP) systems are another common feature of bioprocessing facilities. Automated CIP systems reduce exposure of personnel to hazardous material and assure consistent cleaning. The CIP system design has been treated by Chisti and Moo-Young [56]. For contained facilities, fully closed CIP systems should be used. Attention should be given to cleanability and sterilizability of the CIP system itself, and generation of aerosols should be prevented. In view of the containment and cross-contamination considerations, certain process areas may require dedicated CIP systems.

13.4.5 Personnel Protective Equipment

Certain process operations are diffcult to contain, and even the best designed primary containment can fail. Therefore, use of personnel protective clothing appropriate to risk is essential. Lab coats over street clothing are satisfactory for BL2-LS [33] and lower-rated containment areas. Correct gowning room practice is essential to preventing spread of contamination. Protective clothing – gowns, shoe coverings, head covering, face masks, and gloves – should be removed in the proper fashion before leaving the work area [41].

BL3-LS operations, for example, in hepatitis vaccine production, require that personnel remove all street garments down to underwear and don 'bunny suits', face masks, and gloves, prior to entering the work area [41]. Protective clothing that completely isolates an individual from the environment is sometimes used [13]. Depending on the characteristics of the product, such protective suites may be required even for BL2-LS or lower-rated processes. This type of isolation is provided by positive-pressure suits with battery-operated, forced air supply drawn from the surrounding environment through HEPA filters. Head-and-shoulder half suites with HEPA filtered air supply can also be used over other disposable protective clothing. The suits are equipped with low-battery alarms.

Use of showers is necessary prior to leaving the degowning area of a BL3-LS biohazard containment zone [32]. Lower level containment areas should be equipped with hand washing facilities near the exit from the area. Sinks with automatic or hand- or foot-operated faucets (taps) are used. Suitable germicidal soap should be provided.

Awareness of the biohazard is important to risk reduction. Consequently, BL2-LS and higher containment areas must display the universal biohazard sign (Fig. 13-12) on the entrance to the contained area. Additional information should include the containment level, the specific biohazard agent, any special entry requirements, and emergency contact details of responsible personnel.

Fig. 13-12. Universal biohazard sign.

13.4.6 Personnel Training

Personnel training is an essential part of safe bioprocessing. Even the best designed facilities, equipment, and practices will fail to provide the intended protection if the operators do not have the knowledge, the training, and the right attitude to personal safety, that of colleagues, the product, and the community. Training should be provided in specific processing methods, operation of equipment, use of personnel protective equipment, gowning practices, aseptic and good microbiological technique (see Table 13-5), containment and biosafety measures consistent with the hazard, emergency procedures, and authorized practices. Written operational protocols should identify the specific, actual or potential, hazards. Established practices should be strictly followed [32]. Personnel should be supervised to assure consistent use of prescribed practices [32,40]. Training in Good Manufacturing Practices is also required, and should be a continual process.

13.4.7. Medical Surveillance

Routine medical surveillance appropriate to risk is recommended. For example, electrocardiograms monitoring of individuals working with cardiotoxic substances [13], and seroconversion of individuals handling antigens. Although medical surveillance by itself does not protect against exposure, surveillance is useful in early detection and treatment. Surveillance also helps in identifying procedural or mechanical lapses. Medical surveillance is especially necessary when pathogenic, potentially pathogenic, or new micro-organisms are being investigated, or when the nature of the hazard is unknown. In addition to the viable agent, the bioactivity of the product or any contaminants may pose a health hazard. Thus, for example, a nonpathogenic recombinant species may pose no risk beyond that associated with the corresponding wild strain, but the product of the inserted foreign gene may be highly bioactive, allergenic, or otherwise toxic.

When suitable vaccines are available, the process personnel as well as those providing support services but not directly working in the process areas (e.g., management, administration), should be immunized [41]. A 'wait time' is necessary for development of immunity. Process workers should show resistance to the infective agent before being allowed into the work area [41]. Furthermore, it should be recognized that vaccination may not guarantee protection against a high dose of the etiologic agent [8].

The workplace should be receptive to reporting of illnesses and accidents. Even apparently minor incidents – for example, being scratched while cleaning a process vessel – should be reported and recorded. Personnel who are ill, and those with open sores and cuts, should not be allowed into critical work areas. Immunocompromised individuals, those with diseases such as cancer and diabetes, those undergoing antimicrobial, steroid or immunosuppressive therapy [8] are especially at risk.

Attention to protection of peripheral support staff, for example the maintenance and cleaning personnel, is especially important because they may not have the knowledge or training for the potential risks [3].

13.4.8 Biowaste

All contaminated liquid, solid, and gaseous waste from a facility must be decontaminated prior to release. Solid waste is generally autoclaved or incinerated. Gases are filter sterilized and/or incinerated. Liquid effluent is collected into a containment sump and treated through the biokill system. All decontamination procedures should be validated, and the treated material should be examined for sufficiency of kill before being released. The regulations (OSHA, NIH/CDC, EPA, U.S. Postal Service, etc.) relating to disposal and shipping of biohazardous wastes in the United States have been discussed by Turnberg [63].

Effluent from the containment sump may be chemically disinfected or heat sterilized. Sterilization may be batchwise or continuous using direct steam injection or indirect heating. Good mixing of chemical additives and uniform sterilization temperature must be achieved for defined periods. Waste decontamination areas are generally held at negative pressure which is mandatory when decontaminating BL3-LS effluent. There should be provisions for preventing accidental release of untreated material to sewer. For example, a locked effluent drain valve may be employed; the valve is opened only when a treated batch is released. Use of chemical disinfection prior to thermal sterilization is recommended to reduce the hazard in case of inadvertent effluent release (e.g., rupture disc failure). The decontamination process may be automated.

The sump and the sterilization tanks should be vented only through HEPA filters that may have to be heated to prevent condensation. The filtered exhaust from the waste decontamination tank may have to be treated for odor control if odor is a nuisance [33]. Odor may be controlled by incineration, scrubbing, or absorption.

Effluent decontamination has been discussed further by Wirt et al. [54]. Other relevant effluent management issues have been examined by Court [64]. Miller and

Bergmann [33] have discussed treatment of BL2-LS effluent. Design features of batch thermal biokill systems have been described by Kossik and Miller [65]. In addition, the hygienic design practices noted by Chisti [52,57] for fermentation plant apply also to biokill machinery.

After decontamination, the biohazardous waste is generally disposed of using the same practices that apply to other nonbiological waste. Sometimes special treatment is necessary to destroy nonviable bioactive or otherwise environmentally harmful substances [63,66,67].

13.5 Concluding Remarks

Few bioprocess engineers have experienced a major facility design and construction project; consequently, there is little awareness of industrially relevant biosafety issues. Knowledge of the hazards posed by viable agents and biological materials is essential to safe practice of biotechnology. Industrial processing of any agent must be preceded by assessment of risk and evaluation of the containment needs. Design of the process, process machinery, buildings and the operational practices must be consistent with the biosafety requirements, GMP considerations and other regulatory demands. Process systems and practices should be validated to assure that the intended capability is attained. All this must be complemented with a trained and knowledgeable workforce. Safe bioprocessing rests on a triad of training, procedures, and engineered design [33]. Inadequacies in any of those aspects could lead to safety failures.

The practice and the technology for safe bioprocessing is continually evolving. Guidelines given in some recent literature are already quite dated [5,40,68,69]; therefore, consultation with experts is advised for information on state-of the-art practices. The major containment concepts discussed here are summarized in Table 13-7.

Abbreviations

BLx	Biosafety level x (x = 1–4)
BLx-LS	Biosafety level x (x = 1–3) large scale
CDC	Centers for Disease Control and Prevention
CFU	Colony forming unit
CHO	Chinese hamster ovary
CIP	Clean-in-place
DNA	Deoxyribonucleic acid
EFB	European Federation of Biotechnology
EPA	Environmental Protection Agency
FDA	Food and Drug Administration
GILSP	Good industrial large-scale practice
GLSP	Good large-scale practice

GMO	Genetically modified organisms
GMP	Good manufacturing practices
GRAS	Generally recognized as safe
HEPA	High efficiency particulate air
HIV	Human immunodeficiency virus
HVAC	Heating, ventilation and air conditioning
MCB	Master cell bank
MWCB	Manufacturer's working cell bank
NIH	National Institutes of Health
OECD	Organization for Economic Cooperation and Development
OSHA	Occupational Safety and Health Administration
PVC	Poly(vinyl chloride)
USDA	United States Department of Agriculture
WHO	World Health Organization

References

[1] Chisti, Y., *Trends Biotechnol*, 1993, *11* (6), 265–266. Industrial bioprocess safety.

[2] Stephenson, J. R., Warner, A., *J Chem Technol Biotechnol*, 1996, *65*, 5–14. Release of genetically modified micro-organisms into the environment.

[3] Chisti, Y., in: *Encyclopedia of Bioprocess Technology*: Flickinger, M. C., Drew, S. W. (Eds.), New York: John Wiley, 1999; in press. Solid substrate fermentations, enzyme production, food enrichment.

[4] Angold, R., Beech, G., Taggart, J., *Food Biotechnology,* Cambridge: Cambridge University Press, 1989.

[5] Kearns, M. J., *Pharm Eng,* 1989, *9* (4), 17–21. Containment of biological hazards: Effect of guidelines on the design of pharmaceutical facilities and process equipment.

[6] Winkler, K. C., Parke, J. A. C., in: *Safety in Industrial Microbiology and Biotechnology*: Collins, C. H., Beale, A. J. (Eds.), Oxford: Butterworth-Heinemann, 1992; pp. 34–74. Assessment of risk.

[7] Hacker, J., Ott, M., in: *Safety in Industrial Microbiology and Biotechnology*: Collins, C. H., Beale, A. J. (Eds.), Oxford: Butterworth-Heinemann, 1992; pp. 75–92. Pathogenicity testing.

[8] Liberman, D. F., *Developments in Industrial Microbiology*, 1984, *25*, 69–75. Biosafety in biotechnology: A risk assessment overview.

[9] Lelieveld, H. L. M. et al., *Appl Microbiol Biotechnol*, 1995, *43*, 389–393. Safe biotechnology. Part 6. Safety assessment, in respect of human health, of microorganisms used in biotechnology.

[10] Collins, C. H., *Laboratory-Acquired Infections*, 3rd edition, Oxford: Butterworth-Heinemann, 1993.

[11] Bennett, A. M., in: *Biosafety in Industrial Biotechnology*: Hambleton, P., Melling, J., Salusbury, T. T., (Eds.), London: Chapman Hall, 1994; pp. 109–128. Health hazards in biotechnology.

[12] Chisti, Y., Strategies in downstream processing. In: *Bioseparation and Bioprocessing: Processing Biomolecules and Cell Cultures*, Vol. II: Subramanian, G., (Ed.), Weinheim: Wiley-VCH, 1998; pp. 3–30.

[13] Flickinger, M. C., Sansone, E. B., *Biotechnol Bioeng*, 1984, *26*, 860–870. Pilot- and production-scale containment of cytotoxic and oncogenic fermentation processes.

[14] Richardson, J. H., Barkley, W. E., (Eds.), *Biosafety in Microbiological and Biomedical Laboratories*, 2nd edition, U.S. Department of Health and Human Services, Washington: U.S. Government Printing Office, 1988.

[15] Stentiford, E. I., Dodds, C. M., in: *Solid Substrate Cultivation*: Doelle, H. W., Mitchell, D. A., Rolz, C. E. (Eds.), London: Elsevier, 1992; pp. 211–246. Composting.

[16] Chisti, Y., Moo-Young, M., in. *Biotechnology: The Science and the Business*: Moses, V., Cape, R. E., (Eds.), New York: Harwood Academic Publishers, 1991; pp. 167–209. Fermentation technology, bioprocessing, scale-up and manufacture.

[17] Doran, P. M., *Bioprocess Engineering Principles*, London: Academic Press, 1995.

[18] Vranch, S. P., in: *Bioprocessing Safety: Worker and Community Safety and Health Considerations*: Hyer, Jr., W. C., (Ed.), Philadelphia: American Society for Testing and Materials, 1990; pp. 39–57. Containment and regulations for safe biotechnology.

[19] Hambleton, P., Melling, J., Salusbury, T. T. (Eds.), Biosafety in *Industrial Biotechnology*, London: Chapman Hall, 1994.

[20] Collins, C. H., Beale, A. J., (Eds.), *Safety in Industrial Microbiology and Biotechnology*, Oxford: Butterworth-Heinemann, 1992.

[21] Jones, R. A., Matheson, J. C., *J. Ind. Microbiol.* 1993, *11*, 217–222. Relationship between safety data and biocontainment design in the environmental assessment of fermentation organisms – An FDA perspective.

[22] Özcan, S., Firek, S., Draper, J., *Trends Biotechnol*, 1993, *11* (6), 219. Can elimination of the protein products of selectable marker genes in transgenic plants allay public anxieties?

[23] Kane, J. F., *J Ind Microbiol*, 1993, *11*, 205–208. Environmental assessment of recombinant DNA fermentations.

[24] Frommer, W. et al., *Appl Microbiol Biotechnol,* 1993, *39*, 141–147. Safe biotechnology (5). Recommendations for safe work with animal and human cell cultures concerning potential human pathogens.

[25] Lubiniecki, A. S., (Ed.), *Large-Scale Mammalian Cell Culture Technology*, New York: Marcel Dekker, 1990.

[26] Roberts, P., *J Chem Technol Biotechnol,* 1994, *59*, 110–111. Virus safety in bioproducts.

[27] Rouf, S. A., Moo-Young, M., Chisti, Y., *Biotechnol Adv,* 1996, *14*, 239–266. Tissue-type plasminogen activator: Characteristics, applications and production technology.

[28] Anicetti, V. R., Keyt, B. A., Hancock, W. S., *Trends Biotechnol*, 1989, *7*(12), 342–349. Purity analysis of protein pharmaceuticals produced by recombinant DNA technology.

[29] Garg, V. K., Costello, M. A. C., Czuba, B. A., in: *Purification and Analysis of Recombinant Proteins*: Seetharam, S., Sharma, S. K., (Eds.), New York: Marcel Dekker, 1991; pp. 29–54. Purification and production of therapeutic grade proteins.

[30] Health and Welfare Canada and Medical Research Council of Canada, *Laboratory Biosafety Guidelines*, Ottawa: Minister of Supply and Services Canada, 1990.

[31] Van Houten, J., Fleming, D. O., *J Ind Microbiol*, 1993, *11*, 209–215. Comparative analysis of current US and EC biosafety regulations and their impact on the industry.

[32] Hill, D., Beatrice, M., *Pharm Eng*, 1989, *9* (4), 35–41. Facility requirements for biotech plants.

[33] Miller, S. R., Bergmann, D., *J Ind Microbiol*, 1993, *11*, 223–234. Biocontainment design considerations for biopharmaceutical facilities.

[34] Perkowski, C. A., in: *Bioprocess Engineering: Systems, Equipment and Facilities:* Lydersen, B. K., D'Elia, N. A., Nelson, K. L., (Eds.), New York: John Wiley, 1994; pp. 729–743. Containment regulations affecting the design and operation of biopharmaceutical facilities.

[35] Frommer, W. et al., *Appl Microbiol Biotechnol,* 1989, *30*, 541–552. Safe biotechnology III. Safety precautions for handling microorganisms of different risk classes.

[36] Küenzi, M. et al., *Appl Microbiol Biotechnol,* 1985, *21*, 1–6. Safe biotechnology: General considerations.

[37] Küenzi, M. et al., *Appl Microbiol Biotechnol,* 1987, *27,* 405. Safe biotechnology 2. The classification of microorganisms causing diseases in plants.

[38] Frommer, W. et al., *Appl Microbiol Biotechnol,* 1992, *38,* 139–140. Safe biotechnology (4). Recommendations for safety levels for biotechnological operations with microorganisms that cause diseases in plants.

[39] Lelieveld, H. L. M. et al., *Appl Microbiol Biotechnol,* 1996, *45,* 723–729. Safe biotechnology. 7. Classification of microorganisms on the basis of hazard.

[40] East, D., Stinnett, T., Thoma, R. W., *Developments in Industrial Microbiology* 1984, *25,* 89–105. Reduction of biological risk in fermentation processes by physical containment.

[41] Damm, P. G., in: *Bioprocessing Safety: Worker and Community Safety and Health Considerations:* Hyer, W. C., Jr., (Ed.), Philadelphia: American Society for Testing and Materials, 1990; pp. 58–64. The biological production facility – Design for protection of the worker and the community.

[42] Liberman, D. F., Ducatman, A. M., Fink, R., in: *Bioprocessing Safety: Worker and Community Safety and Health Considerations:* Hyer, W. C., Jr., (Ed.), Philadelphia: American Society for Testing and Materials, 1990; pp. 101–110. Biotechnology: Is there a role for medical surveillance?

[43] Richmond, J. Y., McKinney, R. W., (Eds.), *Primary Containment for Biohazards: Selection, Installation and Use of Biological Safety Cabinets,* U.S. Department of Health and Human Services, Washington: U.S. Government Printing Office, 1995.

[44] First, M. W., *Developments in Industrial Microbiology,* 1984, *25,* 77–87. Ventilation for hazard control.

[45] Van Houten, J., in: *Bioprocessing Safety: Worker and Community Safety and Health Considerations:* Hyer, W. C., Jr., (Ed.), Philadelphia: American Society for Testing and Materials, 1990; pp. 91–100. Safe and effective spill control within biotechnology plants.

[46] Willig, S. H., Stoker, J. R., *Good Manufacturing Practices for Pharmaceuticals: A Plan for Total Quality Control,* 3rd edition, New York: Marcel Dekker, 1992.

[47] Rao, A. K., Adey, H., *Chemtech,* 1989, *October,* 632–637. Designing a bioprocessing facility.

[48] del Valle, M. A., *BioPharm,* 1989, *April,* 26–42. HVAC systems for biopharmaceutical manufacturing plants.

[49] Lee, J. Y., *BioPharm,* 1989, *February,* 42–45. Environmental requirements for clean rooms.

[50] Dobie, D., in: *Bioprocess Engineering: Systems, Equipment and Facilities:* Lydersen, B. K., D'Elia, N. A., Nelson, K. L., (Eds.), New York: John Wiley, 1994; pp. 641–668. Heating, ventilating, and air conditioning (HVAC).

[51] Johnson, H. L., Stutzman, D.A., in: *Bioprocess Engineering: Systems, Equipment and Facilities:* Lydersen, B. K., D'Elia, N. A., Nelson, K. L., (Eds.), New York: John Wiley, 1994; pp. 671–708. Programming and facility design.

[52] Chisti, Y., *Chem Eng Prog,* 1992, *88* (9), 80–85. Assure bioreactor sterility.

[53] Donnelly, R. W., *Pharm Eng,* 1989, *9* (3), 9–12. Design and construction review of one of the first large scale mammalian cell culture facilities.

[54] Wirt, G. D., Orichowskyj, S. T., Wu, J. J., *Chem Eng Prog,* 1991, *87* (1), 49–53. Decontaminate biotech wastes effectively.

[55] Chisti, Y., Moo-Young, M., *Enzyme Microb Technol,* 1986, *8,* 194–204. Disruption of microbial cells for intracellular products.

[56] Chisti, Y., Moo-Young, M., *J Ind Microbiol,* 1994, *13,* 201–207. Clean-in-place systems for industrial bioreactors: Design, validation and operation.

[57] Chisti, Y., *Chem Eng Prog,* 1992, *88* (1), 55–58. Build better industrial bioreactors.

[58] Chisti, Y., *Airlift Bioreactors,* London: Elsevier, 1989.

[59] Weibel, E. K., *Chimia,*1994, *48* (10), 457–459. GMP and biosafety aspects in the production of recombinant IFN alpha-2a.

[60] Deans, J. S., Stewart, I. W., in: *Biosafety in Industrial Biotechnology:* Hambleton, P., Melling, J., Salusbury, T. T., (Eds.), London: Chapman Hall, 1994; pp. 149–177. Containment in downstream processing.

[61] Paul, E. L., in: *Bioprocessing Safety: Worker and Community Safety and Health Considerations:* Hyer, W. C., Jr., (Ed.), Philadelphia: American Society for Testing and Materials, 1990; pp. 65–73. Design criteria for safety in the isolation and purification of antibiotics and biologically active compounds.

[62] Adams, G. D. J., in: *Biosafety in Industrial Biotechnology:* Hambleton, P., Melling, J., Salusbury, T. T., (Eds.), London: Chapman Hall, 1994; pp. 178–212. Freeze-drying of biohazardous products.

[63] Turnberg, W. L., *Biohazardous Waste: Risk Assessment, Policy, and Management,* New York: John Wiley, 1996.

[64] Court, J. R., in: *Biosafety in Industrial Biotechnology:* Hambleton, P., Melling, J., Salusbury, T. T., (Eds.), London: Chapman Hall, 1994; pp. 240–267. Managing the effluent from bio-industrial processes.

[65] Kossik, J. M., Miller, G., *Chem Eng Prog,* 1994, *90* (10), 45–51. Optimize cycle times for batch biokill systems.

[66] Watt, J. C., Wroniewicz, V. S., Ioli, D. F., in: *Environmental Biotechnology:* Omenn, G. S., (Ed.), New York: Plenum Press, 1988; pp. 307–322. Environmental concerns associated with the design of genetic engineering facilities.

[67] Freeman, H. M., (Ed.), *Standard Handbook of Hazardous Waste Treatment and Disposal,* New York: McGraw-Hill, 1989.

[68] Werner, R. G., in: *Safety in Industrial Microbiology and Biotechnology:* Collins, C. H., Beale, A. J., (Eds.), Oxford: Butterworth-Heinemann, 1992; pp. 190–213. Containment in the development and manufacture of recombinant DNA-derived products.

[69] Tuijnenburg Muijs, G., in: *Safety in Industrial Microbiology and Biotechnology*: Collins, C. H., Beale, A. J., (Eds.), Oxford: Butterworth-Heinemann, 1992; pp. 214–238. Monitoring and validation in biotechnological processes.

14 Process Hygiene in Production Chromatography and Bioseparation

Glenwyn D. Kemp

14.1 Introduction

The very nature of industrial bio-chromatography entails the potential for contamination of the equipment with biologically active materials. These materials must not be allowed to reach the final product of the process.

An integral part of any chromatographic step in the production of biologically active materials is the removal of both contaminant and residual material from the chromatography system after each step. The following chapter gives an overview of potential sources of contamination and reviews methods used to remove such contaminants.

The maintenance of cleanliness within a production chromatography plant is considered from the perspective of designing the equipment to both minimize initial infection and to allow the removal of any contamination which may occur. Consideration is also given to methods of sanitization which are applicable to process chromatography.

14.2 General Principles

The preferred method of maintaining hygiene in process-scale chromatography systems is by cleaning-in-place (CIP), whereby the entire contact path is cleaned *in situ*. This is due in large part to the time and effort required to unpack and re-pack very large columns. There is also the additional risk of degeneration of the chromatography media by repeated handling and potential exposure of the operators to bioactive compounds during the pack/unpack cycle. For this reason once a column has been packed and proved to have an acceptable performance it is more economical to clean the column *in situ* for as long as possible, i.e., until the performance is no longer acceptable (due, for example, to loss of capacity or disruption of the gel bed).

In order to get the greatest benefit from effective CIP it is essential to take steps to eliminate infection or contamination of the equipment in the first place. This is best achieved by rigorous cleanliness in handling the chromatography buffers and samples and by strict adherence to current good manufacturing practices (cGMP). Although this adds time and costs to the overall process, the potential for lost time and product due to contamination makes any extra effort in pre-chromatography preparation a worthwhile investment.

It should also be noted that giving adequate consideration to hygiene and CIP from the earliest possible stages of both process and system design will reap great rewards when the final production process is running. One of the best investments which can be made for process reliability is a well-designed chromatograph.

It is frequently the case that the very nature of the feedstock being applied to the column will result in contamination. For example, bacterial cell lysates produced as an initial step in the recovery of recombinant protein will inevitable contain endotoxins and other proteins, DNA and RNA which must be removed from the system after each batch. In such a process there is also a high probability of some unbroken cells being present which can form a potential infection of the system resulting in increasing bio-burden over a period of time. The manufacture of recombinant proteins from the perspective of process validation was the subject of a review by Jungbauer and Boschetti [1].

Some processes require a CIP step regardless of the presence of contaminants. In order to maintain biological activity, many blood products for example are purified chromatographically from unsterilized blood [2]. In accordance with general principles of microbiology, the pooled blood fractions must be assumed to contain pathogenic viruses. Therefore there is a pre-defined requirement to clean the media after each batch, regardless of the actual presence of viral contamination. Because of this it is often the case with very high-value blood products that the gel is simply sterilized and disposed of after each batch. However, the initial sterilization of the gel is usually carried out by an *in situ* step prior to removal of the gel from the column for disposal, to reduce risk to the operators.

In cases such as these, where it is not possible to remove infective contaminants, the equipment being used must be designed for ease of cleaning, with maximum operator safety and the cleaning protocols used must be fully validated. It is important to note that although individual components within a system can be tested and assessed for cleanliness, a full CIP validation can only be made for a complete process under standard operating conditions as a part of the process qualification (PQ).

14.3 Definitions

Sterilization
Sterilization can be defined as 'a method for removing (i.e., killing) all viable organisms'. It should be noted that sterilization may still leave pyrogens within the system, and indeed can be a source of pyrogen increase as the killed organisms break down.

Sanitization
Sanitization can be defined as 'a method of lowering the total content of viable organisms to an acceptable level'. In this case it should be noted that sanitization does not necessarily entail the complete removal of viable organisms, some of

which may remain within the system. The essence here is to keep levels of contamination under control and low enough to represent an acceptable statistical risk. A successful sanitization is usually defined as producing a significant log reduction in the number of viable organisms present. Although sanitization does not necessarily remove pyrogens, it is common for depyrogenation and sanitization to be combined in a single process step.

CIP (Clean-in-place)

CIP can be defined as 'a method of sanitising a system *in situ* without dismantling it'. The aim of CIP is to reduce the amount of viable organisms to an acceptable level while causing the minimum disruption of the system and with the minimum exposure to the operators.

SIP (Sterilize/Steam-in-place)

SIP can be defined as 'a method of sterilising a system *in situ* without dismantling it'. The precise meaning of SIP is still somewhat variable, and can be taken to indicate either steaming in place, or sterilizing in place. It should be noted that steaming in place will usually result in a fully sterile system (provided that the system has been appropriately designed for steaming); however there are alternative methods of sterilization-in-place (such as irradiation) which do not utilize steam.

14.4 Possible Contaminants

Some of the most common contaminants within a system are considered below. Although the list is by no means exhaustive, it will provide an indication of the wide range of potential contaminants present within a bioprocessing environment.

14.4.1 Active Ingredients

By definition, the purification of pharmaceuticals and biologics involves the handling of biologically active materials. Within a given sample there will often be more than one biologically active ingredient. Since only one 'product' is usually being purified, any other active ingredient can be thought of as a contaminant. If any such biologically active materials remain within the system they can be carried over into subsequent batches and cause cross-contamination.

   The extremely high potency of many current and potential therapeutics (such as interferons and peptide hormones) provides an indication of the dramatic effect even trace amounts of biologically active contaminant may have on the safety of the final product. As analytical techniques improve so the limits of detection are increasingly pushed back, placing an ever-increasing onus on the manufacturer to provide purer and purer end products.

14.4.2 Bacteria (Vegetative)

Bacterial infection can be divided into two subgroups, external (adventitious) infections, caused by organisms outside the production process and internal infection by organisms introduces by the production process itself.

14.4.2.1 External Infection

The presence of bacteria in the environment acts as a constant source of potential contamination. All healthy individuals have a microbial flora associated with their skin and mucous membranes. Often, chromatographic operations are not carried out in extreme clean room conditions or in containment areas (although this type of operation is increasingly becoming the norm), and infection from organisms carried by the operators must be avoided. Common commensal organisms include the pathogen *Staphylococcus aureus* and various Pseudomads such as Pseudomonas aeruginosa. These organisms can be introduced into the system during the column packing stage or via buffer tanks.

14.4.2.2 Internal Infection

Bacteria are used extensively in the production of recombinant proteins, the proteins being recovered from bacterial growth broths. Usually there are pre-chromatographic processing steps to harvest and lyse the cells, for intracellular products, or to separate the cells from the growth medium, for extralcluar excretion products. In either case there remains the potential for some viable cells to reach the chromatography stage.

Although the gross presence of cells would inevitably result in the gel bed becoming blocked and manifest as an unacceptable increase in back-pressure, the presence of a few bacterial cells may not cause any obvious problems with the column, but would clearly be enough to infect the system. A further complication is caused by the nature of the sample stream from recombinant products during initial capture steps. In this case, the feedstock is, by definition, a bacterial growth medium and is therefore formulated to encourage the proliferation of any organisms, even if they are in the form of an unwanted infection within the system.

Expanded bed chromatography is a relatively new chromatographic approach to the capture of product from crude growth culture. This method allows the passage of entire growth cultures to be passed through a chromatography resin [3]. Although offering benefits from reduced processing steps, expanded bed chromatography is clearly highly susceptible to contamination by bacteria from the growth medium. Thus, careful consideration must be given to the cleaning and decontamination of expanded bed columns after exposure to crude cell suspensions.

A further source of internal infection can arise from the deliberate infection of the column with bacteria during cleaning validation studies. This will be discussed in more detail below.

14.4.3 Bacteria (Spores)

Spore-forming bacteria pose a particular problem for system sanitation. The spores of bacteria are heavily dehydrated and effectively inert. The are therefore resistant to most methods of sanitation and sterilization. Once again, spores can be introduced accidentally from an external source (the physical nature of spores encourages airborne and aerosol distribution), an internal source (although the conditions for sporulation are normally avoided in the fermenter this is not always possible), or by deliberate infection of the system for validation studies. Clearance of spores forms one of the most rigorous challenges to a CIP protocol.

14.4.4 Fungi/Algae

There is a relatively small, but important, use of fungi in the production of recombinant proteins, and some fungi are used for the production of secondary metabolites as final end products. In general, fungi can be cleaned from a system using similar methods to those used for bacteria. There are, however, the added complications of spore formation and mycelial growth within a system. While *Sacchoromyces* species are only found as single cellular organisms, some other yeasts such as *Candida* species are prone to dimorphism and can switch between single cell and pseudomycelial forms. If this occurs, then the dead organism will not be flushed out of the system after sanitization but will decompose *in situ* potentially giving rise to further problem with decomposition products. Thus, while their larger size make them easier to remove from the process stream before the chromatography step, fungi often prove to be a more formidable problem if infection of the system does occur.

Algae can become a contaminant in poorly maintained process plants. These organisms can proliferate in poorly maintained buffer tanks under conditions of apparently no nutrition. They are also able to live in relatively high pH environments and can thus be particularly difficult to deal with using conventional NaOH solutions. As with fungi, the main principle is to avoid contamination in the first instance.

14.4.5 Viruses

As discussed above, the production of blood products exposes the chromatographic system to viral contamination. The high pathogenicity of the viruses potentially present (hepatitis B/C, HIV) [2] gives extra cause for rigorous cleaning. Chromatographic separation *per se* has been suggested as a virus reduction step [2,4], although this depends upon the nature of the chromatography matrix and separation protocol. It is worth bearing in mind that some degree of viral clearance will accompany most chromatography steps in a manufacturing process.

Due to their parasitic nature, viruses are unable to replicate unless they are within a host cell. For this reason, the presence of a small number of virus particles within a system can be difficult to detect and enumerate. Fortunately, many of the commonly contaminating viruses are relatively fragile outside of the host cell and therefore easy to inactivate. The issue of indirect viral contamination by retention of host cells or microbial vectors should be considered primarily as an exercise in removing the host contaminant.

As with bacteria, viruses may also be used deliberately to infect a column, although this is usually done in a smaller scale-down study due to the cost and hazards of exposing a full-scale production column to potentially pathogenic virus. Viral clearance studies are considered in more detail below.

14.4.6 Endotoxins (Pyrogens)

Endotoxins (or pyrogens) are heat-stable lipopolysaccharides, associated with the outer membranes of Gram-negative bacteria. These molecules initiate a febrile reaction (fever) when injected. The degree of fever varies with different pyrogens and in different patients; in worst cases the reaction may be fatal. Pyrogens are in effect decomposition materials. Elimination of viable bacteria does not necessarily mean the elimination of pyrogens; indeed, the action of bactericides may even produce a novel source of pyrogens as the killed bacteria degrade. The complex nature of pyrogens means they will frequently interact with the chromatographic resin in some way [5] which may hinder their removal.

14.4.7 Cleaning Reagent

The toxic nature of the cleaning reagents themselves is often overlooked as a potential contaminant within the system. It is essential that all traces of CIP solution are removed before the next batch of sample is processed.

14.4.8 Non-specifically Bound Protein

Proteins will bind non-specifically to most surfaces. This is often an area of concern to biochemists, especially when they are dealing with very small amounts of protein. The data in Table 14-1 show that the level of non-specific protein binding is very small for all the materials commonly used in the manufacture of chromatography columns and systems (for example, in a 1000 mm diameter column with a bed height of 250 mm, ~ 240 mg of protein will be bound non-specifically to the glass tube). Given that production systems are usually used in a preparative environment with a large loading of protein in the sample, there is no real problem with the loss of product

Table 14-1. Nonspecific binding of bovine serum albumin to materials commonly used in the manufacture of bioseparations equipment [6].

Material	Protein (mg cm^{-2})
Borosilicate glass	30
Polypropylene	27
TPX (polymethyl pentane)	9
Acetal	30
Stainless steel	< 1[a]
Acrylic (perspex)	12
Santoprene	23
EPDM	32

[a] Below detectable limit.

due to protein binding. However, there is a potential risk of cross-batch contamination from the binding and displacement of proteins from surfaces within the system. A more serious problem can arise from non-specific binding of protein to the chromatography resin due to the intrinsically high surface area available. Thus, this is another area which needs to be addressed by the CIP process. Fortunately, CIP with NaOH solutions is very effective at removing non-specifically bound proteins. Washing the samples referred to in Table 14-1 with 0.1 M NaOH reduced the non-specifically bound protein to less than the level of detection (< 1 mg cm^{-2}) in all cases.

14.5 Design Considerations in System Hygiene

14.5.1 General Materials of Construction

The same materials are utilized almost universally by manufacturers in the construction of columns. The primary consideration in selecting the materials of construction for any system to be used in biotechnology or biopharmaceutical production must be the biocompatibility of the material [7]. However, in order to carry out efficient cleaning, the materials of construction should also be selected to withstand the chemicals used for the process CIP procedures. Since these chemicals are almost always more extreme than those used for the routine separation, the CIP cocktails will often have a strong influence on the materials of construction.

In general, the surface finish is important for all materials; the rougher the surface the greater the area potentially available for contaminants to bind. In extreme cases poor surface finish can provide crevices and 'caves' for contaminants to accumulate. For example, it is important that there are no 'bubbles' within molded component which may be partially exposed during subsequent manufacturing steps.

The relative benefits and shortcomings of the most frequently used materials are discussed below.

14.5.2 Stainless Steel

A variety of grades of stainless steel are available. The grades of stainless steel differ in their content of additive metals (Table 14-2). Although Grade 306 is commonly used for structural elements, such as frames and enclosures, the grade of stainless steel usually used for wetted surfaces is 316L. This has a good compromise between increased corrosion resistance and cost. However, it is important to note that 316L stainless steel is not entirely resistant to salt corrosion. The resistance of stainless steel to corrosion can be significantly decreased by welding, unless appropriate precautions are taken to blanket the metal in an inert gas during welding.

The surface finish of stainless steel is critical to its cleanliness [8]. Finishes are usually quoted as having a certain RA or Grit value (Table 14-3). Typical finishes are illustrated in Fig. 14-1. It can be seen that the smoothest finish is obtained by electropolishing. This entails the electrolytic removal of the surface layer of steel. For the process to be fully effective it requires careful optimization of the polishing solution (electrolyte) used, the current applied and most importantly the design of the electrode 'jig' since the electrode should be positioned near the surface of the component to be electropolished. For small-diameter and complex pipe spools this is often extremely difficult.

The corrosion resistance of stainless steel is due to the presence of an extremely thin oxide layer on its surface. This passive oxide film can only form on clean, uncontaminated surfaces. A typical final treatment for non-electropolished stainless

Table 14-2. Chemical composition (%) of commonly used grades of stainless steel.

Grade	C	Mn	Cr	Ni	Ti	Mo
304	0.08	2.0	18–20	8–10.5	–	–
304L	0.03	2.0	18–20	8–12	–	–
316	0.08	2.0	16–18	10–14	–	2–3
316L	0.03	2.0	16–18	10–14	–	2–3
316Ti	0.08	2.0	16–18	10–14	0.4–0.7	–
317L	0.03	2.0	18–20	11–15	–	3–4
321	0.08	2.0	17–19	9–12	0.4–0.7	–

Table 14-3. Grit and RA (roughness average) equivalents for stainless steel surface finishes.

Grit	Nominal RA (mm)
120	0.8
180	0.4
240	0.3
320	0.23
400	0.15
Bright polish	0.08

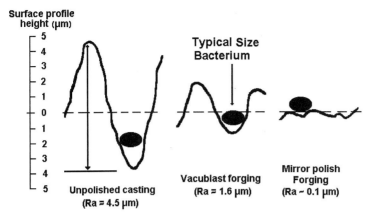

Fig. 14-1. A diagrammatic comparison of the surface profile of stainless steel after it has been subjected to different finishing processes. For scale, a typically sized bacterium is also shown. (After Pfister and Kohler [8].)

steel is passivation or pickling. This entails the exposure of the clean (degreased) stainless steel components to a strongly oxidizing solution which will promote the formation of the protective oxide [9,10]. A typical passivation cocktail would be a solution of 20 % (v/v) nitric acid and 5 % (w/v) potassium dichromate. Electropolishing results in a surface which already has a protective oxide layer, therefore a further chemical passivation step is not required. However, for nonelectropolished components, passivation will increase corrosion resistance and is therefore recommended.

Among the few disadvantages of stainless steel are its weight, making large columns more difficult to handle, and its susceptibility to corrosion by chloride solutions. However, perhaps the greatest disadvantage of stainless steel for the production of chromatography column tubes is the lack of visibility. The ability to observe the gel bed can provide great benefits; void formation, due to bed settling is immediately apparent, as is poor flow distribution if the feedstock is colored.

14.5.3 Borosilicate Glass

Glass is commonly used in laboratory-scale columns. Its advantages are that it is transparent and bio-inert, but its disadvantages are its brittleness and low mechanical strength. While these factors are less important in small columns (where glass can be used successfully within systems operating at relatively high pressures), in production-scale columns the weakness of glass can become a serious disadvantage. Additionally, glass is less suited to a production floor environment where it is more prone to accidental damage. Small scratches or chips in glass tubes will seriously reduce the maximum pressure that the tube is capable of withstanding, and care must be taken when dismantling, cleaning, and assembling glass columns.

14.5.4 Polypropylene

Polypropylene is a cost-effective and easy-to-machine polymer. It is strong and bio-compatible; however, it is not transparent and is only translucent in thin sections. Polypropylene is commonly used to manufacture structural elements within a column or system (end cells, flow cells, etc.) and also for the manufacture of pipework, as it is relatively easy to bend and shape.

14.5.5 Acrylic (Plexiglass)

Acrylic offers the transparency of glass, but with a higher strength. However, it is expensive in large precision-formed sheets or tubes, and has a poor chemical compatibility. The chemical resistance of acrylic to alcohols can often be a problem as ethanolic solutions are used both as part of a CIP procedure and also as a bactericidal storage solution. The relative resistance of acrylic to attack by alcohols depends mainly upon the ethanol concentration, temperature, exposure time, and the manufacturing process of the acrylic tube. In general, column tubes made from cast acrylic show better solvent resistance than tubes formed from annealed rolled sheets.

14.5.6 TPX (Polymethyl Pentane, PMP)

Although TPX is more commonly found in laboratories in the form of beakers and measuring cylinders, the use of this material in chromatography columns has increased significantly in recent years. TPX has good transparency, strength, and chemical resistance, and is biocompatible. Older columns made from TPX had a tendency to turn yellow or cloudy over a period of time as a result of the action of incident UV light. However, modern grades of TPX are now more resistant to UV ageing. The main drawback to TPX is the difficulty in manufacturing large diameter or long column tubes using current moulding techniques.

14.5.7 PVC (Tygon)

Flexible PVC hose is used extensively in food and biopharmaceutical applications for the interconnection of components, especially in coupling columns to chromatographs. PVC hose is available with preformed sanitary (Tri-clover) fittings for extra cleanliness. The flexibility of the hose allows some adjustment of the column without recourse to re-plumbing the supply pipework. However, PVC is not resistant to attack by many organic solvents commonly used in reversed phase chromatography, such as acetonitrile, and may also be weakened by prolonged exposure to high concentrations (> 20 %) of ethanol.

14.5.8 Fluoropolymers

A variety of fluoropolymers are available and are commonly used in pipe work and valve body construction. Although expensive fluoropolymers have outstanding chemical resistance. Careful design and smooth surfaces also help to ensure that fluoropolymer components show very low non-specific binding and are easily cleaned making them suitable for sanitary systems.

14.6 Elastomeric Materials (Seals)

Elastomeric materials are required to seal the columns and pipework. Clearly, this is important both from the point of view of preventing leakage of materials or solutions into the workspace, and to prevent ingress of contaminants into the flow path. Earliest columns and systems commonly used natural rubber seals or their man-made equivalents such as styrenebutadiene (SBR, Buna). Advances in elastomer engineering coupled with more stringent regulatory guidelines have greatly reduced the use of such compounds. The compounds outlined below represent the materials most commonly used for sealing columns.

Each of the polymers listed below has its own benefits and disadvantages. In general, however, it is important to ensure that all sealing materials are biocompatible and nonleaching. Biocompatibility can be determined by implantation (US Pharmacopeia IV) tests [11] and cytotoxicity testing. The USFDA produces a set of guidelines [12] for the composition of elastomeric materials, but it should be borne in mind that these guidelines are in the form of a list of permissible chemical components and the maximum permissible concentrations of those components. Within these guidelines it is possible to have a broad range of ostensibly identical elastomers with widely differing properties.

14.6.1 EPDM (Ethylene Polypropylene)

This is perhaps the most common elastomeric material. It is manufactured from polymerized ethylene propylene with the addition of a controlled amount of plasticizers, stabilizers, and bulking agent (usually carbon black). It is widely available and can be molded (for small components and 'O' rings) or extruded and joined (for large 'O' rings and tubular seals). It is important to be aware that the label EPDM represents a general formulation. Within that specification there is a wide variety of EPDM elastomers available with greatly differing chemical resistance and biocompatibility. It is best to test a sample of the specific elastomer to be used with a variety of process fluids to check for changes in size (usually swelling), hardness, and leachates.

14.6.2 Santoprene (Norprene, Marprene)

Santoprene is similar to EPDM in that it is also based on ethylene polypropylene. However, the bulking agent used in the manufacture of Santoprene is polypropylene rather than carbon black. Santoprene has good chemical resistance and is widely used for peristaltic tubing. Once again, there is a variety of grades and formulations available, and care must be taken to ensure that the grade being used will be acceptable to the regulatory authorities.

14.6.3 Silicone

Silicone is commonly used in flexible hoses, although it is also available as fixed seals. Two varieties of silicone are available; peroxide-cured and the more expensive platinum-cured. Platinum-cured silicone has fewer potential leachables, a smoother surface, and lower levels of protein binding compared with peroxide-cured silicone. The chemical resistance of silicone can be a problem. If it is to have long-term exposure to concentrated acids or in systems stored for long terms in NaOH solutions chemical resistance tests should first, be carried out on samples of the polymer.

14.6.4 Elastomeric PTFE

Elastomeric forms of PTFE are commercially available which have extremely high resistance to solvents and are totally bio-inert. Although used for small seals and components, the very high cost of these compounds makes them unsuitable for use in large columns unless there are no suitable alternatives. This is often the case in preparative HPLC columns where high solvent resistance is necessary.

14.7 Mechanical Construction

14.7.1 Connections and Seals

Laboratory-scale chromatographs are interconnected with flanged or ferruled screw-thread fittings. A similar principle is also applied to process-scale HPLC systems which are usually interconnected with Swagelok connections (or similar). Such systems are not constrained by the requirements for process hygiene and can therefore make use of screw-threaded fittings, which offer benefits in terms of maximum operating pressures, reliability, cost, and ease of use. Unfortunately, a screw thread is also impossible to clean in place. Metal-to-metal contact will allow a leakproof seal, but will not be smooth enough to eliminate small crevices which can harbor contamination.

The most common form of sanitary fitting used is the Tri-clover (also knows as a triclamp or Ladish connector), although Kamlock connectors are also commonly used on large systems. These connectors rely upon an elastomeric seal between the two pieces of pipe or tubing to be connected. The seal is pressed on both faces to form a continuous surface across the joint. It is important to use the correct size seal, and also to be careful not to overtighten the clamp, which may make the seal deform into the flow path. Care should also be taken in selecting the seal material since some elastomers will swell in the presence of commonly used CIP solutions and this swelling will result in extrusion of the seal into the flow path.

As a final note on the question of sanitary connections it has been the all too-frequent experience of the author to witness sanitary triclamp connections attached to flexible hose by means of hose-tail adapters. Whig a hose-tail adapter is undoubtedly the most adaptable and simple method of interconnection available, it is also perhaps the least sanitary, and the use of hose-tail or hose-barb adapters will immediately invalidate any effort on the part of the system designer to produce a sanitary chromatograph! It is also not unknown for some equipment manufacturers to simply mold, weld, or screw a triclamp connector onto an otherwise unchanged component and claim it to be sanitary.

14.7.2 Pipework Spools/Valves

In a small-scale laboratory system it is usual to interconnect components using flexible tubing. However, above the laboratory scale this is impractical and it is unavoidable that fixed pipework spools will be required to interconnect valves, instrumentation, buffer inlet feeds, and fraction collection outlets. For pilot-scale operation, plastic pipework can be used. In order to manufacture a plastic pipework system it is necessary to have separately molded sanitary ends fitted to the pre-shaped spools. Care must be taken when fixing these sanitary end pieces to the pipe, as a poor joint will act as a very effective deadspot and will be liable to accumulate dirt or bio-burden. For larger-scale production systems is it usually more cost-effective to manufacture the pipework spools from electropolished stainless steel. This greatly facilitates the construction of complex spools shapes with welded sanitary fittings (Fig. 14-2). In addition, the overall mechanical strength of the system is better able to cope with a production environment.

If a system is designed to only use two-way valves this will add greatly to the complexity of the pipe Work spools, since manifolds have to be fabricated for diversion of flow (for example, to by-pass a column). Such manifolds will contain T junctions which, by definition, form unswept areas. In general, the construction guidelines applied in these cases is that the length of the 'T' should be no more than six pipe diameters, although it is anticipated that this will be reduced to three pipe diameters. Clearly, a length of three pipe diameters is significant for larger diameter

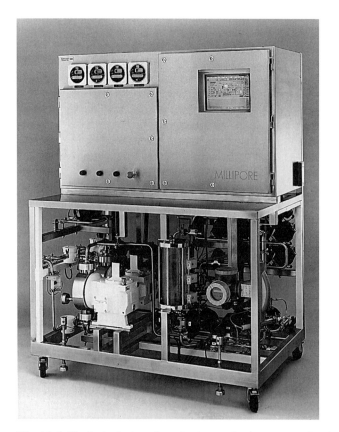

Fig. 14-2. Hygienic design of production-scale chromatographs should take into account the layout and construction of the pipework and valves. Careful consideration at the design stage is essential to the final validation of the system. (Photograph courtesy of Millipore (UK) Ltd.)

pipework and thus this design criterion can be accommodated for high flowrate equipment with wider-bore pipework. However, there is an increasing move towards smaller sanitary systems as very small volume (and very high value) biotechnology products are moving into production. Given that a low flowrate system of say 400 ml min^{-1} would typically have a pipework diameter of 3 mm (1/4 inch), the three pipe-diameter criterion would mean that insufficient pipe may be available to be able to orbitally weld a diaphragm valve in place.

Until recently only two-way valves, such as diaphragm valves or pinch valves, were recognized as being hygienic. Older designs for three-way valves would inevitably incorporate dead spaces and hold-up volumes; neither were such valves easily available with sanitary connections. In recent years several valve manufacturers, often working closely with suppliers of bioseparation equipment, have introduced a variety of sanitary design three-way valves. These will usually have been extensively tested by a variety of clearance studies.

There are several major benefits in using three-way valves, including the elimination of 'T' joints within pipework spools and a decrease in the number of valves required (thus decreasing cost and increasing system reliability).

The initial reluctance to use such valves within a sanitary process is now being overcome as the benefits of three-way valves become apparent. The latest concepts in sanitary valve design incorporate modular multiport valves within a single body. This gives even further potential for a reduction in dead legs and pipework volume. However, careful design is critical to ensure cleanliness in operation.

14.7.3 Column Seal Designs

It is vital within a chromatography column to ensure that all areas in contact with the process fluid are sanitary in nature. The selection of suitable materials is a concern which has been discussed earlier. Although the actual materials utilized are ubiquitous, in the area of seal design there are almost as many variants as there are column manufacturers. The one underlying principle is that the seal should not produce any crevices, or dead spaces. The most basic column seal design is the 'O' ring. This has the advantage of low cost and easy replacement. If standard sizes are used, replacements can be sought from a variety of sources. However, this now places the burden of material validation onto the purchaser and may invalidate the warranty of the column.

The performance of an 'O' ring as a seal is high. Such seals are reliable and not prone to overtightening. However, the nature of an 'O' ring means that there has to be a locating groove into which the seal will fit; this then necessitates a certain degree of dead-space around the seal. This dead-space forms unswept areas which are only cleanable by passive diffusion. 'O' ring seals also require some form of insert for a locating groove. In this region there will inevitably be poor flow distribution for both sample and CIP solution. The efficacy of CIP in the area around the seal can be improved by operating the column in both forward and reverse directions during the cleaning cycle. However, the 'O' ring seal fails one of the first objectives of sanitary design, that no dead spaces be deliberately created.

14.7.3.1 Space-filling Seals

In order to overcome the problem of dead spaces in the sealing area, manufacturers have now produced a variety of space-filling seal designs. These range from the most simple interference fit to complex profile-actuated seals.

Interference fit
Although the most simple to use (and possible the most reliable), these seals are of limited use on all but the smallest columns. The fit of the seal must be tight enough to allow the seal to function at the highest specified operating pressure of the column. Because the seal has to make such a tight fit against the column tube it is

often difficult to adjust the level of the distribution plates on column fitted with interference seals. In addition the seal, by definition, will be made as soon as the adjuster
is inserted into the column tube. This can cause difficulties with entrapment of air
underneath the top distributor plate and seal.

Mechanically Actuated
An alternative to simple interference fit seals are mechanically actuated space filling
seals (Fig. 14-3). These rely on pressure from two actuating plates or rings to force
the seal against the column tube. This mode is analogous to the mode of operation of
conventional 'O' ring seals. However, unlike 'O' ring seals, the shape of the space-
filling seal has been carefully designed to occupy any dead spaces on actuation.
These seals have the benefits of 'O' rings (easy to operate, allow rapid adjustment)
with added cleanliness. There are some disadvantages to mechanically actuated
space-filling seals. When an 'O' ring is actuated, the actuation force is transferred
onto a very small contact area between the column tube and the seal; hence, a strong
reliable seal can be made. In space-filling seals, some of the actuation energy is dissipated in the deformation of the seal to occupy the dead spaces. Hence, these seals
require very careful design to avoid leakage due to under-actuation or poor distribution of the actuating force.

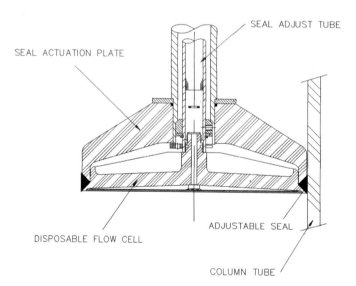

Fig. 14-3. Actuated space-filling seals are an effective and reliable method of minimizing dead-
spaces in laboratory and pilot-scale columns (up to 250 mm diameter). However, in larger columns the mechanical force required to maintain the seal requires tightening mechanisms at multiple points around the seal and is less reliable. (Reproduced with permission from Millipore (UK)
Ltd.)

Dynamically Actuated

On larger column diameters (> 450 mm), simple mechanical actuation of space-filling seals may not give enough compression to allow the seal to operate efficiently. Columns larger than this is usually have dynamically actuated seal mechanisms. These can be pneumatically (Fig. 14-4) or hydraulically operated. In both cases, the principle of action is the same, a hollow elastomeric tube is filled with air or liquid to inflate the seal against the column tube wall. In order to work correctly, the wall of the seal has to be carefully designed to allow selective deformation. The seals are manufactured from extruded lengths which can then be cut and annealed to form seals of the correct size. Pneumatic seals have the advantages of cleanliness of operation and are easy to use and control. However, the elastomers used will have some degree of air permeability and careful formulation is required to ensure that this is kept to a minimum, otherwise the seal can deflate over a period

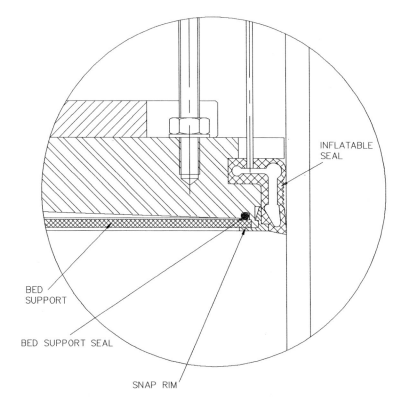

MINIMUM DEAD SPACE INFLATABLE SEAL

Fig. 14-4. Inflatable seals allow reliable sealing of production-scale columns (up to 2000 mm diameter) while maintaining a minimal dead space between the distributor and the column tube. Seals can be inflated by gas (pneumatic) or liquid (hydraulic). (Reproduced with permission from Millipore (UK) Ltd.)

of time. It is common practice to reduce the pressure in the seals before long-term storage (since the gas permeability is proportional to pressure). One less obvious advantage of pneumatic seals is that a breach in the seal integrity is usually immediately apparent and easily traceable by the gas bubbles! Hydraulically actuated seals do not require a clean air supply for operation and are more stable over a long period. However, the liquid within the seal must be regarded as a potential contaminant of the column. Since the liquid within the seal is not subjected to routine CIP procedures it should be sterile and contain a bactericidal agent which should be compatible with the column in case there is a loss in seal integrity.

14.7.4 Distribution Plates

The flow distribution system of a chromatography column end cell is designed to produce plug flow within the column. Thus, the distributor plays a crucial role in ensuring that the feed sample is applied evenly and equally across the face of the gel bed. This is also a requirement for the CIP solution. The design of the distributor system is critical in ensuring that the entire column is exposed to sufficient CIP solution for effective sanitization to occur.

14.7.4.1 Single Port

Single port distributor plates or probably the most common design and have a general design (Fig. 14-5) which is applicable over the entire spectrum of column sizes. Single port distributor plates have an advantage in that they are less susceptible to the presence of air. The conical shape in a typical top end cell can accommodate a relatively large amount of air with only a small loss in chromatographic performance. However, the presence of entrapped air in either top or bottom distributor plates will have a very deleterious effect on a CIP step, since the air bubble will shield part of the distributor surface from the CIP solution.

14.7.4.2 Multi-Port

Multi-port inlets can display a further problem since it is possible for a gradual accumulation of air to form an air-lock in one of the distributor ports. Although this will usually show as a gross reduction in chromatographic performance, it is also a potentially serious problem for CIP steps since a significant proportion of the gel will be underexposed to the CIP regime.

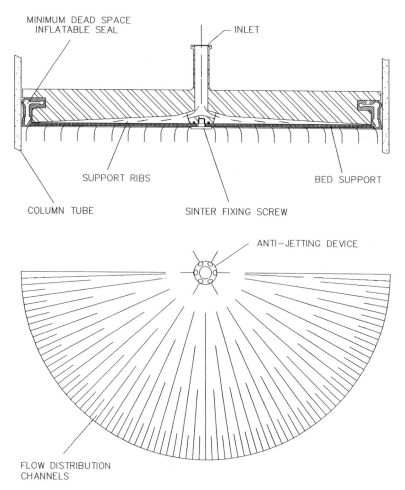

MINIMUM DEAD SPACE
INFLATABLE SEAL

INLET

SUPPORT RIBS

BED SUPPORT

COLUMN TUBE

SINTER FIXING SCREW

ANTI–JETTING DEVICE

FLOW DISTRIBUTION
CHANNELS

Fig. 14-5. The design of the distributor plate is of critical importance in maintaining scalability from pilot to production scale. The distributor plate design shown has been used successfully on columns ranging from 10 mm up to 2000 mm in diameter with excellent scalability. (Reproduced with permission from Millipore (UK) Ltd.)

14.7.5 Expanded Bed Columns

A recent development in process chromatography has been the use of expanded bed resins. The distributor design for expanded bed columns is critical to the performance of the column and the nature of the technique dictates that the distributor consists of a finite and relatively low number of 'holes' within a plate. The use of this method is also aimed specifically at capture of product from crude solutions often containing whole (viable) cells. These two factors clearly combine to form an additional hurdle in CIP procedures. Thus, the CIP of the distributor system needs to be very specifically addressed (and validated) whenever expanded bed chromatography columns are used.

14.7.6 Closed System Columns

Recently, column manufacturers have developed process-scale columns which have been designed to allow packing and unpacking to be carried out without removing the upper distribution plate (Fig. 14-6). These closed system columns offer potential advantages in maintaining good hygiene, since the resin is less exposed to potential

Fig. 14-6. (a) Closed system columns allow the resin to be packed and unpacked *in situ*, reducing the probability of external contamination. Such columns work best with slightly compressible gels and optimization of packing procedures is usually required.

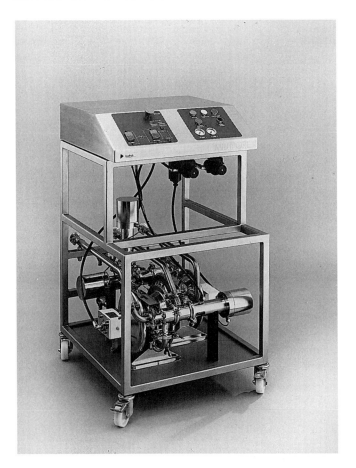

Fig. 14-6. (b) Closed system columns require a separate pumping skid for the gel slurry which must also be designed and built to hygienic standards. (Photographs courtesy of Millipore (UK) Ltd.)

infection. However, it should be noted that such systems require additional pipework and holding tanks which will themselves add to the validation workload for the overall process.

14.7.7 Bubble Traps

Bubble traps are usually fitted either to chromatography systems or to specific columns. The functions of bubble traps are twofold: first (and most important) to protect the resin in the column from air in the process stream; and second, to reduce flow pulsation within the chromatography system and thus avoid pressure shocks to the column.

It is particularly important to protect both the end cell (distributor) and the resin from air. Air in the distribution system may cause problems with CIP as noted above, air in the gel bed will usually require a re-pack. In larger systems, degassing buffers becomes problematic and expensive. Most air will be introduced into the system as a continuous stream of small bubbles within the process line rather than as a single, gross bolus of air (due perhaps to a buffer or feed tank emptying). Thus, a bubble trap is essential and will in turn (if it is performing properly) accumulate air.

The presence of air within the bubble trap makes it both unclean, and also uncleanable, unless specific steps are taken to purge the bubble trap fully with CIP solution to remove all air from inside and ensure complete contact on all surfaces. Once the bubble trap has been through a CIP cycle it must be re-charged with sterile air. In a clean room environment it is usually sufficient to re-charge the bubble trap by using air from the ambient environment; however, for total cleanliness the air used for re-charging should be sterilized (usually by filtration).

14.8 CIP Validation

The most time-consuming part of developing a CIP protocol is inevitably in the validation of the protocol. The well-known and often-quoted definition of validation given by the FDA is: 'establishing documented evidence which provides a high decree of assurance that a specific process will consistently produce a product meeting its predetermined specifications and quality attributes' [11]. In the case of a CIP protocol the 'product' is a chromatography column or system which has a bio-burden reduced to (or below) the 'pre-determined specifications'.

Clearly, this puts the onus on the operator to define the acceptable levels of cleanliness and also to prove that these levels are consistently and reliably achieved.

Much has been written about the validation of separation processes (for recent reviews see [13,14]. However, a vital part of the overall process qualification is the resin regeneration and CIP process. In this area much of the original data are generated by in-house research and are regarded by the companies involved as proprietary, and therefore less likely to be found in the public domain. However, some useful 'discussion documents' have been published in this field [15–17] and some research work published [18]. The CIP part of the process validation may seem to be unduly arduous and protracted; however, it should always be borne in mind that, once in production, product is more likely to be lost or rejected through the presence of unacceptable levels of contaminants than through the failure of the separation chemistry. It is therefore essential that the CIP routine be considered at earliest possible stage in process development and the same rigorous approach used on purification development and scale-up be applied to the CIP cycle [19].

14.8.1 Challenge Testing

One method available to test if a CIP step is effective is to imitate a worst possible case and then carry out the CIP procedure and prove (or otherwise) that it is effective [20]. This is the principle behind challenge testing. Challenge testing can be carried out using either a pure culture of a specified test organism or compound (Table 14-4), or using a sample of normal feedstock which has been 'spiked' to produce an abnormally high load of a specific contaminant. The column and system are exposed (infected) with the contaminant and a standard CIP protocol carried out. On-stream samples can be taken during the process to evaluate the progress of each CIP step.

At the end of the CIP process the efficacy of the sanitization can be assessed by disassembling the system and taking swabs to determine the presence of contaminant, by flushing the system and column with sterile saline solution and collecting a volume for filtration and analysis, or by filling the system with sterile growth media and incubating for a given period before flushing and sampling. The latter is the most rigorous method and will detect the presence of viable organisms to lower levels than the previous two methods.

Although it is most indicative of CIP performance if the challenge tests are carried out using the actual organism or compound most likely to be problematic within the process, this is sometimes not desirable on health and safety grounds. In these cases, standard organisms or compounds are used. The organisms chosen should bear as close a similarity to the most likely contaminant as possible.

Although vegetative bacteria can be used for challenge testing, large column viral challenges are most often carried out on scaled-down columns. The aim of these studies is to validate the chromatography as a virus removal step rather than to validate the cleanliness of the specific production equipment. Some examples of 'standard' organisms used in challenge tests are given in Table 14-4.

Table 14-4. Typical organisms used in challenge test studies for the validation of CIP and virus removal.

Organism	Family	Genome	Enveloped	Size (nm)	Shape	Resistance
Staphylococcus aureus	Gram –ve bacteria	ds DNA	–	~1000	Spherical	Medium
Escherichia coli	Gram –ve bacteria	ds DNA	–	~ 800	Rod	Low
Pseudomonas aeruginosa	Gram –ve bacteria	ds DNA	–	1000	Rod	Medium
Bacillus subtilis	Gram +ve bacteria	ds DNA	–	~1000	Rod	High (spores)
Candida albicans	Yeast	ds DNA	–	~4000	Rod	Medium
Aspergillus niger	Mold	ds DNA	–	–	Mycelial	High (spores)
Poliovirus	Picornaviridae	ss RNA	No	25–30	Icosahedral	Medium
Reovirus	Reoviridae	ds RNA	No	60–80	Spherical	High
Murine leukemia virus (MuLV)	Retroviridae	ss RNA	Yes	80–110	Spherical	Low
Human immunovirus (HIV)	Retroviridae	ss RNA	Yes	80–100	Spherical	Low
Vesicular stomatitis virus	Rhabdoviridae	ss RNA	Yes	80–90	Bullet	Low
Herpessimplex virus	Herpesviridae	ds DNA	Yes	180–200	Spherical	Low
Porcine parvovirus (PPV)	Parvoviridae	ss DNA	No	18–26	Icosahedral	High
Pseudorabies virus (PRV)	Herpesviridae	ds DNA	Yes	150–200	Spherical	Medium
Minute Virus of Mice (MVM)	Parvoviridae	ss DNA	No	18–26	Icosaheral	High
Sindbis virus	Togaviridae	ss RNA	Yes	45–75	Spherical	Low
Simian virus 40	Papovaviridiae	ds DNA	No	45–55	Icosahedral	High

[a] ds, double-stranded; ss, single-stranded.

14.8.2 CIP Conditions

While a variety of CIP solutions are commonly used (see below) some factors need to be taken into account for all CIP processes. The efficacy of a CIP step will depend upon the temperature and exposure time of the system to the CIP cocktail. It is imperative that these factors be validated and written into the CIP protocol. It is an unfortunate fact that decreasing the exposure time or decreasing the concentration of CIP chemical will both lead to a reduction in operating costs. However, if these are carried out without due caution the resulting increase in column contamination will more than wipe out any potential savings. It is the experience of this author that one process ran into problems when the exposure time and the concentration of CIP solution (NaOH) were reduced independently by two separate groups within a plant after process scale-up. While either change alone may have still been sufficient to CIP the columns, the combination of a reduction in both parameters was catastrophic to the process.

14.8.3 CIP Cocktails

While the precise composition of CIP cocktails will vary from application to application, in general there are only a limited set of solutions in common use. Once again, it should be emphasized that a CIP cocktail should be evaluated for the specific process as some of the following solutions will inevitably be incompatible with some processes.

14.8.3.1 Concentrated Salts

Although not bactericidal, concentrated salt is often used as a first step in the regeneration of media. High concentrations (~1.5 M) of salts such as NaCI or KCl can be very effective at removing proteins still bound to ion-exchange gels or proteins bound nonspecifically to gel permeation media. High concentrations (~4 M) of chaotropic agents such as urea or guanidinium hydrochloride are also effective in removing strongly bound proteins and precipitated proteins. Such chaotropic solutions may also be the only available option for some protein-based affinity gels where stronger CIP protocols would damage the ligand.

14.8.3.2 Sodium Hydroxide (NaOH)

Sodium hydroxide is the most common component of process-scale CIP solution, and is used either individually or in combination. The concentration of NaOH used will vary (from 0.1 to 1.0 M) with the degree and type of contamination and the chemical strength of the gel being used. The exposure times will likewise vary (from less than 1 h to several days). A 0.5 M NaOH solution with a minimum exposure time of at least 60 minutes is usually sufficient for most processes. In general, the lowest concentration of NaOH possible should be used to prolong gel life. Affinity gels in particular are prone to damage by excessive exposure to strong NaOH solutions. For affinity gels it is common to calculate CIP solutions in terms of column volumes rather than as exposure times in order to avoid prolonged exposure to alkaline conditions. It should also be noted that some affinity linkages such as divinylsulfone (DVS)-activated gels are not resistant to alkaline conditions. For gels such as these, alternative regeneration/CIP procedures such as acid washes (see below) must be used.

14.8.3.3 Organic Solvents

Organic solvents are used in CIP procedures to assist removal of lipids binding to gels. This type of contamination is especially problematic with columns used in the early stages of a purification. The most common solvent used is ethanol. The addition of acetic acid (~5 %) to the ethanol solution will assist in solubilizing lipid. Longer-chain alcohols can be used in place of ethanol to increase the solubilization of lipid. Longer-chain alcohols also have the added advantage of being more toxic towards micro-organisms (probably due to their greater efficacy in disrupting lipid membranes). Longer-chain alcohols do have a higher viscosity than ethanol, which may cause settling of a packed bed; they can also be more difficult to remove and can, in extreme cases, cause flocculation of the gel.

The inclusion of ethanol in CIP cocktails is commonly intended to act as a sterilization step. However, for efficient sterilization an ethanol concentration of at least 70 % is required. This is much higher than commonly used and, in most cases will raise more problems with chemical resistance and explosion proofing than are justified. Various combinations of ethanol (up to 60 %) and acetic acid (0.5 M) have also been reported as efficient sanitization solutions [21.22]. At concentrations below 70 %, ethanol has little or no sporicidal effect [23] and reduced virucidal effect (a further consideration to note is that ethanol alone will not destroy pyrogens). However, ethanol is a good bacteriostat; therefore ethanol solutions of 10–20 % can be effectively used for the preservation of columns during long-term storage. Ethylene glycol can be used as a chaotropic agent to clean strongly bound species from reversed phase, hydrophobic interaction, or thiophilic gels.

14.8.3.4 Thiomersal (Thimerosal, Merthiolate, Ethyl-mercurithiosalicylate)

Thiomersal is a mercuric compound widely used in the preservation of contact lenses. It has also been used widely in the past as a column sanitization and storage reagent, and as a preservative for vaccines. The concentrations used are small (0.005–0.01 %); however, the mercuric nature of the compound gives rise to the possibility of accumulation in the environment and in exposed organisms. It is vital to ensure that all thiomersal is removed from the column before the feed stock is applied to avoid cross-contamination using a suitably sensitive assay [24]. Further disadvantages of thiomersal are that it can bind nonspecifically to some resins, its use incurs high disposal costs, and it has been reported as a causal agent of cell-mediated immunity [25]. For these reasons the use of thiomersal is becoming less common.

14.8.3.5 Chlorhexidine (Chlorhexidine Digluconate, Hibitane, (1,6-Di(4-chlorophenyl-diguanido)hexane)

Chlorhexidine is widely used for column sanitization. Chlorhexidine itself is almost insoluble in water and is usually used as a 0.5 % solution in 20 % ethanol [26]. Chlorhexidine digluconate cannot be isolated as a solid but is available as a 20 % aqueous solution [27]. Chlorhexidine has good bactericidal properties against vegetative cells, but has no effect on bacterial spores. It is effective against some lipophilic viruses such as herpesvirus and HIV, but is not effective against noncoated viruses such as human Reovirus [28]. The extensive use and safety history of chlorhexidine over the past 30 years makes it the compound of choice for many bioprocesses.

14.8.3.6 Detergents/Solvents

Detergent and solvents used in combination have been found to be very effective in virus inactivation while maintaining mild conditions. A typical virucide cocktail would consist of 0.3 % TNBP (tri-(n-butyl)phosphate) and 1 % Polysorbate (Tween 80) [29].

14.8.3.7 Acidic Conditions

Dilute HCl, low pH glycine buffer, citric acid, and acetic acid are often used in the regeneration of immunoaffinity columns due to the labile nature of these columns in alkaline conditions. Low pH cleaning steps also have the added bonus of virucidal activity [30]. Glycine, citrate, and acetate-based buffers are not recommended for prolonged storage of gels, however, as they form a suitable growth substrate for many micro-organisms.

14.8.3.8 Oxidizing Agents

The sterilization of surfaces and some equipment by the use of strong oxidizing agents such as formalin, sodium hypochlorite, and peracetic acid is widespread within laboratories. Although these compounds are very effective in sterilizing hardware, very few chromatography gels are resistant to them. Some exceptions may be found among newer 'synthetic polymer' gels such as those manufactured in the form of methacrylate polymers (Toyopearl, Macroprep), polystyrene divinylbenzene polymers (Amberchrom), and polystyrenes (HyperD). However, it is strongly advised to seek confirmation from the manufacturers before using oxidizing agents on gels as the resin chemistry may not be compatible, even though the base matrix is. If the resin is compatible with oxidizing agents such as peracetic acid, this can form the basis of a highly effective CIP procedure [31,32].

14.8.4 Storage

The long-term storage of chromatography systems and columns pose further problems. Frequently, the concentrations of chemicals used in CIP cocktails (see above) are too high for long-term storage and would result in damage to either the chromatography hardware (for example over a prolonged period of exposure, 20 % ethanol will attack the acrylic commonly used to manufacture column tubes), or alternatively to the resin itself (many affinity resins will undergo ligand uncoupling after prolonged exposure to NaOH).

Thus, columns should be stored in solutions of lower concentration than those used for CIP. If the column has undergone an efficient CIP step before storage this will not be a problem, since the storage solution merely has to act as a bacteriostat rather than a bactericide. It is also good practice to flush columns in long-term storage at regular intervals with clean buffer and to renew the storage solution. If the column is stored in a solution different from that used for CIP, separate validation of the storage solution must be carried out. In this case, the emphasis should be on bacteriostasis and resin lifetime rather than bactericidal efficiency.

Chromatography pumping systems are usually more resilient to chemical attack. However, storage solutions should also be of lower concentration than CIP cocktails. It is important to remember that even very dilute salt solutions will concentrate during evaporation and can give rise to corrosion, even on stainless steel, while deionized water is very effective at dissolving salts from stainless steels altering the surface chemistry and thus can also be corrosive. When storing systems (and empty columns) for a protracted period the systems should be rinsed in deionized water, drained down, and thoroughly dried. In such cases, pH probes should be removed from the process line and stored in KCl solutions.

14.9 Conclusion

Cleaning is a burden, but unfortunately it is also unavoidable. Effort placed in CIP development and validation may seem unnecessary at the development scale; however, care and consideration at the earliest point in process development will reap great rewards at the production level by increasing process reliability and lifetime whilst decreasing lost production time, lost money and, of course, lost sleep!

References

[1] Jungbauer, A., Boschetti. E., Manufacture of recombinant proteins with safe and validated chromatographic sorbents. *J Chromatogr B,* 1994, *662,* 143–179.

[2] Sofer, G., Preparative chromatographic separations in parmaceutical, diagnostic, and biotechnology industries: current and future trends. *J Chromatogr A,* 1995, *707,* 23–28.

[3] Burnouf, T., Chromatography in plasma fractionation: benefits and future trends. *J Chromatogr B,* 1995, *664,* 3–15.

[4] Darling. A.J., Spalto J.J., Process validation for virus removal. *Biopharm,* 1996, *9,* 42–50.

[5] Anspach, F.B.. Hilbeck. O., Removal of endotoxins by affinity sorbents. *J Chromatogr A,* 1995, *711,* 1–92.

[6] Kemp. G. D., *Applications group internal report LR 02029.* Amicon Ltd, Stonehouse, Y.K., 1995.

[7] Flavell, P., Selecting the correct Medical-grade polymer. *Medical Device Technology,* 1996, *November,* 16–22.

[8] Pfister, M., Kohlzr, W. G., Fittings and components for aseptic processes in the chemical and pharmaceutical industry. *Chemical Plants and Processing,* 1993, *March,* 24–26.

[9] White, P. E., Stainless steel for food and beverage processing-advantages in hygiene and cleanability. *Stainless Steel Industry,* 1988, *September,* 2–4.

[10] Roessell, T., Swain. J., Effective chemical cleaning of stainless steel fabrications. *Stainless Steel Industry,* 1990, Vol. 18, no. 102, 1–3.

[11] Center for Drugs and Biologics and Center for Devices and Radiological Health, guidelines on general principles of process validation; Food and Drug Administration, May, 1987.

[12] US Department of Health and Human services. Code of Federal Regulations, Title 21, part 177; Food and Drug Administration, .

[13] Eckman, B., Validation, in: *Handbook of Downstream processing;* Goldberg, E. (Ed.), Blackie Academic Professional, 1996.

[14] Sofer, G., Hagel, L., *Process chromatography,* London: Academic Press, 1997.

[15] Agalloco, J., 'Points to consider' in the validation of equipment cleaning procedures. *J Parenteral Sci Technol,* 1992, *46,* 81–86.

[16] Parenteral Drug Asociation Industry perspective on the validation of column-based separation processes for the purification of proteins. *J Parenteral Sci Technol,* 1992, *46,* 87–97.

[17] Zeller, A. O., Cleaning validation and residue limits: a contribution to current discussions. *Pharmaceutical Technology Europe,* 1993, *November,* 18–27.

[18] Boschetti, E., Pouradier Duteil, X., Nguyen, C., Moroux, Y., Concerns and solutions for a proper decontamination of chromatographic packings. *Chimicaoggi,* 1993, *March/April,* 29–35.

[19] Edwards, J., Large-scale -column chromatography – a GMP manufacturing perspective, in: *Handbook of Downstream processing;* Goldberg, E. (Ed.), Blackie Academic Professional, 1996.

[20] Berube, R., Oxborrow, G. S., methods of testing sanitizers and bacteriostatic substances, in: *Disinfection sterilization and preservation,* 4th edition; Block, S. S. (Ed.), Philadelphia: Lea Febiger, 1991, pp. 1058–1068.

[21] Boschetti, E., Girot, P., Guerrier, L., Silica-dextran sorbent composites and their cleaning in place. *J Chromatogr,* 1990, *523,* 35–42.

[22] Girot, P., Moroux, Y., Pouradier Duteil, X., Nguyen, C., Boschetti, E., Composite affinity sorbents and their cleaning in place. *J Chromatogr,* 1990, *510,* 213–223.

[23] Larson, E. L., Morton, H. E., Alcohols, in: *Disinfection sterilization and preservation,* 4th edition; Block, S. S. (Ed.), Philadelphia: Lea Febiger, 1991, pp. 191–203.

[24] Shrivastaw K. P., Singly S., A new method for spectrophotometric determination of thiomersal in biologicals. *Biologicals,* 1995, *23,* 65–69.

[25] Seal, D., Ficker. L., Wright, P., Andrews, V., The case against thiomersal. *The lancet*, 1991, *338*, 315–316.

[26] Adner, N., Sofer, G., Biotechnology product validation, part III: chromatography cleaning validation. *Pharm Tech Eur*, 1994, *April*, 21–28.

[27] Denton, G. W., Chlorhexidine, in: *Disinfection sterilization and preservation*, 4th edition; Block, S. S. (Ed.), Philadelphia, Lea Febiger, 1991, pp. 274–289.

[28] Springthorpe, V. S., Grenier, J. L., Lloyd-Evans, N., Sattar, S. A., Chemical disinfection of human rotaviruses: Efficacy of commercially available products in suspension tests. *J Hygeine*, 1986, *97*, 139–161.

[29] Horowitz, M. S., Bolmer, S. D., Horowitz, B. Elimination of disease-transmitting enveloped viruses from human blood plasma and mammalian cell culture products. *Bioseparation*, 1991, *1*, 409–417.

[30] Burstyn, D. G., Hageman, T. C. Startegies for viral removal and inactivation, in: *Viral safety and evaluation of viral clearance from biopharmaceutical products;* Brown, F., Lubiniecki, A. S. (Eds.), Developmental Biological Standards, Karger, 1996, vol. 88, pp. 73–79.

[31] Jungbauer, A., Lettner H., Chemical disinfection of chromatography resins, part I: preliminary studies and microbial kinetics. *Biopharm*, 1994, *7*, 46–56.

[32] Jungbauer, A., Lettner, H., Guerrier, L., Boschetti, E., Chemical disinfection of chromatography resins, part II: *in situ* treatment of packed columns and long term stability of resins. *Biopharm*, 1994, *7*, 37–42.

15 Strategies and Considerations for Advanced Economy in Downstream Processing of Biopharmaceutical Proteins

Joachim K. Walter

15.1 Introduction

Downstream processing (DSP), that is the recovery and purification of biotechnically derived proteins, as well as proteins derived from fluids or tissues of biological origin, represents a sequential arrangement of individual unit operations which form a rather complex synthesis of chromatographic and filtration methodologies. Each single unit operation of DSP is designated to contribute selectively and significantly to the removal of impurities, including contaminating peptides, proteins, and lipids. Devoted care needs to be taken of drug safety regarding potential harmful infectious contaminants, hence the removal of DNA and (potential) virus or virus-like-particles (VLP) is crucial and requires respective efforts in validation.

15.2 Economic Potential in Downstream Processing

The yield of product at each purification step strongly depends on the potency and resolution of the separation media chosen, as well as on the design of the operational procedures around that purification step. At an economic basis, typical step yields are in the range of 92–98 %; significant lower yields down to 80 % are accepted in the light of an unique purification effect of such an operation, typically being located at the initial phase of a downstream process.

The number of purification steps required, and their individual step yields determine the overall yield. On this account the design of the downstream process contributes considerably to the economy of the manufacturing process.

Currently the overall yield of typical purification processes composed of about 8–10 unit operations is in the range of 50–80 %, a yield that – at its lower end – can diminish minor process achievements in fermentation processes. This is one of the main driving forces to accomplish yield improvement during downstream processing. Since typical yields are already in the range of about 95 % or higher for an individual purification step, potential improvements on single-step yields are to be considered as marginal. Nevertheless, the use of appropriate technologies might contribute to reduce hands-on time and investment in hardware, e.g., the application

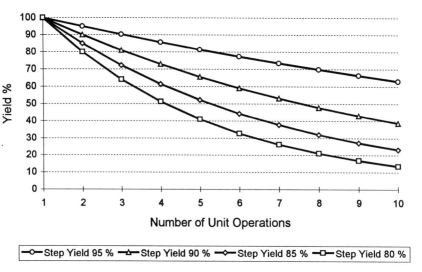

Fig. 15-1. Influence of the number of unit operations and their step yields on the overall yield of bulk product.

of membrane-based chromatography, in the mode of negative chromatography, even being used as disposable in-line filters.

Due to the large volumes of several thousand liters used in fermentation processes and to significant improvements in cell productivity resulting in product titers in mammalian cell culture of currently up to 1000 mg l^{-1}, the development of downstream processes might be faced with severe technical or economic limitations.

Despite the fact that a further increase of the dynamic loading capacity of the chromatography matrices for the product would solve certain difficulties, the steric hindrance in case of large protein molecules and the aggregation of proteins at high concentrations during elution due to limited solubility rarely allows the usage of protein loads significantly beyond 100 mg ml^{-1} of matrix. The application of high linear flowrates in a chromatographic process is clearly another approach to shorten process time. However, adsorption kinetics of the protein product to the matrix and appropriate access of the product molecule to the inner surfaces of typical chromatographic beads remain as limiting factors.

Highly expensive separation media are of impact on production economy at large scale. As a practical approach, the protein purification can result in a split of an individual fermentation batch. By preference, such a split is limited to those steps with limited capacity and implies re-combination of the resulting pools, a strategy that helps to tighten the time frame of manufacturing and avoid the generation of several batches of bulk product.

A significant contribution to process amelioration should be expected by the optimization of those steps that show yields lower than 95 %. At times, an even more dramatic loss of product might proceed with the cell harvest process using microfiltration or related techniques. Much effort has been undertaken to reduce product loss during the harvest procedure, i.e., the separation of the crude product solution from cells.

Beside a continuous optimization of conventional cell harvest equipment, e.g., microfiltration and efforts to master clogging and membrane fouling respectively, or centrifugation and respective efforts to conquer the generation of cell debris but still guarantee equivalent cell removal, the merger of cell removal and product-specific adsorption to a chromatographic matrix – known as fluidized or expanded bed chromatography – is an attractive approach both to accomplish loss of product and reduce the number of steps [1].

The other option for process amelioration – a reduction in the number of steps – is even more conducive to a distinct increase in total yield as well as process economy. However, such a reduction is limited with respect to product safety by the need for a validation of removal factors for potential DNA and viral contaminants [2–5].

The effectiveness of methods and technologies for the inactivation and removal of potentially harmful contaminants such as viral particles and cellular DNA is crucial with regard to the number of labor-intensive procedures which are necessary in order to obtain the required reduction factors and to satisfy validation needs [6].

15.3 Strategic Development of Unit Operations

The development of a complex downstream process for biopharmaceuticals is oriented according to the application profile of the protein product: cGMP manufacturing which yields a highly pure product featuring drug safety and state-of-the-art process validation is devoted to therapy/*in vivo* diagnosis. Thus, the extent of unit operations and measures to obtain product quality suitable for an *in vitro* diagnostic is significantly reduced [7].

Hence, prior to the start-up of any process development, premises must be formulated which are considered as guidelines for the duration of the development: technical basis, raw materials, process design, and process validation.

15.3.1 Technical Basis

The fundamental technical pre-requisite for the development in order to achieve a maximum benefit and economy from the process is the identification and knowledge of the prospective manufacturing scale, plant, and equipment. Certainly only those unit operations which can be scaled to the intended scale are allowed for development, and with due regard to all technical contingencies and conditions the transfer to production scale and thence to the production plant is improved considerably.

The design and construction of equipment and piping must allow for a simple, complete and easily validated sanitization as cleaning-in-place (CIP). In general, all process components must be planned from the very beginning with respect to the feasibility of validation measures.

15.3.2 Raw Materials and Equipment

A focal point of the development is on the selection of raw materials and process media, and their impact on suitability and functionality at the intended technical scale. Each of the raw materials must be available in appropriate quality and quantity: salts for buffer preparation, solutions and solvents must be specified according to applicable pharmacopoiea, e.g., DAB *(Deutsches Arzneimittel Buch)* or USP *(United States Pharmacopoiea)*. Process media such as chromatographic matrices and membranes for the different types and modes of filtration are selected according to their potency and commercial availability. (Table 15-1).

Table 15-1. Selection criteria for media applied to DSP.

● Suitability and functionality at technical scale
  – separation performance, regeneration, life time

● Reproducibility of all relevant process parameters
  – selectivity
  – resolution
  – yield
  – dynamic binding capacity
  – chromatographic profiles, filtration profiles
  – flowrates
  – system pressures
  – conductivity, pH-value

● Commercial availability
  – batch size, batch consistency, deliverability

● Authorities' acceptance
  – Drug Master File (DMF)
  – GLP/GMP documentation

The influence of process media on process economy is fundamental: economically, the efficacy of the purification steps of the initial phase in DSP ('recovery') determines the total number of unit operations which are necessary to obtain the required product quality. The individual step yields contribute essentially to the total product yield, and contribute substantially to the overall economy of the downstream process. Accordingly, the selection for raw materials focuses on their commercial availability in a reproducible and reliable quality. Batch sizes might become crucial at larger scales, as the substantial analysis for release of the raw material including functional testing can be expensive. For the display of their resolution capability, excellent separation media require an equipment of optimal design and functionality. For large-scale operations, advanced chromatographic columns are constructed for an automated, fully controlled *in situ* filling. Although extremely expensive, such columns allow an excellent return on investment due to both their outstanding handling and reproducibility. Even columns with volumes of up to several hundred liters of matrix can easily be handled by a single person, as no heavy parts need to be moved. Table 15-2 shows data on the preparation of columns for gel permeation

Table 15-2. Preparative gel permeation chromatography.

Column size	Pressure (mPa)	Flowrate (cm h^{-1})	Slurry (%)	Theoretical plate number N (m^{-1})	Asymmetry factor (A$_F$)	HETP (cm)	Packing time (min)
Diameter 180 mm Height 900 mm	0.44	34*	48	13 680	1.43	0.0073	115
Diameter 180 mm Height 900 mm	0.45	44	45	13 515	1.12	0.0074	125
Diameter 180 mm Height 900 mm	0.42	38	68	13 930	1.14	0.0072	72
Diameter 600 mm Height 900 mm	0.40	32.5	62	16 010	1.14	0.0063	69
Diameter 600 mm Height 900 mm	0.46	40	65	16 807	1.14	0.0060	66

Scale-up of preparative gel permeation chromatography was performed using Chromaflow™ Columns (Euroflow, U.K.) and Superdex 200 pg (Pharmacia, Sweden). Total volume of the packed gel bed was 30 L for the 180-mm diameter and 255 L for the 600-mm diameter column. The packing direction was top-down except for the marked 180/900 (*) column, which was prepared bottom-up and resulted in a less favorable asymmetry factor of 1.43. This difference was reproducible. Otherwise, the columns allow the reliable *in-situ* preparation of columns at large technical scale even under varying conditions as shown in Table 15-2. The obtained resolutions are comparable with those which can be achieved with laboratory-scale columns for the identical matrix. The 600-mm diameter column could be optimized to $N > 16\,000$.

chromatography, which is a nearly inextricable task for conventional columns at the featured scale.

Furthermore, the performance of resolution is an invaluable fundamental for a reliable scale-up. However, as with other large-scale equipment, the time consumption for design, specification, construction, and qualification is not to be underestimated, and will easily be in the range of 6–12 months. Due to such lead times, already results from early stages in development must be highly reliable, as they might urgently be requested for decision finding in an investment of great consequence.

15.3.3 Process Design

DSP is an integral part of biopharmaceutical manufacturing. Although most operational procedures within the DSP might be dealt with as individual unit operations from a technical point of view, the impact and influence of the adjacent scientific and technological disciplines need to be considered (Fig. 15-2).

The process design comprises the sequence of various unit operations, employment of personnel capacities and process validation.

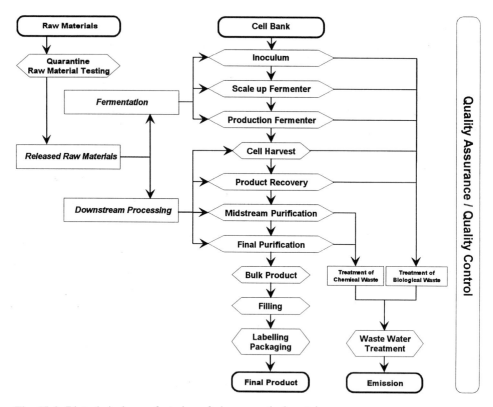

Fig. 15-2. Biotechnical manufacturing of pharmaceutical proteins.

The suitable combination of efficient separation technologies is one key function for the realization of utmost process economy. The sequential application of individual unit operations promotes the strategy of a modular process design.

Along a general arrangement of the downstream process (Table 15-3), individual operations might be shifted up- and downstream within the process in order to achieve a maximum benefit regarding product quality and process economy. The skilful arrangement of different types of chromatography is an illustrious example: ion-exchange chromatography typically applies a low-conductive buffer for loading in order to maximize the dynamic binding capacity, while elution is promoted by increasing the conductivity. With hydrophobic interaction chromatography it is exactly the reverse: binding is enhanced at a high conductivity while elution occurs at low conductivities – hence a promising succession of two powerful separation tools is feasible, excluding an intermediate ultra/diafiltration step as an additional operation.

A graphic description for different options which can be found in a chromatographic separation is gel permeation chromatography (GPC) (Fig. 15-3). The product load determines the separation efficacy of a GPC column and, depending on the separation qualifications, the total load, i.e., the combination of product concentration and volumetric load, needs to be fixed in a narrow range. This range is crucial

Table 15-3. Downstream processing of biopharmaceutical proteins.

DSP section	Unit Operations	Process volume (%)	Product concentration (mg mL^{-1})
Cell Harvest ⇓	TFF Microfiltration	100	0.2
Capture ⇓	Ultra/diafiltration Chromatography	100–10	0.2–2
Midstream ⇓	Chromatography Ultra/diafiltration Virus removal Virus inactivation	10–2	2–10
Polishing ⇓	Chromatography Ultra/diafiltration	1–0.2	20–100
Filling	Sterile filtration	20–0.2	1–100

and, due to the unprofitable relation of product load to column size and volume, GPC can become a delicate downstream operation with respect to economy. The economic benefit of the elimination of an entire process step is evident, but the ingenious selection of buffer components and the total number of different buffers is also of broad economic and logistic impact. The availability of buffer make-up tanks and storage tanks might be limited, but it is even more the turnover procedure including CIP and SIP. Buffer make-up and regeneration procedures under GMP are time-consuming and labor-intensive, as they comprise beside the actual handling appropriate logistics for raw material testing, release and storage, LAL testing of the prepared buffer, and buffer storage validation. Therefore the limitation to simple buffer compositions of inexpensive ingredients, e.g., physiological buffers such as phosphate, glycine/NaOH, acetate or citrate buffers, and highly effective regeneration solutions, e.g., 0.5–1.0 M NaOH, is a valuable task for basic process development. Thus, the resistance of separation media in chromatography and filtration is stringent.

The application of highly selective but unfortunately vastly expensive matrices, e.g., affinity chromatography of antibodies on Protein A or G, is of impact on long-term production economy at large scale. This leads to considerations regarding batch versus cycle mode of processing. There is no general answer to this question, as the advantages of a batch mode (short process time, limited sampling and intermediate analysis, no need for definition and validation of pooling criteria, clear lot designation) need to be balanced against the preferences of a cycle mode (reduced initial investment in equipment and separation matrix, space-saving reduced size of equipment). The lifetime of the chromatographic matrix is another criterion which must be thoroughly determined, and which will definitely be one decisive factor for the mode selection. With respect to regulatory concerns a defined single batch strategy might be recommended; however, in the case of manufacturing campaigns limited to only a few product batches, cycling will be – and will remain – a viable option.

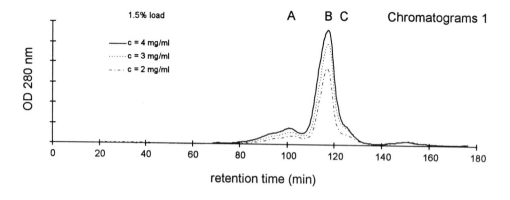

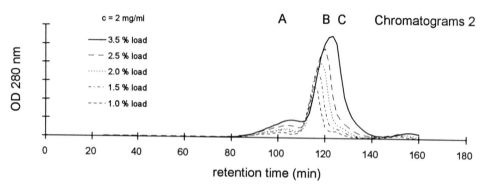

Fig. 15-3. Exploitation of gel permeation chromatography. The chromatograms show GPC separation of dimeric (A) from monomeric (B) monoclonal antibody (Mab, $IgG2_a$). The chromatograms illustrate two different strategies in maximizing the column load. Chromatograms 1 feature an identical volumetric load (1.5 % of the packed gel bed volume) at different protein concentrations and look quite comparable regarding resolution. For chromatograms 2 the volumetric load varied from 1.0 to 3.5 % at a constant product concentration. Resolution clearly decreases with an increasing volumetric load: there is no resolution left towards the lower-molecular weight species (C), but – depending on the requested performance – still an acceptable separation of the dimer fraction from the monomers.

Technical data: column diameter 5.0 cm   linear flow rate: 25.0 cm h^{-1}
                column height   86.0 cm   matrix: Superdax 200 pg (Pharmacia)
                column volume   1687 ml   buffer: 50 mM citrate pH 6.25

15.3.4 Process Validation

The modular design of DSP supports all aspects of process validation. In particular, drug safety is improved early on in development due to the application of proven measures and technologies for the removal and inactivation of potential contaminating DNA and virus. Although a number of preventive measures can be taken to minimize the extent of contamination, the DSP for any clinical-grade protein derived from biological sources must be validated accordingly (Table 15-4).

Table 15-4. Measures for the prevention and clearance of potential contaminants in biopharmaceutical manufacturing.

- Virus prevention/clearance
    - virus-free producer cell in animal cell culture
    - virus-free sera and raw materials
    - validation of respective unit operations in DSP for virus inactivation and virus removal

- DNA prevention/clearance
    - gentle cell harvest to minimize cell rupture
    - anion-exchange chromatography, preferably no product binding
    - product-specific adsorption chromatography with no DNA binding

- Pyrogen prevention/clearance
    - sanitary environment
    - pyrogen-free media and buffers in fermentation and DSP
    - ultrafiltration of media and buffer at a membrane cut-off < 10 kDa
    - anion-exchange chromatography, preferably no product binding
    - product-specific adsorption chromatography with no pyrogen binding

The use of filtration technologies, e.g., ultrafiltration and nanofiltration, for the mechanical removal of viral particles can easily be implemented to a large extent independent of the process stage. The simple but highly efficient removal of contaminating residual cellular DNA and DNA fragments by means of anion-exchange chromatography provides for a valuable and time-saving approach. Chromatographic removal of DNA might be based upon two competing methodologies: the use of chromatographic beads or membranes (Fig. 15-4).

Assuming a position in the midstream section of DSP, the total load of DNA is typically fairly low at 50–500 pg per mg protein (Fig. 15-5). Consequently, the overall process time which is correlated with the effective volumetric flowrate gains importance. Membranes feature a superior pressure/flow relation; therefore, the total membrane area can be reduced to a dimension which ensures a reliable dynamic capacity for DNA clearance. In contrast, the size of a conventional chromatographic column containing beads is most likely balanced on the applicable volumetric flowrate. Hence, the application of strong anion-exchange membranes can diminish a loss of product based on nonspecific adsorption in consequence of a considerable reduction of total surface compared with conventional chromatographic beads. In addition, the application of adsorptive membranes for the purpose of DNA removal is of practical impact on the performance and validation work: otherwise than labor-intensive and hardware-expensive column chromatography, the membranes are easily handled, installed as in-line filters, and can be disposed after use, not requiring comprehensive regeneration procedures.

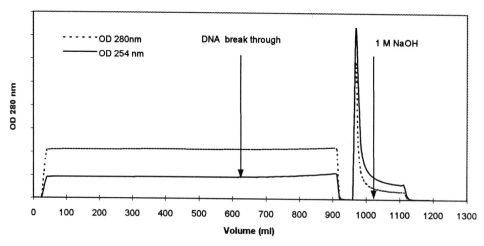

Fig. 15-4. Anion-exchange chromatography for DNA removal. A monoclonal antibody, dissolved in a 10 mM phosphate/250 mM NaCI buffer pH 7.25, 25 mS cm^{-1} at a concentation of 0.63 mg ml^{-1}, was spiked with 25 µg ml^{-1} DNA isolated from CHO-cells. A total of 900 ml was filtered through a Sartobind Q100 anion-exchange membrane (Sartorius, Göttingen, Germany) at a flowrate of 30 ml min^{-1}. The effluent was monitored photometrically parallel at 254 nm (DNA) and 280 nm (protein). Beside photometrical determination, DNA was analytically measured at 50-ml fractions. A slight continuous increase of the 254 nm line starting at 620 ml effluent indicates the DNA breakthrough. The dynamic loading capacity for the Sartobind Q membrane was determined by analytical DNA measurement with 156 µg DNA cm^{-2}. In a further experiment the dynamic binding capacity was determined for a bead-type anion-exchange matrix: Q Sepharose FF (Pharmacia, Uppsala, Sweden). A chromatographic column ($h = 11$ cm, $d = 1.0$ cm, $v = 8.6$ ml) was loaded comparable with the Sartobind Q membrane. However, a linear flowrate of $F_{lin} = 100$ cm h^{-1} equivalent to 78 ml h^{-1}, was applied to the column, which is a magnitude below the flowrate applied to the membrane (1800 ml h^{-1}). Due to the vastly increased residence time, the respective dynamic binding capacity for Q Sepharose FF amounted to 581 µg DNA ml^{-1}. Under these coditions, a membrane area of 3.75 cm^2 is equivalent to 1 ml bead-type matrix. The process time is reduced 23-fold, excluding the benefit in handling for set-up.

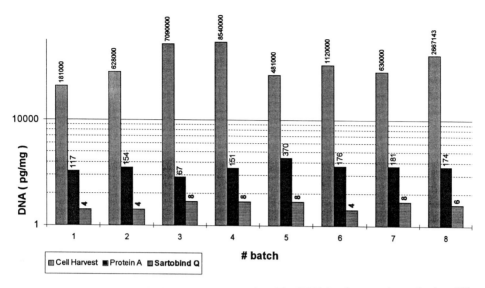

Fig. 15-5. DNA clearance in downstream processing. The DNA burden was determined at different stages in the DSP of a monoclonal antibody (MAb). The variation of DNA load within the different cell harvests reflects the variability of a biological process: it is due to cell number and cell viability in the fermentation fluid as well as cell rupture during the harvest procedure Already, Mab-specific affinity chromatography on Protein A reduces the DNA to a very consistent level. Filtration through an anion-exchange membrane (Sartobind Q) clears the DNA below the detection limit.

15.4 Economy in Process Completion

Process automation and online data acquisition allow the transfer of noncritical but time-consuming process sections to periods with reduced personnel employment, e.g., night shift or weekend. Automation might make works accessible which can not or not fully be performed during such times. The interconnection of process sections might help to reduce the overall process time. The warranty of a suitable process control, including continuous documentation of relevant process parameters, already during early stages in development helps to limitate the number of scale-up experiments, assuming that a proper linear scale-up is followed. Thus, an initial risk analysis can be performed at a very early stage in development, which helps to identify weak points and to improve the process robustness.

The course of the process development of a protein drug may be reflected by the stages in product development (Table 15-5). Early studies in the product development (Toxicology, Clinical Phase I and IIa) are content with a limited amount of the desired protein product: the purpose of these stages in product development is the evidence of safety, tolerance, and efficacy. The protein drug is produced at smaller scale, but needs to feature appropriate quality and safety. The magic term is 'dedi-

Table 15-5. Course of development for a biotechnical protein drug.

- Genetechnical development
  - DNA gene sequence
  - transfection
  - cell screening, amplification
  - cell banking

- Assay development analytics
  - physical assays
  - functional bioassays
  - immunoassays

- Feasibility study
  - fermentation
  - DSP

- Development of the basic process
  - cell biology: MCB, MWCB
  - fermentation: media optimisation, process operation
  - DSP: process design
  - Toxicology studies

- Process consolidation/validation
  - DSP: validation of DNA/virus clearance

- IND filing
  - Phase I and IIa clinical studies

- Process optimization/scale-up
  - process transfer to designated manufacturing scale and site

- Validation of the manufacturing process
  - Phase IIb and III clinical studies

- BLA filing/licensing

cated equipment', which should be a guarantee to exclude the cross-contamination with another product. This principle is fundamental for large-scale manufacturing, but sounds rather hindering for a production campaign limited to a single or few batches only in a developmental phase, or in the manufacturing of a drug with a limited market need. Chromatographic matrices as well as filters and membranes applied to ultra/diafiltration, nanofiltration, and microfiltration represent a vast surface which comes into close contact with the processed fluids. The regeneration procedures for such materials are key operations in the cGMP manufacturing, guaranteeing a most reliable product quality. Nevertheless, the use of filters, membranes, and chromatographic matrices is undoubtly dedicated to a distinctive product: the required efforts in validation in order to eliminate any potential cross-contamination would easily exceed a potential financial benefit given by matrix re-use for different protein products. However, respective cleaning and regeneration operations are to be applied to all other surfaces exposed to process fluids: piping, probes, pumps as well as chromatography columns, filter housings, and vessels. Cleaning fluids based on caustic and acidic solutions and detergents are feasible for stainless steel equipment (the recommended grades are SS316L and DIN1.4435, both electropolished with an interior finish of Ra < 0.8 μm). Their application as a CIP procedure and its performance must be re-validated accordingly. The time requirements and financial impact are significant due to labor and analytical expenditures. The analytical tools need to be developed product-specifically and in place at the right time. The feasibility of all such efforts is given for a manufacturing process which goes into operation after a considerable development time: the product has successfully passed the various clinical phases and has been shown to provide both efficiency/safety and product quality in an extensive range of analytical investigations.

At the time when a designated protein product is at the beginning of its clinical trials, i.e., after development at laboratory scale and first scale-up in order to obtain reasonable amounts of product for a clinical start-up, not all of the described validation and analytical tools might be available. Nevertheless a high through-put of different protein products in a pilot plant designated for clinical use requires an appropriate validation of the product changeover. While generic regeneration methods applied to product-dedicated chromatography and filtration media can be used as 'process modules', the cleaning of equipment (tanks, piping, etc.) which will be used for different products does not allow for compromises. An exciting solution both to technology and economy is the use of disposable hardware – piping as well as containers – which can remove a significant burden from pilot-scale manufacturing of clinical-grade biopharmaceuticals. Appropriate disposables feature a technical solution even for the cubic meter scale. Attention is drawn to materials, compatibility, leachables, temperature resistance, and flexibility. Design and sizes of process containers are different and dependent on the purpose of their application – either as a simple storage device or as designated containment for mixing/adjustment of process fluids. Polypropylene or polyethylene tanks are used widely as inexpensive hardware for storage of process fluids. Although they are dedicated to distinctive process buffers or product solutions, they do not really represent ‚disposables' as their repeated use within a production campaign again calls for significant efforts in their cleaning and sanitization. In addition, such containers occupy valuable storage space when not in use, and their supply and storage prior to a scheduled campaign may be problematic. Flexible bags such as Flexboy™, BioPharm™, or BioPharm XL™ (Stedim, Aubagne, France) are manufactured from inert polymers and eradicate the obvious shortcomings of plastic tanks. One excellent advantage is that the bags are delivered sterile and pyrogen-free. They are available in sizes from 50 mL to 800 L and are equipped with customer-specified piping (tubing), including sterile filter(s), sample ports, and connectors. Equipped in this way 'bag technology' effectively supports any need for operations in closed systems under full cGMP. Beside their application as storage devices for samples, buffers, product intermediates, and bulk product, designed bags provide an excellent mixing behavior, and thus can be applied conveniently to ultra/diafiltration operations (Fig. 15-6).

With regard to process logistics, such 'bag technology' facilitates the availability of required containers at any time. Moreover, such bags utilize a minimum storage space and handling; thus the cost–benefit relation-ship in comparison with plastic tanks is undoubted in favor of the bags.

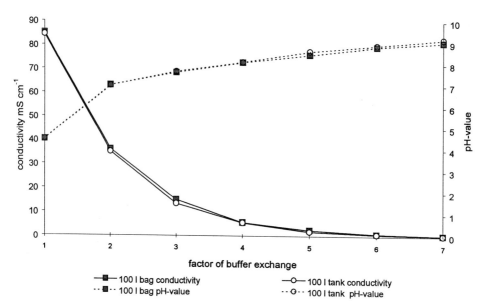

Fig. 15-6. Bag technology in ultra/diafiltration. Ultra/diafiltration was performed using a stirred 100-L PP tank in comparison with a non-stirred 100-L BioPharm™ bag. In the experiment on conductivity, a 1 M NaCI solution was diafiltered against deionized water. In the experiment on pH, a 0.1 M phosphate buffer of different pH was diafiltered (pH 4.46 versus pH 9.1). Performance in the efficiency of mixing was good in both types of hardware.

References

[1] Born, C., Thömmes, J., Biselli, M., Wandrey, C., Kula, M.-R., *Bioprocess Eng,* 1996, *15,* 21– 29.
[2] Walter, J. K., Werz, W., Berthold, W., *Biotech Forum Europe,* 1992, *9,* 560–564.
[3] Walter, J. K., Werz, W., Berthold, W., in: *Viral Safety and Evaluation of Viral Clearance from Biopharmaceutical Products;* Brown, F., Lubiniecki, A. S. (Eds.), Dev Biol Stand, Karger: Basel, 1996, vol. 88, pp. 99–108.
[4] Walter, J. K., Nothelfer, F., Werz, W., in: *ACS Symposium Series, Validation of Biopharmaceutical Manufacturing Processes,* 1998.
[5] Walter, J. K., Nothelfer, F., Werz, W., in: *Bioseparation and Bioprocessing,.Vol. I;* Subramanian, G. (Ed.), Weinheim, Wiley-VCH: 1998.
[6] Walter, J. K., Allgaier, H., in: *Manipulation of Mammalian Cells;* Wagner, R., Hauser, H. J. (Eds.), GBF Braunschweig, Berlin-New York: Walter de Gruyter, 1997, pp. 453–482.
[7] Berthold, W., Walter, J. K., *Biologicals,* 1994, *22,* 135–150.

Index

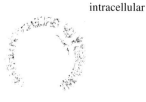